ALGEBRA: FORM AND FUNCTION

Draft Edition 2.0

BICENTENNIAL
1807
⊕WILEY
2007
BICENTENNIAL

THE WILEY BICENTENNIAL—KNOWLEDGE FOR GENERATIONS

*E*ach generation has its unique needs and aspirations. When Charles Wiley first opened his small printing shop in lower Manhattan in 1807, it was a generation of boundless potential searching for an identity. And we were there, helping to define a new American literary tradition. Over half a century later, in the midst of the Second Industrial Revolution, it was a generation focused on building the future. Once again, we were there, supplying the critical scientific, technical, and engineering knowledge that helped frame the world. Throughout the 20th Century, and into the new millennium, nations began to reach out beyond their own borders and a new international community was born. Wiley was there, expanding its operations around the world to enable a global exchange of ideas, opinions, and know-how.

For 200 years, Wiley has been an integral part of each generation's journey, enabling the flow of information and understanding necessary to meet their needs and fulfill their aspirations. Today, bold new technologies are changing the way we live and learn. Wiley will be there, providing you the must-have knowledge you need to imagine new worlds, new possibilities, and new opportunities.

Generations come and go, but you can always count on Wiley to provide you the knowledge you need, when and where you need it!

WILLIAM J. PESCE
PRESIDENT AND CHIEF EXECUTIVE OFFICER

PETER BOOTH WILEY
CHAIRMAN OF THE BOARD

ALGEBRA: FORM AND FUNCTION

Draft Edition 2.0

Produced by the Calculus Consortium and initially funded by a National Science Foundation Grant.

William G. McCallum
University of Arizona

Eric Connally
Harvard University Extension

Deborah Hughes-Hallett
Univers ity of Arizona

Philip Cheifetz
Nassau Community College

Ann Davidian
Gen. Douglas MacArthur HS

David Lovelock
University of Arizona

Patti Frazer Lock
St. Lawrence University

Pat Shure
University of Michigan

Ellen Schmierer
Nassau Community College

Elliot J. Marks

with the assistance of

Andrew M. Gleason
Harvard University

Gregory Hartman
University of Arizona

Pallavi Jayawant
Bates College

David Lomen
University of Arizona

Karen Rhea
University of Michigan

John Wiley & Sons, Inc.

PUBLISHER	Laurie Rosatone
SENIOR ACQUISITIONS EDITOR	Angela Y. Battle
ASSOCIATE EDITOR	Shannon Corliss
MARKETING MANAGER	Amy Sell
MEDIA EDITOR	Stefanie Liebman
FREELANCE DEVELOPMENTAL EDITOR	Anne Scanlan-Rohrer
SENIOR PRODUCTION EDITOR	Ken Santor
EDITORIAL ASSISTANT	Jeffrey Benson
MARKETING ASSISTANT	Tara Martinho

This book was set in Times Roman by the Consortium using TeX, Mathematica, and the package AsTeX, which was written by Alex Kasman. It was printed and bound by Courier Westford. The cover was printed by Courier Westford. The process was managed by Elliot Marks.

This book is printed on acid-free paper.

The paper in this book was manufactured by a mill whose forest management programs include sustained yield harvesting of its timberlands. Sustained yield harvesting principles ensure that the numbers of trees cut each year does not exceed the amount of new growth.

This material is based upon work supported by the National Science Foundation under Grant No. DUE-9352905. Opinions expressed are those of the authors and not necessarily those of the Foundation.

ISBN-13 978-0-471-27175-8

Printed in the United States of America

10 9 8 7 6 5 4 3 2 1

PREFACE

Algebra is fundamental to the working of modern society, yet its origins are as old as the beginnings of civilization. Algebraic equations describe the laws of science, the principles of engineering, and the rules of business. Algebraic expressions are programmed into almost every modern electronic device.

The power of algebra lies in its efficient symbolic representation of complex ideas, and this also presents the main difficulty in learning it. It is easy to forget the underlying structure of algebra and rely instead on a surface knowledge of algebraic manipulations.

A Balanced Approach: Form, Function, and Fluency

Form is related to function. An airplane wing has the form it does because of its lifting function. The pillars of the Parthenon and the girders of a skyscraper are shaped to the purpose of supporting their massive structures. Similarly, the form of an algebraic expression or equation reflects its function. To use algebra in later courses, students need not only procedural fluency, but an ability to recognize algebraic form and an understanding of the purpose of different forms. For example, a physics student needs to be able to see that the expression $L_0\sqrt{1 - v^2/c^2}$ is zero when $v = c$; a finance student should notice that $P(1 + r/12)^{12n}$ is linear in P; a statistics student should recognize that $\sigma/\sqrt{n}$ halves every time n is multiplied by 4. Knowing which manipulation to perform for a certain purpose, or whether to perform one at all, is as important as being able to carry it out.

Restoring Meaning to Expressions and Equations

The profound ideas of expression and equation are basic to everything else in algebra. Therefore, after introducing each type of function—linear, power, quadratic, exponential, polynomial—we study the basic forms of expressions for that function and which of its properties those expressions reveal. We also study the types of equations that arise from each function. We encourage students to pause and examine the basic forms, see how they are constructed, and consider the role of each component. We pay attention to the meaning of expressions in various contexts and the purposes to which they are put. On these foundations we proceed to study how each type of function is used in modeling data.

Maintaining Manipulative Skills: Tools Sections

To play a sport, it is not enough to recognize your position on the field—you also need to be able to execute a play. Similarly, acquiring the skills to perform basic algebraic manipulations is as important as recognizing algebraic forms. Throughout the book, there are Tools sections placed at the ends of chapters to which they are particularly pertinent. In most cases, students will recognize the material in these sections from previous courses. It can be used as a review for students who need it. In courses with students who are not familiar with the material, the Tools sections can be taught before the main text of the chapter in which they appear.

Expectations for Students

Students using this book should have completed high school algebra. The book is thought-provoking for well-prepared students while still accessible to students with weaker backgrounds, making it understandable to students of all ability levels. By emphasizing the basic ideas of algebra, the book provides a conceptual basis to help students master the material.

After completing this course, students will be well-prepared for Precalculus, Calculus, and other subsequent courses in mathematics and other disciplines. The focus on interpreting algebraic form, supported by graphical and numerical representations, enables students to obtain a deeper understanding of the mathematics. Our goal is to help bring students' understanding of mathematics to a higher level. Whether students go on to use "reform" texts, or more traditional ones, this knowledge will form a basis for their future studies.

Content

This content represents our vision of how algebra can be taught.

Chapter 0: The Basic Ideas of Algebra

In this chapter we review the ideas of expression and equation, as well as the difference between them. We discuss the basic underlying principles for transforming expressions, and we construct, read, and analyze simple expressions and equations.

Chapter 1: Functions, Expressions, and Equations

In this chapter we consider functions defined by algebraic expressions and how equations arise from functions. We consider other ways of describing functions—graphs, tables, and verbal descriptions—that are useful in analyzing and understanding the expressions for them. We conclude the chapter with a discussion of proportionality as a prototype of modeling with functions.

Chapter 2: Linear Functions, Expressions, and Equations

In this chapter we introduce functions that represent change at a constant rate. We consider different forms for linear expressions and what each form reveals about the function it expresses. We see how linear equations in one variable arise in considering where two linear functions attain the same value and how linear equations in two variables arise in describing straight lines. We use linear functions to model data and applications. We conclude the chapter with a discussion of systems of linear equations.

Chapter 3: Powers and Power Functions

Power functions express relationships in which one quantity is proportional to a power of another. We relate the basic graphical properties of a power function to the properties of the exponent and use the laws of exponents to put functions in a form where the exponent can be clearly recognized. We consider equations involving power functions and conclude with applications that can be modeled by power functions. In the Tools sections we review the properties of exponents, showing how they follow from the rules of arithmetic.

Chapter 4: More on Functions

In this chapter we use what he have learned about analyzing algebraic expressions to study the inner workings of functions. We consider the possible inputs and outputs of expressions (domain and range), see how functions can be built up from and decomposed into simpler functions, and consider how to construct inverse functions by reversing the operations from which they are made up.

Chapter 5: Quadratic Functions, Expressions, and Equations

We start this chapter by looking at quadratic functions and their graphs. We then consider the different forms of quadratic expressions—standard, factored, and vertex form—and show how each form reveals a different property of the function it defines. We consider two important techniques for solving quadratic equations: factoring and completing the square. The second technique leads to the quadratic formula. The Tools section reviews factoring techniques.

Chapter 6: Exponential Functions and Expressions

In this chapter we consider exponential functions such as 2^x and 3^x, in which the base is a constant and the variable occurs in the exponent. We show how to interpret different forms of exponential functions. For example, we see how to interpret the base to give the growth rate and how to interpret offset and scaled exponents in terms of growth over different time periods. We then look at exponential equations. Although they cannot be solved using the basic operations introduced so far, we show how to find qualitative information about solutions and how to estimate solutions to exponential equations graphically and numerically. We conclude with a section on modeling with exponential functions.

Chapter 7: Logarithms

In this chapter, we develop the properties of logarithms from the properties of exponents and use them to solve exponential equations. We explain that logarithms do for exponential equations what taking roots does for equations involving powers: They provide a way of isolating the variable so that the equation can be solved.

Chapter 8: Polynomials

We start by introducing polynomials as expressions that can be built up out of basic operations of arithmetic. Then, as in the chapter on quadratics, we consider the form of polynomial expressions, including the factored form, and study what form reveals about different properties of polynomial functions. This and the following chapter could be taught immediately after Chapter 4.

Chapter 9: Polynomials and Rational Functions

In this chapter we look at the graphical and numerical behavior of polynomials and rational functions on both large and small scales.

Supplementary Materials

The following supplementary materials are available for the Draft Edition 2.0:
- **Instructor's Manual** contains information on planning and creating lessons and organizing in-class activities. There are focus points as well as suggested exercises, problems and enrichment problems to be assigned to students. This can serve as a guide and check list for teachers who are using the text for the first time.
- **Instructor's Solution Manual** with complete solutions to all problems.
- **Student's Solution Manual** with complete solutions to half the odd-numbered problems.

About the Calculus Consortium

The Calculus Consortium was formed in 1988 in response to the call for change at the "Lean and Lively Calculus" and "Calculus for a New Century" conferences. These conferences urged mathematicians to re-design the content and pedagogy used in calculus. The Consortium brought together mathematics faculty from Harvard, Stanford, the University of Arizona, Southern Mississippi, Colgate, Haverford, Suffolk Community College and Chelmsford High School to address the issue. Finding surprising agreement among their diverse institutions, the Consortium was awarded funding from the National Science Foundation to design a new calculus course. A subsequent NSF grant supported the development of a precalculus and multivariable calculus curriculum

The Consortium's work has produced innovative course materials. Five books have been published. The first edition Calculus was the most widely used of any first edition calculus text ever; the precalculus book is currently the most widely used college text in the field. Books by the Consortium have been translated into Spanish, Portuguese, French, Chinese, Japanese, and Korean. They have been used in Australia, South Africa, Turkey, and Germany.

During the 1990s, the Consortium gave more than 100 workshops for college and high school faculty, in addition to numerous talks. These workshops drew a large number of mathematicians into the discussion on the teaching of mathematics. Rare before the 1990s, such discussions are now part of the everyday discourse of almost every university mathematics department. By playing a major role in shaping the national debate, the Consortium's philosophy has had widespread influence on the teaching of mathematics throughout the US and around the world.

During the 1990s, about 15 additional mathematics faculty joined the Consortium. The proceeds from royalties earned under NSF funding were put into a non-profit foundation, which supported efforts to improve the teaching of mathematics.

Since its inception, the Calculus Consortium has consisted of members of high school, college, and university faculty, all working together toward a common goal. The collegiality of such a disparate group of instructors is one of the strengths of the Consortium.

Acknowledgments

We would like to thank the many people who made this book possible. First, we would like to thank the National Science Foundation for their trust and their support; we are particularly grateful to Jim Lightbourne and Spud Bradley.

Working with Laurie Rosatone, Angela Battle, Amy Sell, Anne Scanlan-Rohrer, Ken Santor, Shannon Corliss, and Jeffrey Benson at John Wiley is a pleasure. We appreciate their patience and imagination.

Without testing of the Draft Edition in classes across the country, this book would not have been possible. We thank Victor Akasta, Jacob Amidon, Catherine Aust, Irene Gaither, Tony Giaquinto, Berri Hsiao, Pallivi Ketkar, Richard Lucas, Katherine Nyman, Daniel Pinzon, Katherine Pinzon, Abolhassan Taghavy, Chris Wetzel, Charles Widener, Andrew Wilson, Alan Yang, and all of their students for their willingness to experiment and for their many helpful suggestions.

Many people have contributed significantly to this text. They include: Charlotte Bonner, Pierre Cressant, Tina Deemer, Valentina Dimitrova, Carolyn Edmund, David Halstead, Robert Indik, Donna Krawczyk, Ernani Maia, Gowri Meda, Hideo Nagahashi, Igor Padure, Alex Perlis, Seung Hye Song, Elias Toubassi, Laurie Varecka, Mariamma Varghese, and Katherine Yoshiwara.

Special thanks are owed to "Suds" Sudholz for administering the project, to Alex Kasman for his software support, and to Kyle Niedzwiecki for his help with the computers.

William G. McCallum	Ann Davidian	Ellen Schmierer
Eric Connally	Patti Frazer Lock	Pat Shure
Deborah Hughes-Hallett	David Lovelock	
Philip Cheifetz	Elliot J. Marks	

To Students: How to Learn from this Book

- This book may be different from other mathematics textbooks that you have used, so it may be helpful to know about some of the differences in advance. At every stage, this book emphasizes the *meaning* (in algebraic, practical, graphical or numerical terms) of the symbols you are using. There is more emphasis on the interpretation of expressions and equations than you may expect. You will often be asked to explain your ideas in words.

   Why does the book have this emphasis? Because *understanding* is the key to being able to remember and use your knowledge in other courses and other fields. Much of the book is designed to help you gain such an understanding.

- The book contains the main ideas of algebra written in plain English. It was meant to be read by students like yourself. Success in using this book will depend on reading, questioning, and thinking hard about the ideas presented. It will be helpful to read the text in detail, not just the worked examples.

- There are few examples in the text that are exactly like the homework problems, so homework problems can't be done by searching for similar-looking "worked out" examples. Success with the homework will come by grappling with the ideas of algebra.

- Many of the problems in the book are open-ended. This means that there is more than one correct approach and more than one correct solution. Sometimes, solving a problem relies on common sense ideas that are not stated in the problem explicitly but which you know from everyday life.

- The following quote from a student may help you understand how some students feel. "I find this course more interesting, yet more difficult. Some math books are like cookbooks, with recipes on how to do the problems. This math requires more thinking and I do get frustrated at times. It requires you to figure out problems on your own. But, then again, life doesn't come with a cookbook."

- This book attempts to give equal weight to three skills you need to use algebra successfully: interpreting form and structure, choosing the right form for a given application, and transforming an expression or equation into the right form. There are many situations where it is useful to look at the symbols and develop a strategy before going ahead and "doing the math."

- Students using this book have found discussing these problems in small groups helpful. There are a great many problems which are not cut-and-dried; it can help to attack them with the other perspectives your classmates can provide. Sometimes your teacher may organize the class into groups to work together on solving some of the problems. It might also be helpful to work with other students when doing your homework or preparing for exams.

- You are probably wondering what you'll get from the book. The answer is, if you put in a solid effort, you will get a real understanding of algebra as well as a real sense of how mathematics is used in the age of technology.

Table of Contents

Chapter Zero

The Basic Ideas of Algebra

Contents

0.1 EXPRESSIONS

An *algebraic expression* is a way of representing a calculation, using letters to stand for some of the numbers in the calculation.

Example 1   (a) Describe a method for calculating a 20% tip on a restaurant bill, and use it to calculate the tip on a bill of $8.95 and a bill of $23.70.
(b) Choosing the letter B to stand for the bill amount, represent your method in symbols.

Solution   (a) Taking 20% of a number is the same as multiplying it by 0.2, so

$$\text{Tip on } \$8.95 = 0.2 \times 8.95 = 1.79 \text{ dollars}$$
$$\text{Tip on } \$23.70 = 0.2 \times 23.70 = 4.74 \text{ dollars}.$$

(b) The tip on a bill of B dollars is $0.2 \times B$ dollars. Usually in algebra we leave out the multiplication sign or represent it with a dot, so we write

$$0.2B \quad \text{or} \quad 0.2 \cdot B.$$

We call B a *variable*, because it can stand for various different numbers, such as 8.95 or 23.70.

Example 2   (a) Pick two numbers, find their sum and product, and then the average of their sum and product.
(b) Using the variables x and y to stand for the two numbers, write an algebraic expression that represents this calculation.

Solution   (a) Take 3 and 5, for example. Their sum is 8 and their product is 15, so the average of the sum and product is

$$\frac{(3+5)+3 \cdot 5}{2} = \frac{8+15}{2} = \frac{23}{2} = 11.5.$$

(b) If x and y are any numbers, instead of 3 and 5, then the sum is $x + y$ and the product is xy, so

$$\text{Average of sum and product} = \frac{(x+y)+xy}{2}.$$

Evaluating Expressions

If we assign particular values to the variables in an expression, then we can calculate the corresponding value of the expression itself. We call this process *evaluating the expression.*

Example 3   Evaluate each expression using (i) $x = 2$, $y = -5$ and (ii) $x = -2$, $y = 3$.
(a) $3x - 4y$ (b) $4x^2 + 9x + 7y$

Solution   (a) (i) If $x = 2$ and $y = -5$, then

$$3x - 4y = 3 \cdot 2 - 4 \cdot (-5) = 6 + 4 \cdot 5 = 26.$$

(ii) If $x = -2$ and $y = 3$, then

$$3x - 4y = 3 \cdot (-2) - 4 \cdot 3 = -6 - 12 = -18.$$

(b) (i) If $x = 2$ and $y = -5$, then

$$4x^2 + 9x + 7y = 4 \cdot 2^2 + 9 \cdot 2 + 7(-5) = -1.$$

(ii) If $x = -2$ and $y = 3$, then

$$4x^2 + 9x + 7y = 4 \cdot (-2)^2 + 9 \cdot (-2) + 7 \cdot 3 = 4 \cdot 4 - 9 \cdot 2 + 7 \cdot 3 = 19.$$

Example 4   Juan wants to buy chips and soda for a party. He estimates that he needs 5 bags of chips and 10 bottles of soda. His total cost for these items is given by

$$\text{Total cost} = 5c + 10s,$$

where c is the price of a bag of chips and s is the price of a bottle of soda. If chips cost \$2.99 per bag and soda costs \$1.29 per bottle, find the total cost.

Solution   We have $c = 2.99$ and $s = 1.29$, so

$$\text{Total cost} = 5 \cdot c + 10 \cdot s = 5 \cdot 2.99 + 10 \cdot 1.29 = 27.85 \text{ dollars.}$$

Example 5   Noah wants to buy CDs and DVDs. He knows that CDs cost \$16.99 each and DVDs cost \$29.95 each. His total cost for these items is given by

$$\text{Total cost} = 16.99m + 29.95n,$$

where m is the number of CDs purchased and n is the number of DVDs purchased. If he buys 5 CDs and 3 DVDs, find his total cost.

Solution   We have $m = 5$ and $n = 3$ so

$$\text{Total cost} = 16.99 \cdot m + 29.95 \cdot n = 16.99 \cdot 5 + 29.95 \cdot 3 = 174.80 \text{ dollars.}$$

Example 6   The surface area of a cylinder of radius r and height h is given by the formula

$$\text{Surface area} = 2\pi r^2 + 2\pi rh.$$

(a) Describe in words how the surface area is computed.
(b) Find the surface area of the cylinder of radius 3 inches and height 4 inches.

Solution   (a) Take the square of the radius and multiply the result by 2π, then take the product of the radius and the height and multiply the result by 2π, then add the results of these two calculations.
(b) If $r = 3$ and $h = 4$ then

$$\text{Surface area} = 2\pi(3)^2 + 2\pi(3)(4) = 18\pi + 24\pi = 42\pi \text{ in}^2.$$

Recognizing the Form of an Expression

Representing calculations by algebraic expressions enables us to see patterns that different calculations have in common.

Example 7   Each of the following is the result of evaluating the expression

$$\frac{(x + y) + xy}{2}.$$

Give possible values of x and y that make the expression have the given value.

(a) $\dfrac{(3 + 4) + 3 \cdot 4}{2}$ (b) $\dfrac{10 + 21}{2}$ (c) $\dfrac{(2r + 3s) + 6rs}{2}$

Solution   (a) If $x = 3$ and $y = 4$, then

$$\frac{(x + y) + xy}{2} = \frac{(3 + 4) + 3 \cdot 4}{2}.$$

(b) Here we want
$$\frac{(x + y) + xy}{2} = \frac{10 + 21}{2},$$
so we want $x + y = 10$ and $xy = 21$. Possible values are $x = 7$ and $y = 3$.

(c) Here we want $x + y = 2r + 3s$ and $xy = 6rs$. Since $(2r)(3s) = 6rs$, we can choose $x = 2r$ and $y = 3s$. Notice that in this case the "value" we have given to x and y involves other variables, rather than specific numbers.

Sometimes small differences between two algebraic expressions can radically change the calculation they represent.

Example 8   In each of the following pairs, the second expression differs from the first by the placement of parentheses. Explain what difference this makes to the calculation the expressions represent and give values of the variables that illustrate the difference.

(a) $2x^2$ and $(2x)^2$ (b) $2l + w$ and $2(l + w)$ (c) $3 - x + y$ and $3 - (x + y)$

Solution   In each case the parentheses change the order in which we do the calculation, which changes the value of the expression.

(a) In the expression $2x^2$, we square the number x and multiply the result by 2. In $(2x)^2$, we multiply the number x by 2 then square the product. For example, if $x = 3$, the first calculation gives $2 \cdot 9 = 18$, whereas the second gives $6^2 = 36$.

(b) In $2l + w$, we begin with two numbers, l and w. We double l and add w to it. In $2(l + w)$, we add l to w, then double the sum. If $l = 3$ and $w = 4$, then the first calculation gives $6 + 4 = 10$, whereas the second gives $2 \cdot 7 = 14$.

(c) In $3 - x + y$, we subtract x from 3, then add y. In $3 - (x + y)$ we add x and y, then subtract the sum from 3. If $x = 1$ and $y = 2$, then the first calculation gives $2 + 2 = 4$, whereas the second gives $3 - 3 = 0$.

Longer expressions can be built up from simpler ones. To read a complicated expression, we can sometimes break the expression down as a sum or product of parts and analyze each part. The parts of a sum are called *terms*, and the parts of a product are called *factors*.

Example 9   Describe how each expression breaks down into parts.

(a) $\frac{1}{2}h(a + b)$ (b) $3(x - y) + 4(x + y)$ (c) $\frac{R + S}{RS}.$

Solution   (a) This expression is the product of three factors, $1/2$, h, and $a + b$. The last factor is the sum of two terms, a and b.

(b) This expression is the sum of two terms, $3(x - y)$ and $4(x + y)$. The first term is the product of two factors, 3 and $x - y$, and the second is the product of 4 and $x + y$. The factors $x - y$ and $x + y$ are each sums of two terms.

(c) This expression is a quotient of two expressions. The numerator is the sum of two terms, R and S, and the denominator is the product of two factors, R and S.

Example 10   Decide whether the expressions can be put in the form
$$\frac{R + S}{RS}.$$
For those that are of this form, identify R and S.

(a) $\frac{2 + 4w}{8w}$ (b) $\frac{(x + 1) + (x - 1)}{x^2 + 1}$ (c) $\frac{6p - 2}{-12p}$

Solution
(a) The numerator of this expression is the sum of 2 and $4w$. If we multiply these terms, the result is $8w$, which is the denominator. Therefore we can take $R = 2$ and $S = 4w$.
(b) The numerator of this expression is the sum of $x + 1$ and $x - 1$. If we multiply these terms, the result is $x^2 - 1$. Since $x^2 - 1$ is not the denominator, this expression does not fit the pattern.
(c) The numerator of this expression can be written as $6p + (-2)$. If we multiply $6p$ by -2, the result is $-12p$, which is the denominator. Therefore we can take $R = 6p$ and $S = -2$.

Constructing Expressions

Given a verbal description of a calculation, we can write an expression that represents it by identifying the variables and the mathematical operations associated with them.

Example 11   You buy a corporate bond for p dollars, an amount called the face value. The interest each year is 5% of the face value, and after t years the total interest is the product of the number of years and the interest. The balance remaining is the sum of the face value and the total interest.

(a) Write an expression for the total interest after t years.
(b) Write an expression for the balance after t years.

Solution
(a) The variables are the number of years, t, and the face value, p. Because $5\% = 0.05$, the interest is given by $0.05 \cdot$ (Face value), so the total interest is given by

$$\text{Total interest} = (\text{Number of years}) \cdot 0.05 \cdot (\text{Face value}) = t0.05p = 0.05tp.$$

If an expression is the product of a constant and a variable, it is customary to put the constant at the beginning of the expression, which is why we rewrote $t0.05p$ as $0.05tp$.
(b) The balance is the sum of the face value and the total interest and is given by

$$\text{Balance} = p + 0.05tp.$$

Example 12   A student's grade in a course depends on a homework grade, h, three test grades, t_1, t_2, and t_3, and a final exam grade f. The course grade is the sum of 10% of the homework grade, 60% of the average of the three test grades, and 30% the final exam grade. Write an expression for the course grade.

Solution
The primary mathematical operation is the addition of three terms. The first term is 10% of the homework grade. The second term is 60% of the average of the three test grades. The third term 30% of the final grade. The course grade is given by

$$\text{Course grade} = 0.10h + 0.60\left(\frac{t_1 + t_2 + t_3}{3}\right) + 0.30f.$$

Example 13   The surface area of a right circular cone (not including the base) is the product of π, the radius, and the square root of the sum of the squares of the radius and the height.

(a) Write an expression for the surface area of a cone of radius r and height h.
(b) Find the surface area of a cone of radius 3 and height 4.

Solution
(a) The surface area of the cone is the product of three quantities:

$$\text{Surface area} = \pi r \sqrt{r^2 + h^2}.$$

(b) If $r = 3$ and $h = 4$, we have

$$\text{Surface area} = \pi r \sqrt{r^2 + h^2} = (\pi)(3)\sqrt{3^2 + 4^2} = 15\pi.$$

Exercises and Problems for Section 0.1

Exercises

In Exercises 1–4, write an expression for the sales tax on a car.

1. Tax rate is 7%, price is p.

2. Tax rate is r, price is $20,000.

3. Tax rate is 6%, price is $1000 off the sticker price, p.

4. Tax rate is r, price is 10% off the sticker price, p.

In Exercises 5–8, describe the sequence of operations that produces the expression.

5. $2(u+1)$ **6.** $2u+1$

7. $1-3(B/2+4)$ **8.** $3-2(s+5)$

In Exercises 9–12, write an expression for the sequence of operations.

9. Subtract x from 1, double, add 3.

10. Subtract 1 from x, double, add 3.

11. Add 3 to x, subtract the result from 1, double.

12. Add 3 to x, double, subtract 1 from the result.

In Exercises 13–19, write an expression for the quantity described.

13. Twelve x's

14. Six more than a number, n

15. Six less than a number, n

16. Seven less than twice the radius, r

17. Four years later than the present year, t

18. Four years earlier than the present year, t

19. Two more than five q's

In Exercises 20–26, evaluate the expression using the values given.

20. $3x^2 - 2y^3$, $x=3$, $y=-1$

21. $-16t^2 + 64t + 128$, $t=3$

22. $(0.5z + 0.1w)/t$, $z=10$, $w=-100$, $t=-10$

23. $(1/4)(x+3)^2 - 1$, $x=-7$

24. $(a+b)^2$, $a=-5$, $b=3$

25. $(1/2)h(B+b)$, $h=10$, $B=6$, $b=8$

26. $((b-x)^2 + 3y)/2 - 6/(x-1)$, $b=-1$, $x=2$, $y=1$

In Exercises 27–30, evaluate the expressions given that $u = -2$, $v=3$, $w=2/3$.

27. $uv - vw$ **28.** $v + \dfrac{u}{w}$

29. $u^2 + v^2 - (u-v)^2$ **30.** $u^v + w^v$

31. A caterer for a party buys 75 cans of soda and 15 bags of chips. Write an expression for the total cost if soda costs s dollars per can and chips cost c dollars per bag.

32. Apples are 99 cents a pound, and pears are $1.25 a pound. Write an expression that represents the total cost, in dollars, of a pounds of apples and p pounds of pears.

33. A person's body mass index is his or her weight in pounds, w, multiplied by 704.5, and divided by the square of his or her height in inches, h. Write an expression for the body mass index.

Problems

In Problems 34–35, you have p pennies, n nickels, d dimes, and q quarters.

34. Write an expression for the dollar value of these coins.

35. Write an expression for the total number of coins if you change your quarters into nickels and your dimes into pennies.

Oil well number 1 produces r_1 barrels per day, and oil well number 2 produces r_2 barrels per day. Each expression in Problems 36–41 describes the production at another well. What does the expression tell you about the well?

36. $r_1 + r_2$ **37.** $2r_1$ **38.** $\dfrac{1}{2}r_2$

39. $r_1 - 80$ **40.** $\dfrac{r_1 + r_2}{2}$ **41.** $3(r_1 + r_2)$

In Problems 42–45, describe the calculation given by each expression and explain how they are different.

42. $p + q/3$ and $(p+q)/3$ **43.** $(2/3)/x$ and $2/(3/x)$

44. $a-(b-x)$ and $a-b-x$ **45.** $3a + 4a^2$ and $12a^3$

In Problems 46–50, decide whether the expressions can be put in the form

$$\frac{ax}{a+x}.$$

For those that are of this form, identify a and x.

46. $\dfrac{3y}{y+3}$

47. $\dfrac{b^2\theta^2}{b^2+\theta^2}$

48. $z(k+z)^{-1}k$

49. $\dfrac{8y}{4y+2}$

50. $\dfrac{5(y^2+3)}{y^2+8}$

The surface area of a cylinder with radius r feet and height h feet is $2\pi r^2 + 2\pi rh$ square feet. In Problems 51–54, find the surface area of the cylinder with the given radius and height.

51. Radius 5 feet and height 10 feet.

52. Radius 10 feet and height 5 feet.

53. Radius 6 feet and height half the radius.

54. Radius b feet and height half the radius.

55. Two streams feed a reservoir, one at a rate of r_1 m³/day and the other at a rate of r_2 m³/day. Write an expression for the amount of water in the reservoir after t days if at time $t=0$ it holds V_0 m³ of water.

56. A car travels from Tucson to San Francisco, a distance of 870 miles. It has rest stops totaling 5 hours. While driving, it maintains a speed of v mph. Give an expression for the time it takes. What is the difference in time taken between a car that travels 65 mph and a car that travels 75 mph?

57. A mathematics professor uses the following scheme to determine the course grade: he takes the sum of the four semester grades g_1, g_2, g_3, and g_4, then adds twice the final grade, f, then divides the total by 6. Write an expression for calculating the course grade.

58. The number of registered voters in Town A is x and the number in Town B is y. If 42% of the voters in Town A and 53% of those in Town B are registered Republicans, write an expression for the total number of Republican voters in both towns.

59. Write an expression that is the sum of two terms, the first being the quotient of x and z, the second being the product of $L+1$ with the sum of y and $2k$.

60. A rectangle with base b and height h has area bh. A triangle with the same base and height has area $(1/2)bh$. Write a brief explanation of where the $1/2$ in this expression comes from by comparing the area of the triangle to the area of the rectangle.

61. The perimeter of a rectangle of length l and width w is $2l+2w$. Write a brief explanation of where the constants 2 in this expression come from.

62. The volume of a cylinder of height h and radius r is the product of the area of the base and the height. Find the volume.

63. For what value(s) of w, if any, does the expression $2w^2+1$ equal 19?

0.2 TRANSFORMING EXPRESSIONS

In Example 8 on page 4, we saw that a small change in the form of an expression could change its value. On the other hand, sometimes two expressions look quite different, but turn out to have the same value.

> **Equivalent Expressions**
>
> Two expressions are equivalent if they have the same value no matter what the value of the variables in them.

It is useful to mentally replace the variables with some actual numbers to help decide if two expressions are equivalent, although the only way to be sure is to use general properties of number operations.

Example 1   Is the expression $\dfrac{a}{2}$ equivalent to the expression $\left(\dfrac{1}{2}\right) a$?

Solution   Imagine choosing any particular value for a, say $a = 10$, and evaluating both expressions:

$$\frac{a}{2} = \frac{10}{2} = 5$$

$$\left(\frac{1}{2}\right) a = \left(\frac{1}{2}\right) 10 = 5.$$

No matter what value of a we choose, we get the same value for both expressions. This is because dividing by 2 is the same as multiplying $1/2$. For any number a,

$$\left(\frac{1}{2}a\right) = \left(\frac{1}{2}\right)\left(\frac{a}{1}\right) = \frac{a}{2}.$$

Because the two expressions have the same value for all values of a, they are equivalent.

Two expressions are not equivalent if there are any values of the variables that give the expressions different values.

Example 2   Is the expression $(x + y)^2$ equivalent to $x^2 + y^2$?

Solution   Here, as in Example 8 on page 4, the parentheses change the order in which we perform the operations of squaring and adding. If, for example, $x = 4$ and $y = 3$, the expressions have different values:

$$(x+y)^2 :\quad (4+3)^2 = \quad 7^2 \quad = 49 \quad \text{add first, then square}$$
$$x^2 + y^2 :\quad 4^2 + 3^2 = 16 + 9 = 25 \quad \text{square first, then add.}$$

Since 25 is not equal to 49, it follows that $(x + y)^2$ is not equivalent to $x^2 + y^2$.

When two expressions are not equivalent it does not mean they never have the same value, only that they do not have the same value for all values of the variable.

Example 3   Show that the expressions $2x^2 - 5x + 3$ and $x^2 - 2x + 1$ have the same values at $x = 1, 2$ but not at $x = 3$.

Solution   We have

$$2x^2 - 5x + 3 \qquad\qquad x^2 - 2x + 1$$
$$2 \cdot 1^2 - 5 \cdot 1 + 3 = 0 \qquad 1^2 - 2 \cdot 1 + 1 = 0 \quad \text{equal values at } x = 1$$
$$2 \cdot 2^2 - 5 \cdot 2 + 3 = 1 \qquad 2^2 - 2 \cdot 2 + 1 = 1 \quad \text{equal values at } x = 2$$
$$2 \cdot 3^2 - 5 \cdot 3 + 3 = 6 \qquad 3^2 - 2 \cdot 3 + 1 = 4 \quad \text{different values at } x = 3.$$

Since there is a value ($x = 3$) at which the expressions are not equal, they are not equivalent, even though they have equal values at $x = 1$ and at $x = 2$.

What Changes Can We Make to an Expression?

Since $x + 1$ and $1 + x$ are equivalent, we can say

$$x + 1 = 1 + x, \quad \text{for all values of } x.$$

Thus we can rewrite $x + 1$ as $1 + x$ in any expression without changing its value. On the other hand, we cannot rewrite $x - 1$ as $1 - x$, because they have different values (except when $x = 1$). In general, we can make any change to an expression that follows the rules of arithmetic.

Reordering and Regrouping

We can reorder and regroup addition, and reorder and regroup multiplication, because these actions do not change the value of a numerical expression. For example, reordering allows us to write

$$3 \cdot 5 = 5 \cdot 3 \quad \text{and} \quad 3 + 5 = 5 + 3,$$

while regrouping allows us to write

$$2 \cdot (3 \cdot 5) = (2 \cdot 3) \cdot 5 \text{ and } 2 + (3 + 5) = (2 + 3) + 5.$$

These same principles are valid for algebraic expressions, not just numbers:

$$ab = ba \quad \text{and} \quad a + b = b + a, \qquad \text{for all values of } a \text{ and } b$$
$$a(bc) = (ab)c \quad \text{and} \quad a + (b + c) = (a + b) + c \quad \text{for all values of } a, b, \text{ and } c.$$

Example 4 In each of the following, an expression is changed into an equivalent expression by reordering addition, reordering multiplication, regrouping addition, regrouping multiplication, or a combination of these. Which principles are used?

(a) $(x + 2)(3 + y) = (3 + y)(x + 2)$ (b) $(2x)x = 2x^2$

(c) $(2c)d = c(2d)$

Solution (a) The product of the two factors on the left is reversed on the right, so multiplication is reordered.

(b) We have a product of two terms, $2x$ and x, on the left. On the right, the two xs are grouped together to make an x^2. The multiplication is regrouped.

(c) Here both regrouping and reordering of multiplication are used. First, $2c$ is rewritten as $c \cdot 2$, and then the product is regrouped. In symbols, we have

$$(2c)d = (c \cdot 2)d = c(2d).$$

Example 5 During a normal month a bike shop sells q bicycles at a price of p dollars each, and its gross revenue in dollars is given by the expression qp. During a sale month it halves the selling price and the number of bicycles sold triples. Write an expression for the revenue during a sale month. Does the revenue change, and if so, how?

Solution Since the bike shop sells q bicycles during a normal month, and since that number triples during a sale month, it sells $3q$ bicycles during a sale month. Similarly, the price is $(1/2)p$ during a sale month. Thus,

$$\text{Revenue during sale month} = (3q)\left(\frac{1}{2}p\right).$$

By regrouping and reordering the factors in this product we get

$$\text{Revenue during sale month} = (3q)\left(\frac{1}{2}p\right) = 3 \cdot \frac{1}{2}qp = \frac{3}{2}qp.$$

Since the original revenue was qp the new revenue is $3/2$ or 1.5 times the original revenue.

Rewriting Subtraction and Division

We can rewrite $3 - 5$ as $3 + (-5)$, and $12/6$ as $12 \cdot \frac{1}{6}$. Just as the principles of regrouping and reordering are valid for general algebraic expressions, so are the principles used to rewrite subtraction and division:

$$a - b = a + (-b) \quad \text{rewrite subtraction as addition}$$

$$a/b = a \cdot \frac{1}{b} \quad \text{rewrite division as multiplication}$$

If $b \neq 0$, the number $1/b$ is called the *reciprocal* of b.

Example 6   Are the following pairs of expressions algebraically equivalent?

(a) $a + (2 - d)$ and $(a + 2) - d$

(b) $6 + r/2$ and $3 + 0.5r$

Solution   (a) Yes. We have

$$
\begin{aligned}
a + (2 - d) &= a + (2 + (-d)) && \text{rewriting subtraction as addition} \\
&= (a + 2) + (-d) && \text{regrouping addition} \\
&= (a + 2) - d && \text{rewriting addition as subtraction}
\end{aligned}
$$

(b) No. We have

$$
\begin{aligned}
6 + r/2 &= 6 + r \cdot \frac{1}{2} && \text{rewriting division as multiplication} \\
&= 6 + \frac{1}{2} \cdot r && \text{reordering multiplication} \\
&= 6 + 0.5r && \text{rewriting } \tfrac{1}{2} \text{ as } 0.5.
\end{aligned}
$$

Notice that because addition follows multiplication in the order of operations, the factor of 0.5 does not apply to the six; it applies only to r.

Example 7   If $x + y + z = 1$ find the value of $(x + 10) + (y - 8) + (z + 3)$.

Solution   By the regrouping principle, the parentheses are unnecessary for determining the value of the expression. We can reorder the terms in the expression to get

$$x + 10 + y + (-8) + z + 3 = x + y + z + 10 + (-8) + 3.$$

Then, grouping the first and last three terms together, we find

$$(x + y + z) + (10 - 8 + 3) = (x + y + z) + 5.$$

Since $x + y + z = 1$, it follows that $(x + 10) + (y - 8) + (z + 3) = 6$.

Combining Like Terms

Regrouping and reordering allows us to write expressions more concisely. For example,

$$2x^2 + 5x^2 = (x^2 + x^2) + (x^2 + x^2 + x^2 + x^2 + x^2) = 7x^2.$$

We say $2x^2$ and $5x^2$ are *like terms* because they have the same variables raised to the same powers. Terms in $3x + 4x^2$ may not be combined because the variables are raised to different powers.

Example 8   Combine like terms in each of the following expressions:

(a) $3x^2 + 6x + 9x - x^2$

(b) $z^3 + 5z^3 - 3z$.

Solution   (a) We begin by reordering:

$$3x^2 + 6x + 9x - x^2 = 3x^2 - x^2 + 6x + 9x.$$

We now regroup and combine like terms:

$$3x^2 - x^2 + 6x + 9x = (3x^2 - x^2) + (6x + 9x) = (2x^2) + (15x) = 2x^2 + 15x.$$

(b) We can add together only the z^3 terms here, since the term $-3z$ is not like the terms with z^3:

$$z^3 + 5z^3 - 3z = 6z^3 - 3z.$$

Exercises and Problems for Section 0.2

Exercises

In Exercises 1–3,

(a) Write an algebraic expression representing each of the given operations on a number b.

(b) Are the expressions equivalent? Explain what this tells you.

1. "Multiply by one fifth"
   "Divide by five"

2. "Multiply by 0.4"
   "Divide by five-halves"

3. "Multiply by eighty percent"
   "Divide by eight-tenths"

For Exercises 4–9, are the two expressions equivalent?

4. $3(z + w)$ and $3z + 3w$ **5.** $(a - b)^2$ and $a^2 - b^2$

6. $\sqrt{a^2 + b^2}$ and $a + b$ **7.** $-a + 2$ and $-(a + 2)$

8. $(3 - 4t)/2$ and $3 - 2t$ **9.** $bc - cd$ and $c(b - d)$

Simplify and then combine like terms in each of the expressions in Exercises 10–15.

10. $3p^2 - 2q^2 + 6pq - p^2$

11. $y^3 + 2xy - 4y^3 + x - 2xy$

12. $(1/2)A + (1/4)A - (1/3)A$

13. $(a + 4)/3 + (2a - 4)/3$

14. $3(2t - 4) - t(3t - 2) + 16$

15. $5z^4 + 5z^3 - 3z^4$

In Exercises 16–20, complete the table. Which, if any, of the expressions in the left-hand column are equivalent to each other? Justify your answer algebraically.

16.

x	-11	-7	0	7	11
$2x$					
$3x$					
$2x + 3x$					
$5x$					

17.

t	-11	-7	0	7	11
$2t$					
$-3t$					
$2t - 3t$					
$-t$					

18.

m	-1	0	1
m^4			
$2m^2$			
$2m^2 + 2m^2$			
$4m^4$			
$4m^2$			

19.

I	-2	-1	0	1	2
$-I$					
$-(-I)$					

20.

x	-2	-1	0	1	2
$x + 3$					
$-(x + 3)$					
$-x$					
$-x + 3$					
$-x - 3$					

For Exercises 21–24, is the attempt to combine like terms correct?

21. $2x^2 + 3x^3 = 5x^5$

22. $2AB^2 + 3A^2B = 5A^3B^3$

23. $3h^2 + 2h^2 = 5h^2$

24. $3b + 2b^2 = 5b^3$

25. Write a sentence explaining what it means for two expressions to be equivalent.

Problems

26. To convert from miles to kilometers, Abby takes the number of miles, m, doubles it, takes 20% of the result, then subtracts it. Renato first divides the number of miles by 5, and then multiplies the result by 8.

   (a) Write an algebraic expression for each method.

   (b) Use your answer to part (a) to decide if the two methods give the same answer.

27. The area of a triangle is often expressed as $A = (1/2)bh$. Is the expression $bh/2$ equivalent to the expression $(1/2)bh$?

28. Show that the following expressions have the same value at $x = -8$:

$$\frac{12 + 2x}{4 + x} \quad \text{and} \quad \frac{12}{4 - x}.$$

Does this mean these expressions are algebraically equivalent? Explain your reasoning.

29. If $pqr = 17$ find the value of $p(2qr + 3r) + 3r(pq - p)$.

In Problems 30–32, each row of the table is obtained by performing the same operation on all the entries of the previous row. Fill in the blanks in the table.

30.

-2	-1	0	1	2		a
-1	0	1	2	3		
$-1/2$	0	$1/2$	1	$3/2$		
$1/2$	0	$-1/2$	-1	$-3/2$		
$3/2$	1	$1/2$	0	$-1/2$		

31.

-2	-1	0	1	2		a
-1	$-1/2$	0	$1/2$	1		
1	$1/2$	0	$-1/2$	-1		
$3/2$	1	$1/2$	0	$-1/2$		

32.

1	2	3	4	5	x	$x-1$	
2		6	8	10			B
3	5	7		11			

33. On Figure 1, indicate an interval of length $1 - x$, and then use this to indicate an interval of length $1 - (1 - x)$. What two expressions does this suggest are equivalent?

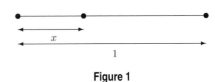

Figure 1

34. On Figure 2, indicate intervals of length

   (a) $x + 1$ **(b)** $2(x + 1)$

   (c) $2x$ **(d)** $2x + 1$

   What does your answer tell you about whether $2x + 1$ and $2(x + 1)$ are equivalent?

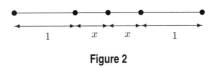

Figure 2

35. Rewrite the expression $2x^3 + 3x^2 + 4x$ using only the variable x and the operations of addition and multiplication. Your answer should not involve constants like $2, 3,$ and 4, nor should it involve powers.

In Problems 36–37, use the fact that an even number is one that can be written in the form $2n$, and an odd number is one that can be written in the form $2n + 1$, where n is an integer.

36. Is the sum of an odd and an even integer always odd, always even, or sometimes one and sometimes the other? Use algebraic expressions to justify your answer.

37. Is the product of an odd and an even integer always odd, always even, or sometimes one and sometimes the other? Use algebraic expressions to justify your answer.

0.3 THE DISTRIBUTIVE LAW

In the previous section we saw that simple rules of arithmetic allow us to rewrite expressions in different forms. One rule that is behind many of the transformations we use is the distributive law.

Example 1 Suppose Abby's bill at a restaurant is B dollars and Renato's is b dollars. If they each pay a 20% tip, what is the total tip?

Solution   There are two ways to think about this:

- The total bill is $B + b$, so the total tip is 20% of $B + b$, or $0.2(B + b)$.
- Alternatively, Abby's tip is $0.2B$ and Renato's tip is $0.2b$, so the total tip is $0.2B + 0.2b$.

In general, if we multiply a number (like 0.2) by a sum (like $B + b$), the result is the same as multiplying the number by each term in the sum and then adding the results.

Distributive Law

For any numbers a, b, and c,

$$a(b + c) = ab + ac.$$

Example 2   Simplify $3(x + 3) - x - 14$.

Solution   Distributing 3 across the sum $(x + 3)$ gives

$$3(x + 3) = 3x + 9.$$

Replacing $3(x + 3)$ with $3x + 9$, then reordering and combining like terms, we obtain

$$3(x + 3) - x - 14 = 3x + 9 - x - 14 = \overbrace{3x - x}^{\text{Combine}} + \overbrace{9 - 14}^{\text{Combine}} = 2x - 5.$$

Example 3   Use the distributive law to write two equivalent expressions for the perimeter of a rectangle, and interpret each expression in words.

Solution   Let l be the length and w the width of the rectangle. The perimeter is the sum of the lengths of the 4 sides. Since there are 2 sides of length l and 2 of length w, the perimeter is given by the expression $2l + 2w$. By the distributive law, this is equivalent to $2(l + w)$. The expression $2l + 2w$ says "double the length, double the width, and add." The expression $2(l + w)$ says "add the length to the width and double."

Distributing a Minus Sign

Since $-n = (-1)n$, the distributive law tells us how to distribute a minus sign:

$$-(n + m) = \underbrace{(-1)}_{a}\underbrace{(n + m)}_{(b+c)} \qquad \text{Let } a = -1, b = n, c = m$$

$$= \underbrace{(-1)n}_{ab} + \underbrace{(-1)m}_{ac} \qquad \text{Because } a(b + c) = ab + ac$$

$$= -n - m.$$

Example 4   Let n and k be positive integers. Show that $(n - (n - k))^3$ does not depend on the value of n, but only on the value of k.

Solution   Simplifying inside the cube gives

$$n - (n - k) = n - n + k = k.$$

Thus our expression is

$$(n - (n - k))^3 = (k)^3 = k^3,$$

which depends only on the value of k, not n.

Factoring: Using the Distributive Law in Both Directions

Sometimes we use the distributive law from left to right:

$$a(b+c) \longrightarrow ab + ac.$$

We call this distributing a over $b+c$. Other times, we use the distributive law from right to left:

$$a(b+c) \longleftarrow ab + ac.$$

We call this *taking out a common factor* of a from $ab + ac$.

Example 5   Take out a common factor in   (a) $2x + xy$ (b) $-wx - 2w$ (c) $12lm^2 - m^2$.

Solution   (a) We factor out an x, giving $2x + xy = x(2 + y)$.
(b) We factor out the $-w$, giving $-w(x + 2)$.
(c) Here there is a common factor of m^2. Factoring it out gives $m^2(12l - 1)$.

The Distributive Law and Algebraic Fractions

Often, algebraic expressions take the form of one expression divided by another.

Example 6   Suppose a car's fuel efficiency is 25 mpg in the city and 30 mpg on the highway. What is the car's average fuel efficiency if it drives:

(a) 150 miles in the city and 300 miles on the highway?
(b) c miles in the city and h miles on the highway?

Solution   (a) The car uses $150/25 = 6$ gallons in the city, and $300/30 = 10$ gallons on the highway. We have

$$\text{Average fuel efficiency} = \frac{\text{Total miles driven}}{\text{Total gallons gas}} = \frac{\text{City miles} + \text{Highway miles}}{\text{City gallons} + \text{Highway gallons}}$$

$$= \frac{150 + 300}{6 + 10} = \frac{450}{16} = 28.125 \text{ mpg}.$$

(b) Following the pattern from part (a), we have

$$\text{Average fuel efficiency} = \frac{\overbrace{\text{City miles}}^{c} + \overbrace{\text{Highway miles}}^{h}}{\underbrace{\text{City gallons}}_{c/25} + \underbrace{\text{Highway gallons}}_{h/30}} = \frac{c + h}{\dfrac{c}{25} + \dfrac{h}{30}}.$$

To check, notice that if we let $c = 150$ and $h = 300$, we have

$$\frac{150 + 300}{\dfrac{150}{25} + \dfrac{300}{30}} = 28.125,$$

which is the same answer that we got in part (a).

The distributive law tells us how to add fractions with the same denominator.

Example 7   Rewrite $\dfrac{j}{7} + \dfrac{2j}{7}$ as a single fraction.

Solution   We have

$$
\begin{aligned}
\frac{j}{7} + \frac{2j}{7} &= \frac{1}{7} \cdot j + \frac{1}{7} \cdot 2j && \text{rewriting division as multiplication} \\
&= \frac{1}{7} \cdot (j + 2j) && \text{factoring out 1/7} \\
&= \frac{1}{7} \, (3j) && \text{collecting like terms} \\
&= \frac{3j}{7} && \text{rewriting multiplication as division.}
\end{aligned}
$$

This is the same answer we get by adding numerators.

Example 8   Rewrite $\dfrac{-x + y}{-3}$ as $\dfrac{x}{3} - \dfrac{y}{3}$.

Solution   Since dividing $(-x + y)$ by -3 is the same as multiplying it by $-1/3$, we have

$$
\begin{aligned}
\frac{-x + y}{-3} &= -\frac{1}{3}(-x + y) \\
&= -\frac{1}{3} \cdot (-x + y) \\
&= -\frac{1}{3} \cdot (-x) + \left(-\frac{1}{3}\right) \cdot y && \text{distributing } -1/3 \\
&= \frac{1}{3} \cdot x - \frac{1}{3} \cdot y && \text{rewriting addition as subtraction} \\
&= \frac{x}{3} - \frac{y}{3} && \text{rewriting multiplication as division.}
\end{aligned}
$$

The previous example shows that a fraction with a sum in the numerator can be expressed as a sum of two fractions. On the other hand, there is not much that you can do with a sum in the denominator.

Example 9   Show

(a) $\dfrac{2}{x + y}$ is not equivalent to $\dfrac{1}{x} + \dfrac{1}{y}$.

(b) $2/(x + y)$ is not equivalent to $2/x + y$.

Solution   (a) One way to see this is to let $x = 1$ and $y = 1$ in each expression. We obtain

$$
\frac{1}{x} + \frac{1}{y} = \frac{1}{1} + \frac{1}{1} = 1 + 1 = 2,
$$

$$
\frac{2}{x + y} = \frac{2}{1 + 1} = \frac{2}{2} = 1.
$$

Since $2 \neq 1$, the expression $1/x + 1/y$ is not equivalent to $2/(x + y)$.

(b) We do not expect the two expressions to be equivalent since they represent quite different calculations:

$$2/(x + y) = \frac{2}{x + y} \quad \text{divide 2 by the sum } (x + y)$$

$$2/x + y = \frac{2}{x} + y \quad \text{divide 2 by } x \text{ then add } y.$$

For example, if $x = 1$ and $y = 1$, the first calculation gives 1 and the second gives 3.

To add fractions without the same denominator, we put them over a common denominator.

Example 10   Simplify the expression

$$\frac{c + h}{\dfrac{c}{25} + \dfrac{h}{30}}$$

from part (b) of Example 6 by writing it as a single fraction.

Solution   First we write the denominator as a simple fraction:

$$\frac{c}{25} + \frac{h}{30} = \frac{30c}{25 \cdot 30} + \frac{25h}{25 \cdot 30} = \frac{30c + 25h}{25 \cdot 30} = \frac{30c + 25h}{750}.$$

Then we use the rule for division of fractions to rewrite the entire expression:

$$\frac{c + h}{\dfrac{c}{25} + \dfrac{h}{30}} = = \frac{c + h}{\dfrac{30c + 25h}{750}} = \frac{750(c + h)}{30c + 25h}.$$

Rules for Operations on Algebraic Fractions

Example 10 illustrates some of the general rules for adding and multiplying algebraic fractions, which are the same as the rules for ordinary fractions.

Addition and Subtraction

$$\frac{a}{b} + \frac{c}{d} = \frac{ad + cb}{bd} \quad \text{and} \quad \frac{a}{b} - \frac{c}{d} = \frac{ad - cb}{bd}, \quad \text{provided } b \text{ and } d \text{ are not zero.}$$

We call bd the common denominator of the two fractions.

Multiplication

$$\frac{a}{b} \cdot \frac{c}{d} = \frac{ac}{bd}, \quad \text{provided } b \text{ and } d \text{ are not zero.}$$

Division

$$\frac{a/b}{c/d} = \frac{a}{b} \cdot \frac{d}{c} = \frac{ad}{bc}, \quad \text{provided } b, c, \text{ and } d \text{ are not zero.}$$

These rules hold if a, b, c, d are numbers, variables, or more complicated expressions.

Example 11   Express as a single fraction (a) $\dfrac{3}{v+w}\cdot\dfrac{vw^2}{7}$ (b) $\dfrac{3/(v+w)}{vw^2/7}$

Solution   (a) Using the rule $\dfrac{a}{b}\cdot\dfrac{c}{d}=\dfrac{ac}{bd}$, we have

$$\frac{3}{v+w}\cdot\frac{vw^2}{7}=\underbrace{\overset{a}{\overbrace{3}}}_{b\;\underbrace{v+w}}\cdot\underbrace{\overset{c}{\overbrace{vw^2}}}_{d\;\underbrace{7}}=\underbrace{\overset{ac}{\overbrace{3\cdot vw^2}}}_{bd\;\underbrace{(v+w)\cdot 7}}=\frac{3vw^2}{7(v+w)}.$$

(b) Using the rule $\dfrac{a/b}{c/d}=\dfrac{a}{b}\cdot\dfrac{d}{c}$, we have

$$\frac{3/(v+w)}{vw^2/7}=\frac{\overset{a}{\overbrace{3}}/\overset{b}{\overbrace{(v+w)}}}{\underbrace{vw^2}_{c}/\underbrace{7}_{d}}=\frac{3}{v+w}\cdot\frac{7}{vw^2}$$

$$=\frac{21}{(v+w)vw^2}.$$

Cancelation

Sometimes we can simplify an algebraic fraction by canceling a common factor in the numerator and denominator.

Example 12   Simplify $\dfrac{6x-6}{3}$.

Solution   The numerator has a common factor of 6:

$$6x-6=6(x-1).$$

Now, dividing both the numerator and denominator by 3, we obtain

$$\frac{6x-6}{3}=\frac{{}^2\cancel{6}(x-1)}{{}^1\cancel{3}}=2(x-1).$$

We can cancel the 3 because it amounts to re-expressing the fraction as a product with the number 1, where 1 has been written in a useful way:

$$\frac{6(x-1)}{3}=\frac{3\cdot 2(x-1)}{3}=\overset{1}{\overbrace{\frac{3}{3}}}\cdot\frac{2(x-1)}{1}=2(x-1).$$

Example 13   Simplify the answer to Example 10.

Solution   The fraction on the right can be simplified by canceling a 5 from the numerator and denominator:

$$\frac{750(c+h)}{30c+25h}=\frac{{}^{150}\cancel{750}(c+h)}{{}^6\cancel{30}c+{}^5\cancel{25}h}=\frac{150(c+h)}{6c+5h}.$$

Notice that we must cancel the common factor of 5 from each term in the denominator, since we use the distributive law to take it out.

Canceling Expressions

We can cancel expressions from the numerator and denominator of a fraction, but it is important to remember that cancelation is valid only when the factor being canceled is not zero.

Example 14   Simplify $\dfrac{5a^3 + 10a}{10a^2 + 20}$.

Solution   We factor $5a$ out of the numerator, giving

$$5a^3 + 10a = 5a(a^2 + 2),$$

and we factor 10 out of the denominator, giving

$$10a^2 + 20 = 10(a^2 + 2).$$

Thus, our original fraction becomes

$$\frac{5a^3 + 10a}{10a^2 + 20} = \frac{5a(a^2 + 2)}{10(a^2 + 2)}.$$

We simplify by dividing both the numerator and denominator by 5 and $(a^2 + 2)$.

$$\frac{5a^3 + 10a}{10a^2 + 20} = \frac{{}^1\!\cancel{5}a\cancel{(a^2 + 2)}}{{}^2\!\cancel{10}\cancel{(a^2 + 2)}} = \frac{a}{2}.$$

Since the quantity $(a^2 + 2)$ is never 0, the division is legal for all values of a.

Example 15   Factor the numerator and denominator to simplify $\dfrac{3 - x}{2x - 6}$.

Solution   Notice that we can factor out a -1 from the numerator:

$$3 - x = -1(-3 + x) = -1(x - 3).$$

We can also factor a 2 from the denominator:

$$2x - 6 = 2(x - 3).$$

Both numerator and denominator have $x - 3$ as a factor. So

$$\frac{3 - x}{2x - 6} = \frac{-1\cancel{(x - 3)}}{2\cancel{(x - 3)}} = -\frac{1}{2}, \quad \text{provided } x \neq 3.$$

Exercises and Problems for Section 0.3

Exercises

1. Which of the following expressions is equivalent to $3(x^2 + 2) - 3x(1 - x)$?

   (i) $6 + 3x$ (ii) $-3x + 6x^2 + 6$

   (iii) $3x^2 + 6 - 3x - 3x^2$ (iv) $3x^2 + 6 - 3x$

2. Rewrite the expression $a + 2(b - a) - 3(c + b)$ without using parentheses. Simplify your answer.

3. Is the fraction $\dfrac{x + 3}{x}$ equivalent to $1 + \dfrac{3}{x}$?

4. Is the fraction $\dfrac{x}{x + 3}$ equivalent to $1 + \dfrac{x}{3}$?

Write each of the expressions in Exercises 5–11 as a single fraction.

5. $\dfrac{m}{2} + \dfrac{m}{3}$

6. $z + \dfrac{z}{2} + \dfrac{2}{z}$

7. $1 + \dfrac{1}{1 + \frac{1}{x}}$

8. $\dfrac{-1}{x} - \dfrac{1}{-x} + \dfrac{-1}{-x}$

9. $\frac{1}{4}(e/2)$

10. $\dfrac{\frac{1}{3}r + r/4}{2r/5 - \frac{1}{11}(3r)}$

11. $j/(j/(3/j))$

For Exercises 12–17, combine the two terms into one by using a common denominator.

12. $2 + \dfrac{3}{x}$

13. $\dfrac{1}{a} + \dfrac{1}{b}$

14. $\dfrac{3}{x} + \dfrac{4}{x - 1}$

15. $\dfrac{1}{x - 2} - \dfrac{1}{x - 3}$

16. $\dfrac{1}{x - a} - \dfrac{1}{x - b}$

17. $\dfrac{1}{a - b} + \dfrac{1}{a + b}$

In Exercises 18–29, multiply and simplify. Assume any factors you cancel are not zero.

18. $\dfrac{5p}{6q^2} \cdot \dfrac{3pq}{5p}$

19. $\dfrac{3xy^2}{4x^2z} \cdot \dfrac{8xy^3z}{6xy^5}$

20. $\dfrac{2ab}{5b} \cdot \dfrac{10a^2b^2}{6a}$

21. $\dfrac{4}{6x + 12y} \cdot \dfrac{3x + 6y}{10}$

22. $\dfrac{x + 3}{x + 4} \cdot \dfrac{2x + 8}{4x + 12}$

23. $\dfrac{1}{ab + abc} \cdot (c(a + b) + (a + b))$

24. $\dfrac{2r + 3s}{4s} \cdot \dfrac{6r}{6r + 9s}$

25. $\dfrac{p^2 + 4p}{p^2 - 2p} \cdot \dfrac{3p - 6}{3p + 12}$

26. $\dfrac{w^2r + 4wr}{2r^2w + 2wr} \cdot \dfrac{r + r^2}{4w + 16}$

27. $\dfrac{cd + c}{cd - 8d} \cdot \dfrac{16 - 2c}{4 + 4d}$

28. $\dfrac{5v^2 + 15v}{vw - v} \cdot \dfrac{3w + 3}{5v + 15}$

29. $\dfrac{ab + b}{2b^2 + 6b} \cdot \dfrac{3a^2 + 6a}{a + a^2}$

Problems

30. To determine the number of tiles needed to cover A square feet of wall, a tile layer multiplies A by the number of tiles in a square foot and then adds 5% to the result to allow for breakage. If each tile is a square with side length 4 inches, write an expression for the number of tiles.

31. Consider the following sequence of operations on a number n: "Add four, double the result, add the original number, subtract five, divide by three."

    (a) Write an expression in n giving the result of the operations.

    (b) Show that the result is always one more than the number you start with.

32. Consider the following sequence of operations: "Pick a number, add 3, double the result, subtract 4 from that answer and triple what you get."

    (a) If you pick 10, what number do you get?

    (b) If you pick 5, what number do you get?

    (c) If you pick a number and end up with 30, what number did you pick?

    (d) Explain how to find someone's number if you are told what the answer is after doing all the steps.

    (e) Maris has a shortcut for finding someone's number. "Take the answer, divide by 6, and subtract 1." Is Maris right?

33. Complete the table. Are the two expressions in the left column equivalent? Justify your answer.

a	-2	-1	0	1	2
$-(1/2)(a+1)+1$					
$-(1/2)a+(1/2)$					

34. (a) Show the following expression is equivalent to y:
$$\frac{(3(y-2)+6)-y}{2}.$$

(b) Use your answer to part (a) to write a sequence of operations on a number y that always yields the number you started with.

In Problems 35–38, the distributive law $a(b+c)=ab+ac$ is used in the algebraic transformations shown. For each question, identify possible values for a and b.

35. $2(x+u)=2x+2u$ **36.** $3z(2z+7)=6z^2+21z$

37. $\dfrac{r^2s^3-rs^2}{r}=rs^3-s^2$

38. $(2x+3)(3x+1)=6x^2+9x+2x+3$

In Problems 39–44, find an equivalent expression (a)–(f), if possible.

(a) $2x$ **(b)** $\dfrac{1}{2x}$ **(c)** $\dfrac{2}{x}$

(d) $\dfrac{4}{x+2}$ **(e)** $\dfrac{1}{1-x}$ **(f)** $\dfrac{1}{x+1}$

39. $\dfrac{1}{x}+\dfrac{1}{x}$ **40.** $\dfrac{x}{0.5}$ **41.** $\dfrac{1/x}{2}$

42. $-\dfrac{1}{-1-x}$ **43.** $\dfrac{3}{x+1}$ **44.** $-\dfrac{1}{x-1}$

45. Assuming $A>0$ and $x\neq 0$, which of the following expressions are equivalent to A/x?

(a) $\dfrac{1}{2}\cdot\dfrac{2A}{2x}$ **(b)** $\dfrac{\sqrt{A^2}}{\left(\sqrt{x}\right)^2}$ **(c)** $\dfrac{1}{\frac{1}{A}\cdot\frac{1}{1/x}}$

(d) $\dfrac{Ax+x}{Ax+A}$ **(e)** $\dfrac{A^2+A}{Ax+x}$ **(f)** $\dfrac{2}{\frac{x}{A}+\frac{1}{A/x}}$

46. It takes Tiffany 3 hours to mow the lawn. It takes Lily 4 hours to mow the same lawn.

(a) If they work together, what portion of the lawn does each of them mow in one hour? Do they finish mowing the lawn?

(b) If they work together, does it take them more than two hours to mow the lawn?

(c) If they work for h hours, what portion of the lawn does each of them mow?

(d) Express your answer to part (c) as a single fraction.

47. In electronics, when two resistors, with resistances A and B, are connected in a parallel circuit, the total resistance is
$$\frac{1}{1/A+1/B}.$$

Rewrite this so there are no fractions in the numerator or denominator.

In Problems 48–50, rewrite the expression $3(n+f)$.

48. Without using parentheses.

49. Without using parentheses or the operation of multiplication.

50. Without using the operation of addition. [Hint: Try using the operation of subtraction.]

0.4 EQUATIONS

If we have only \$4200 to spend on a car, we might want to know the maximum sticker price we can afford if taxes are included. Since the expression $p+0.05p$ represents the result of adding 5% tax to a sticker price p, the maximum price p satisfies the condition

$$p+0.05p=4200.$$

This sort of statement, one that asserts two expressions are equal, is called an *equation*. When we assign numerical values to the variables in an expression, we get a numerical result. But when we assign numerical values to the variables in an equation, the equation does not yield a numerical result—it becomes true or false. For instance, the equation $p+0.05p=4200$ is true for some values of p but not others:

Let $p=4000$: $4000+0.05(4000)=4200$ a true statement.

Let $p=2000$: $2000+0.05(2000)=4200$ a false statement.

The first statement is true because the value of the expression on the left equals the value of the expression on the right. The second statement is false because these values are not equal. A value of the variable that makes an equation true is called a *solution of the equation*, and we say that the solution *solves* or *satisfies* the equation. In the next example, both sides of the equation include a variable.

Example 1

You have a $10 discount certificate for a pair of pants. When you go the store you discover that there is a 25% off sale on all pants, but no further discounts apply. For what tag price will you end up paying the same amount with each discount method?

Solution

Let p be the tag price, in dollars, of a pair of pants. With the discount certificate you pay $p - 10$, and with the store discount you pay 75% of the price, or $0.75p$. You want to know what values of p make the following statement true:

$$p - 10 = 0.75p.$$

By trying several different values of p, we find that $p = 40$ does the job (see Table 1). You pay the same price with either discount method when the tag price is $40. Thus 40 is a solution to the equation $p - 10 = 0.75p$.

Table 1   *Comparison of two discount methods*

Tag price, p (dollars)	20.00	30.00	40.00	50.00	60.00
Discount certificate, $p - 10$	10.00	20.00	30.00	40.00	50.00
Store discount, $0.75p$	15.00	22.50	30.00	37.50	45.00

In Table 1, we see that the equation $p - 10 = 0.75p$ is true for $p = 40$ but not for $p = 20, 30, 50$, or 60. Thus we see that $p = 40$ is a solution to the equation and that the other values of p are not solutions.

An expression by itself is not an equation: An equation is an assertion about the value of two different expressions, so it must contain an equal sign.

Example 2

Which of the following are not equations?

(a) $3(x - 5) + 6 - x$
(c) $ax^2 + bx + c$
(b) $3(x - 5) + 6 = x$
(d) $at^2 + t = bt^2 - 2$

Solution

An equation must include an equal sign, so (a) and (c) are not equations.

Example 3

Suppose you have $10.00 to spend on n bottles of soda, each of which costs $1.50. Is each of the following an expression? An equation? Give an interpretation.

(a) $1.50n$
(c) $10 - 1.50n$
(b) $1.50n = 6.00$
(d) $10 - 1.50n = 2.50$

Solution

(a) Since there is no equal sign, $1.50n$ is an expression that represents the cost for n bottles of soda when each bottle costs $1.50.
(b) There is an equal sign between the two expressions, so this is an equation. The solution to the equation $1.5n = 6.00$ tells us the number of bottles that can be purchased for $6.00.
(c) This is an expression. It represents the amount of money remaining after buying n bottles of soda.
(d) This is an equation, whose solution is the number of bottles you bought if the change you received was $2.50.

Solutions of Equations

We have seen that an equation may be a true statement for some values of the variable and a false statement for others. To test whether a value of the variable is a solution to an equation, you evaluate each side of the equation and see if you obtain a true statement.

Example 4   For each of the following equations, which of the given values is a solution?

(a) $3 - 4t = 5 - (2 + t)$, for the values $t = -3, 0$
(b) $3x^2 + 5 = 8$, for the values $x = -1, 0, 1$.

Solution   For each equation, we test the given values of the variable to see if the two sides are equal.

(a) We have

$$3 - 4t = 5 - (2 + t)$$
$$t = -3: \quad 3 - 4(-3) = 5 - (2 + (-3)) \quad \longrightarrow \quad 15 = 6 \quad \text{not a solution}$$
$$t = 0: \quad 3 - 4 \cdot 0 = 5 - (2 + 0) \quad \longrightarrow \quad 3 = 3 \quad \text{a solution.}$$

The two sides agree when $t = 0$ and disagree when $t = -3$. Thus, $t = 0$ is a solution, and $t = -3$ is not a solution.

(b) We have

$$3x^2 + 5 = 8$$
$$x = -1: \quad 3(-1)^2 + 5 = 8 \quad \longrightarrow \quad 8 = 8 \quad \text{a solution}$$
$$x = 0: \quad 3 \cdot 0^2 + 5 = 8 \quad \longrightarrow \quad 5 = 8 \quad \text{not a solution}$$
$$x = 1: \quad 3 \cdot 1^2 + 5 = 8 \quad \longrightarrow \quad 8 = 8 \quad \text{a solution.}$$

Both $x = -1$ and $x = 1$ satisfy the equation and are solutions, but $x = 0$ is not a solution.

In Example 4, we saw that an equation can have more than one solution. We cannot be sure we have all the solutions to an equation by trial and error. If an equation is simple enough, we can often reason out the solutions using arithmetic.

Example 5   Find all the solutions to the following equations.

(a) $x + 5 = 17$ (b) $20 = 4a$ (c) $s/3 = 22$ (d) $g^2 = 49$

Solution   (a) There is only one number that gives 17 when you add 5, so the only solution is $x = 12$.
(b) Here the product of 4 and a is 20, so $a = 5$ is the only solution.
(c) Here 22 must be one third of s, so s must be $3 \cdot 22 = 66$.
(d) The only numbers whose squares are 49 are 7 and -7, so $g = 7$ and $g = -7$ are the solutions.

For more complicated equations, we may not be able to find all solutions. Even so, it is sometimes possible to see at least one of the solutions by looking at the structure of the expressions on either side.

Example 6   Give a solution to each equation.

(a) $\dfrac{x + 1}{5} = 1$

(b) $\dfrac{x - 7}{3} = 0$

(c) $9 - z = z - 9$

(d) $t^3 - 5t^2 + 5t - 1 = 0$

Solution   (a) The value of the left side of the equation $\dfrac{x+1}{5} = 1$ must be one. The only way a fraction can be equal to one is for the numerator to equal the denominator. Therefore, $x + 1$ must be equal to five. So, $x = 4$ is the only solution.

(b) The only way a fraction can be equal to zero is if its numerator is equal to zero. Therefore $x = 7$ is the only solution.

(c) The left side of the equation is the negative of right side of the equation. The only number that is equal to its negative is zero. This means that $9 - z$ and $z - 9$ must be zero, so $z = 9$.

(d) Careful inspection of the left side of the equation reveals that its coefficients sum to zero. Therefore, if $t = 1$,

$$t^3 - 5t^2 + 5t - 1 = 1 - 5 + 5 - 1 = 0,$$

so $t = 1$ is a solution. There could be other solutions as well.

With some equations, it is possible to see from their structure that there is no solution.

Example 7   For each of the following equations, why is there no solution?[1]

(a) $x^2 = -4$ (b) $t = t + 1$ (c) $\dfrac{3x+1}{3x+2} = 1$ (d) $\sqrt{w + 4} = -3$

Solution   (a) Since the square of any number is positive, this equation has no solutions.

(b) The letter t represents the same number on each side of the equation. However, no number can be equal to one more than itself. Therefore this equation has no solutions.

(c) A fraction can equal one only when its numerator and denominator are equal. Since the denominator is one larger than the numerator, the numerator and denominator can never be equal.

(d) The square root of a number must be positive. Therefore, this equation has no solutions.

Writing Equations

A problem given verbally can often be solved by translating it into symbols and solving an equation.

Example 8   In order to receive a grade of C in a course, your average on 4 exams must be between 70 and 79. Suppose your scores on the first three exams were 50, 78, and 84. Write an equation that would allow you to determine the lowest score you have to attain on the fourth exam in order to receive a C for the course. What is the lowest score?

Solution   To find the average, we add the four test scores and divide by 4. If we let g represent the unknown score, then we want

$$\frac{50 + 78 + 84 + g}{4} = \frac{212 + g}{4} = 70.$$

Since the numerator must be divided by 4 to obtain 70, the numerator has to be 280. So $g = 68$.

----

[1] Unless otherwise specified, in this book we will be concerned only with solutions that are real numbers.

Example 9   A drought in Central America causes the price of coffee to rise 25% from last year's price. You have a $3 discount coupon and spend $17.00 for two pounds of coffee. What did coffee cost last year?

Solution   If we let c represent the price of one pound of coffee last year, then this year's price would be 25% more, which is
$$c + 0.25c = 1.25c.$$
Two pounds of coffee would cost $2(1.25c)$. When we subtract the discount coupon, the cost would be $2(1.25c) - 3$. Since we spent $17.00, our equation is
$$2(1.25c) - 3 = 17,$$
or
$$2.5c - 3 = 17.$$
The value $c = 8$ satisfies this equation, so the price of coffee last year was $8 per pound.

Equations and Identities

When we solve an equation such as $3x + 2 = 5$, we are considering the question: "Is there a number x which makes $3x + 2 = 5$?" On the other hand, when we say two expressions are equivalent, such as $x + x$ and $2x$, we are really saying: "For all numbers x, we have $x + x = 2x$." In order to distinguish this use of equations, we refer to
$$x + x = 2x$$
as an *identity*. An identity is really a special equation, one that is true for all values of the variables.

Inequalities

Whereas an equation is an assertion that two quantities are equal, an inequality is an assertion that one quantity is greater than or less than another.

Example 10   Water freezes if the temperature drops to 0°C or below, and boils if the water rises to 100°C or above. If H stands for the temperature in °C of a quantity of water, write an inequality describing H given that
(a)   The water is frozen.   (b)   The water is boiling.
(c)   The water is liquid and is not boiling.

Solution   (a) The value of H must be less than or equal to 0°C, so $H \leq 0$.
(b) The value of H must be greater than or equal to 100°C, so $H \geq 100$.
(c) The value of H must be greater than 0°C and less than 100°C, so $0 < H < 100$.

Solutions to Inequalities

A value of a variable that makes an inequality a true statement is a solution to the inequality.

Example 11   For each of the following inequalities, which of the given values is a solution?
(a) $20 - t > 12, t = -10, 10$.
(b) $16 - t^2 < 0, t = -5, 0, 5$.

Solution   (a) We have:

For $t = -10$: $\underbrace{20 - (-10)}_{30} > 12$ a true statement.

For $t = 10$: $\underbrace{20 - 10}_{10} > 12$ a false statement.

So $t = -10$ is a solution but $t = 10$ is not.

(b) We have:

For $t = -5$: $\underbrace{16 - (-5)^2}_{-9} < 0$ a true statement.

For $t = 0$: $\underbrace{16 - 0^2}_{16} < 0$ a false statement.

For $t = 5$: $\underbrace{16 - 5^2}_{-9} < 0$ a true statement.

So $t = -5$ and $t = 5$ are solutions but $t = 0$ is not.

Determining the Sign of an Expression from its Factors

We are often interested in determining whether a quantity N is positive, negative, or zero:

N is positive: $N > 0$

N is negative: $N < 0$

N is zero: $N = 0$.

In practice, determining the sign of an expression often amounts to inspecting its factors.

Example 12   For what values of x is the quantity $5x - 105$ positive?

Solution   Factoring $5x - 105$, we obtain $5(x - 21)$. This shows that if x is greater than 21, the quantity $5x - 105$ is positive. For a review of factoring, see page 223.

Example 13   What determines the sign of $a^2 x - a^2 y$ if $a \neq 0$?

Solution   Factoring, we obtain
$$a^2 x - a^2 y = a^2(x - y).$$
Since a^2 is always positive, this result shows that

- If $x > y$, then $a^2(x - y)$ is positive.
- If $x = y$, then $a^2(x - y)$ is zero.
- If $x < y$, then $a^2(x - y)$ is negative.

Example 14   The quantity H takes on values between 0 and 1, inclusive, that is, $0 \le H \le 1$. For positive A, what are the possible values of

(a) $1 + H$ (b) $1 - H$ (c) $A + AH$ (d) $3A - 2HA$?

Solution   (a) Since H takes values between 0 and 1, the value of $1 + H$ varies between a low of $1 + 0 = 1$ and a high of $1 + 1 = 2$. So
$$1 \le 1 + H \le 2.$$

(b) The value of $1 - H$ varies between a low of $1 - 1 = 0$ and a high of $1 - 0 = 1$. So
$$0 \le 1 - H \le 1.$$

(c) The value of $A + AH$ can be investigated directly, or by factoring out an A and using the results of part (a).
$$A + AH = A(1 + H).$$
Since A is positive,
$$A \cdot 1 \le A(1 + H) \le A \cdot 2.$$
So
$$A \le A + AH \le 2A.$$

(d) Factoring out an A gives
$$3A - 2HA = A(3 - 2H).$$
Since H varies between 0 and 1, the value of $3 - 2H$ varies between a low of $3 - 2 \cdot 1 = 1$ and a high of $3 - 2 \cdot 0 = 3$. Thus
$$A \cdot 1 \le A(3 - 2H) \le A \cdot 3$$
$$A \le 3A - 2AH \le 3A.$$

Exercises and Problems for Section 0.4

Exercises

In Exercises 1–6, is the value of the variable a solution to the equation?

1. $t + 3 = t^2 + 9, t = 3$

2. $x + 3 = x^2 - 9, x = -3$

3. $x + 3 = x^2 - 9, x = 4$

4. $\dfrac{a + 3}{a - 3} = 1, a = 0$

5. $\dfrac{3 + a}{3 - a} = 1, a = 0$

6. $4(r - 3) = 4r - 3, r = 1$

7. Which of the following numbers is a solution to the following equation?
$$\frac{-3x^2 + 3x + 8}{2} = 3x(x + 1) + 1$$

(a) 1   (b) 0   (c) -1 (d) 2

Write an equation for each situation presented in Exercises 8–12, letting x stand for the unknown number.

8. Twice a number is 24.

9. A number increased by 2 is twice the number.

10. Six more than a number is the negative of the number.

11. Six less than a number is 30.

12. A number is doubled and then added to itself. The result is 99.

In Exercises 13–17, write in words the statement represented by the equation.

13. $2x = 16$

14. $x = x^2$

15. $s + 10 = 2s$

16. $0.5t = 250$

17. $y - 4 = 3y$

Solve each equation in Exercises 18–22.

18. $5x = 20$

19. $t + 5 = 20$

20. $w/5 = 20$

21. $y - 5 = 20$

22. $20 = 5 - x$

23. **(a)** Construct a table showing the values of the expression $1 + 5x$ for $x = 0, 1, 2, 3, 4$.
   (b) For what value of x does $1 + 5x = 16$?

24. **(a)** Construct a table showing the values of the expression $3 - a^2$ for $a = 0, 1, 2, 3, 4$.
   (b) For what value of a does $3 - a^2 = -6$?

Which of the equations in Exercises 25–28 are identities?

25. $x^2 + 2 = 3x$

26. $2x^2 + 3x^2 = 5x^2$

27. $2u^2 + 3u^3 = 5u^5$

28. $t + 1/(t^2 + 1) = (t + 1)/(t^2 + 1)$

In Exercises 29–34, write an inequality describing the given quantity. *Example*: An mp3 player can hold up to 120 songs. *Solution*: The number of songs is n where $0 \leq n \leq 120$, n an integer.

29. Chain can be purchased in one-inch lengths from one inch to twenty feet.

30. Water is a liquid above $32°$F and below $212°$F.

31. A 200-gallon holding tank fills automatically when its level drops to 30 gallons.

32. Normal resting heart rate ranges from 40 to 100 beats per minute.

33. Minimum class size at a certain school is 16 students, and state law requires fewer than 24 students per class.

34. An insurance policy covers losses of more than $1000 but not more than $20,000.

Problems

A ball thrown vertically upward at a speed of v ft/sec rises a distance d feet in t seconds, given by $d = 6 + vt - 16t^2$. In Problems 35–38, write an equation whose solution is the given value. Do not solve the equation.

35. The time it takes a ball thrown at a speed of 88 ft/sec to rise 20 feet.

36. The time it takes a ball thrown at a speed of 40 ft/sec to rise 15 feet.

37. The speed with which the ball must be thrown to rise 20 feet in 2 seconds.

38. The speed with which the ball must be thrown to rise 90 feet in 5 seconds.

39. The value of a computer t years after it is purchased is given by

$$\text{Value of the computer} = 3500 - 700t.$$

   (a) What is value of the computer when it is purchased?
   (b) Write an equation whose solution is the time when the computer is worth nothing.

40. Write a short explanation describing the difference between an expression and an equation.

Is $t = 0$ a solution to the equations in Problems 41–44?

41. $20 - t = 20 + t$

42. $t^3 + 7t + 5 = 5 - \dfrac{rt}{n}$

43. $t + 1 = \dfrac{1 - t}{t}$

44. $t(1 + t(1 + t(1 + t))) = 1$

45. Verify that $t = 1, 2, 3$ are solutions to the equation

$$t(t - 1)(t - 2)(t - 3)(t - 4) = 0.$$

   Can you find any other solutions?

In Problems 46–49, explain how you can tell from the form of the equation that it has no solution.

46. $1 + 3a = 3a + 2$

47. $\dfrac{3x^2}{3x^2 - 1} = 1$

48. $\dfrac{1}{4z^2} = -3$

49. $\dfrac{a + 1}{2a} = \dfrac{1}{2}$

In Problems 50–52, determine whether the sentence describes an identity.

50. Eight more than a number n is the same as two less than six times the number.

51. Twice the combined income of Carlos and Jesse equals the sum of double Carlos' income and double Jesse's income.

52. In a store, N bottles of one brand of bottled water, at $1.19 a bottle, plus twice that many bottles of another brand at $1.09 a bottle, cost $6.74.

Table 2 shows values of three unspecified expressions in x for various different values of x. Give as many solutions of the equations in Problems 53–55 as you can find from the table.

Table 2

x	-1	0	1	2
Expression 1	1	2	-1	0
Expression 2	1	0	-1	0
Expression 3	0	2	-1	-1

53. Expression 1 = Expression 2
54. Expression 1 = Expression 3
55. Expression 2 = Expression 3
56. The number of houses remaining unsold after t days in a given market is given by the expression $30e^{-0.1t}$, where e is a positive constant. It can be shown that the only solution to the equation

$$30e^{-0.1t} = 10$$

is 11. What does this solution tell you about the number of houses remaining unsold?

57. A number of bids are entered for a government contract. Let p_1 be the highest bid (in dollars) and p_2 be the second highest. Explain what the following equation tells you about the bids:

$$p_1 = 2p_2 + 10,000.$$

58. For what values of x is the expression $x(x^2+4)-5(x^2+4)$ positive?

59. For what values of x is the expression $4x^2 - 28x$ negative?

60. Consider the equation

$$\frac{1}{a} = \frac{1}{b} + \frac{1}{c}.$$

(a) An incorrect technique for solving this equation would be to invert each fraction individually to get $a = b + c$. Demonstrate the mistake by using the values $a = 6/5$, $b = 2$, and $c = 3$. Show that they satisfy the original equation, but $a \neq b + c$.
(b) Solve for a in terms of b and c.

61. Two resistors, with resistance R_1 and R_2, in ohms, are connected in parallel. The total resistance R_T is given by

$$\frac{1}{R_T} = \frac{1}{R_1} + \frac{1}{R_2}.$$

(a) Solve for R_T.
(b) Two resistors of 2 and 5 ohms are connected in parallel. What is the total resistance?

0.5   SOLVING EQUATIONS

In the previous section we saw that we can sometimes guess a solution to an equation. But how can we be sure of finding a solution to an equation, and how can we be sure that we have found all the solutions? In Example 5 on page 22 the equations were simple enough to solve by direct reasoning about numbers. For a more complicated equation, one we cannot solve by direct reasoning, we try to find a simpler equation having the same solutions.

Equivalent Equations

We say two equations are *equivalent* if they have exactly the same solutions.

How can we tell when two equations are equivalent? We can think of an equation like a scale on which things are weighed. When the two sides are equal, the scale balances, and when they are different it is unbalanced. If we change the weight on one side of the scale, we must change the other side in exactly the same way to be sure that the scale remains in the same state as before, balanced or unbalanced. Similarly, in order to transform an equation into an equivalent one, we must perform an operation on both sides of the equal sign that keeps the relationship between the two sides the same, either equal or unequal. In that way we can be sure that the new equation has the same solutions—the same values that make the scale balance—as the original one.

The simplest form of equation that we can aim for is one of the form

$$x = \text{Number},$$

which directly tells us the solution.

Example 1   Solve the equation $16.8x = 84$.

Solution   We divide both sides of the equation by 16.8:

$$16.8x = 84$$
$$\frac{16.8x}{16.8} = \frac{84}{16.8} \quad \text{divide both sides by 16.8}$$
$$x = 5.$$

We have transformed our equation into one which shows the solution $x = 5$. This equation is equivalent to the original one, because dividing the two sides by 16.8 does not change whether they are equal or not.

In the previous example, we solved the equation by dividing both sides by 16.8. We chose this operation to isolate the x on one side. Being able to anticipate the effect of an operation is an important skill in solving equations.

Example 2   What operation transforms the first equation into the second equation? Check to see that the solutions of the second equation are also solutions of the first equation.

(a) $3t + 58.5 = 94.5$ $\qquad\qquad$ (b) $a^2/1.6 = 40$
$\qquad\quad 3t = 36$ $\qquad\qquad\qquad\qquad\quad a^2 = 64$

Solution   (a) We subtract 58.5 from both sides of the first equation to get

$$3t + 58.5 - 58.5 = 94.5 - 58.5$$
$$3t = 36.$$

The solution to the last equation is 12. Substituting $t = 12$ into the first equation gives $3(12) + 58.5 = 36 + 58.5 = 94.5$ so 12 is also a solution of the first equation.

(b) We multiply both sides of the first equation by 1.6 to get

$$1.6(a^2/1.6) = 40(1.6)$$
$$a^2 = 64.$$

There are two solutions to the second equation, 8 and -8. Substituting these values into the first equation gives $8^2/1.6 = 64/1.6 = 40$ and $(-8)^2/1.6 = 64/1.6 = 40$, so 8 and -8 are both solutions to the first equation as well.

What Operations Can We Perform on an Equation?

In Examples 1 and 2 we transformed equations by performing the same operation on both sides. What operations can we perform that produce an equivalent equation?

> We can transform an equation into an equivalent equation using any operation that does not change the balance between the two sides. This includes:
> - Adding or subtracting the same number to both sides
> - Multiplying or dividing both sides by the same number, provided it is not zero
> - Replacing any expression in an equation by an equivalent expression.

These operations ensure that the original equation has the same solutions as the new equation, even though the expressions on each side might change. We explore other operations in later chapters.

Example 3   Without solving, explain why the equations in each pair have the same solution.

$$\text{(a)} \quad 2.4(v - 2.1)^2 = 15 \qquad\qquad \text{(b)} \quad y^3 + 4y = y^3 + 2y + 7$$
$$(v - 2.1)^2 = 6.25 \qquad\qquad\qquad 4y = 2y + 7$$

Solution   (a) We divide both sides of the first equation by 2.4 to obtain the second equation.
(b) We subtract y^3 from both sides of the first equation to get the second.

Not every operation that we can perform on both sides of an equation leads to an equivalent equation.

Example 4   What operation transforms the first equation into the second equation? Explain why this operation is not a valid step for solving the equation.

$$x^2 = 3x$$
$$x = 3.$$

Solution   We divide both sides of $x^2 = 3x$ by x to obtain $x = 3$. This step is not valid because we must not divide by zero, and x might be equal to zero. In fact, both $x = 3$ and $x = 0$ are solutions to the first equation, but only $x = 3$ is a solution of the second equation. When we divide by x, we lose one of the solutions of the original equation.

Deciding Which Operation to Use

When we evaluate the expression $2x + 5$, we first multiply the x by 2 then add the 5. When we solve the equation

$$2x + 5 = 13$$

we first subtract the 5 from both sides, then divide by 2. Notice that when solving the equation, we undo in reverse order the operations used to evaluate the expression.

Example 5   Which operation should we use to solve each equation?

(a) $x + 5 = 20$ 　　　　　　　(b) $5x = 20$ 　　　　　　　(c) $x/5 = 20$

Solution   (a) In this equation, 5 is added to x, so we should subtract 5 from both sides of the equation.
(b) Because x is multiplied by 5, we should divide both sides of the equation by 5.
(c) Here x is divided by 5, so we should multiply both sides of the equation by 5.

Example 6   Solve each equation.

(a) $\sqrt[3]{z} = 8$ 　　　　　　　　　　　　　　　(b) $\dfrac{1}{z} = 2.5$

Solution   (a) To undo the operation of taking the cube root, we cube both sides of the equation. Since cubing the two sides does not change whether they are equal or not, this produces an equivalent equation.

$$\sqrt[3]{z} = 8$$
$$(\sqrt[3]{z})^3 = 8^3 \qquad \text{cube both sides}$$
$$z = 512.$$

We check to see that 512 is a solution: $\sqrt[3]{512} = 8$.

(b) To undo the operation of dividing by z, we multiply both sides by z. This is allowed because z cannot be equal to zero in the original equation, since the left-hand side would be undefined:

$$\frac{1}{z} = 2.5$$
$$1 = z \cdot 2.5 \qquad \text{multiply both sides by } z$$
$$\frac{1}{2.5} = z \qquad \text{divide both sides by 2.5}$$
$$0.4 = z.$$

We check to see that 0.4 is a solution: $1/0.4 = 2.5$. Note that we could also take the reciprocal of both sides to arrive at the same answer.

Why are these Operations Legal?

In Example 6 we performed some new operations on both sides of an equation: cubing and multiplying by z. Cubing two numbers does not change whether they are equal or not: different numbers have different cubes. So this operation produces an equivalent equation. Multiplying both sides by z could be a problem if $z = 0$. But in Example 6(b), the value $z = 0$ is not a solution to either

$$\frac{1}{z} = 2.5 \quad \text{or} \quad 1 = z \cdot 2.5,$$

so the two equations are equivalent.

Transforming Equations Versus Transforming Expressions

Let's compare what we have learned about expressions and equations. To transform an expression into an equivalent expression, we can use any operation that does not change the value of the expression. These operations include reordering, regrouping, and using the distributive law. To transform an equation, we can use any operation that does not change the equality of the two sides. However, when we solve an equation, the expressions on either side of the equal sign may change at each step. Consider the equation $2x + 5 = 13$, whose solution is 4. If we subtract 5 from both sides of the equation, we get

$$2x + 5 = 13$$
$$2x = 8 \quad \text{subtracting 5 from both sides,}$$

whose solution is also 4. But the expression $2x$ is not equivalent to the expression $2x + 5$. Although we can subtract 5 from both sides of an equation, we cannot subtract 5 from an expression without changing the value of the expression. Thus, the operations we can use to solve an equation includes some that we cannot use to create an equivalent expression.

Example 7   (a) Does $\dfrac{x}{2} + \dfrac{3}{4} = 2x$ have the same solution as $2x + 3 = 8x$?

(b) If we multiply $\dfrac{x}{2} + \dfrac{3}{4}$ by 4, we get $2x + 3$. Is $2x + 3$ an equivalent expression to $\dfrac{x}{2} + \dfrac{3}{4}$?

Solution   (a) If we multiply both sides of the first equation by 4, we get

$$\frac{x}{2} + \frac{3}{4} = 2x$$
$$4\left(\frac{x}{2} + \frac{3}{4}\right) = 4(2x) \quad \text{multiplying both sides by 4}$$
$$4\left(\frac{x}{2}\right) + 4\left(\frac{3}{4}\right) = 4(2x) \quad \text{applying the distributive law}$$
$$2x + 3 = 8x.$$

Therefore, the two equations have the same solution. You can verify that $x = 1/2$ is the solution to both equations.

(b) The expression $x/2 + 3/4$ is not equivalent to $2x + 3$. This can be seen by substituting $x = 0$ into each expression. The first expression becomes

$$\frac{x}{2} + \frac{3}{4} = \frac{0}{2} + \frac{3}{4} = \frac{3}{4},$$

but the second expression becomes

$$2x + 3 = 2(0) + 3 = 3.$$

Since 3 is not equal to $3/4$, the expressions are not equivalent. We cannot multiply this expression by 4 without changing its value.

Solving Inequalities

When solving an inequality, the same principle applies as when solving an equation: At any step in the process, the resulting inequality must have exactly the same solutions as the original.

Example 8   Solve the equation $3x - 12 = 0$ and the inequality $3x - 12 < 0$.

Solution   The techniques used for solving the equation and the inequality are the same:

$$
\begin{array}{l|ll}
3x - 12 = 0 & 3x - 12 < 0 & \\
3x = 12 & 3x < 12 & \text{add 12 to both sides} \\
x = 4 & x < 4 & \text{divide both sides by 4.}
\end{array}
$$

Some operations that are legal for equations are not legal for inequalities. In Example 8, it was legal to divide both sides of the inequality by 3, because if 3 times a number is less than 12, then the number must be less than 4. But we have to be careful when dividing or multiplying both sides of an inequality by a negative number. For example, although $2 > 1$, this does not mean that $(-3) \cdot 2 > (-3) \cdot 1$. In fact, it is the other way around: $-6 < -3$.

Example 9   Solve the equation $30 - x = 50$ and the inequality $30 - x > 50$.

Solution   Solving the equation gives

$$
\begin{array}{ll}
30 - x = 50 & \\
-x = 20 & \text{subtract 30 from both sides} \\
x = -20 & \text{multiply both sides by } -1.
\end{array}
$$

To solve the inequality, we start out the same way:

$$
\begin{array}{l}
30 - x > 50 \\
-x > 20
\end{array}
$$

The last inequality states that, when x is multiplied by -1, the result must be greater than 20. This is true for numbers *less than* -20. We conclude that the correct solution is

$$x < -20.$$

Notice that multiplying both sides by -1 without thinking could have led to the incorrect solution $x > -20$.

Another way to solve Example 9 is to write

$$
\begin{array}{ll}
30 - x > 50 & \\
30 > 50 + x & \text{add } x \text{ to both sides} \\
30 - 50 > x & \text{subtract 50 from both sides} \\
-20 > x. &
\end{array}
$$

The statement $-20 > x$ is equivalent to the statement $x < -20$.

Exercises and Problems for Section 0.5

Exercises

Each of the equations in Exercises 1–4 can be solved by performing a single operation on both sides. State the operation and solve the equation.

1. $x + 7 = 10$

2. $3x = 12$

3. $\dfrac{x}{9} = 17$

4. $x^3 = 64$

Each of the equations in Exercises 5–8 can be solved by performing two operations on both sides. State the operations in order of use and solve the equation.

5. $2x + 3 = 13$

6. $2(x + 3) = 13$

7. $\dfrac{x}{3} + 5 = 20$

8. $\dfrac{x + 5}{3} = 20$

In Exercises 9–16, is the second equation the result of a valid operation on the first? If so, what is the operation?

9. $3 + 5x = 1 - 2x$
$\ \ \ \ 3 + 7x = 1$

10. $3 + 2x = 5$
$\ \ \ \ \ \ \ \ \ 3 = 2x + 5$

11. $1 - 2x^2 + x = 1$
$\ \ \ \ \ \ x - 2x^2 = 0$

12. $\dfrac{x}{3} - \dfrac{3}{4} = 0$
$\ \ \ 4x - 9 = 0$

13. $9x - 3x^2 = 5x$
$\ \ \ 9 - 3x = 5$

14. $\dfrac{3}{4} - \dfrac{x}{3} = 2$
$\ \ \ 9 - 3x = 2$

15. $5x^2 - 20x = 90$
$\ \ \ x^2 - 4x = 18$

16. $\dfrac{x + 2}{5} = 1 - 3x$
$\ \ \ x + 2 = \dfrac{1 - 3x}{5}$

In Exercises 17–24, what operation on both sides of the equation isolates the variable on one side? Give the solution of the equation.

17. $0.1 + t = -0.1$

18. $-10 = 3 + r$

19. $-t + 8 = 0$

20. $5y = 19$

21. $-x = -4$

22. $7x = 6x - 6$

23. $\dfrac{-x}{5} = 4$

24. $0.5x = 3$

25. You can verify that $t = 2$ is a solution to the two equations $t^2 = 4$ and $t^3 = 8$. Are these equations equivalent?

Problems

In Problems 26–33, which of the following operations on both sides transforms the equation into one whose solution is easiest to see?

(a) Add 5   **(b)** Add x **(c)** Collect like terms
(d) Multiply by 3   **(e)** Divide by 2

26. $x - 5 = 6$

27. $2x = 2$

28. $\dfrac{x}{3} - \dfrac{1}{3} = 0$

29. $5 - x = 0$

30. $2x - 7 - x = 3$

31. $2 - 2x = 2 - x$

32. $\dfrac{2x}{3} + \dfrac{x}{3} = 2$

33. $5 - 2x = 0$

In Problems 34–37, the solution to the equation depends on the constant a. Assuming a is positive, what is the effect of increasing a on the value of the solution? Does the solution increase, decrease, or remain unchanged? Give a reason for your answer that can be understood without solving the equation.

34. $x - a = 0$

35. $ax = 1$

36. $ax = a$

37. $\dfrac{x}{a} = 1$

38. A town's total allocation for firemen's wage and benefits in a new budget is $600,000. If wages are calculated at $40,000 per fireman and benefits at $20,000 per fireman, write an equation whose solution is the maximum number of firemen the town can employ, and solve the equation.

39. (a) Does $x/3 + 1/2 = 4x$ have the same solution as $2x + 3 = 24x$?
(b) Is $x/3 + 1/2$ equivalent to $2x + 3$?

40. (a) Does $8x - 4 = 12$ have the same solution as $2x - 1 = 3$?
(b) Is $8x - 4$ equivalent to $2x - 1$?

41. The equation
$$x^2 - 5x + 6 = 0$$
has solutions $x = 2$ and $x = 3$. Is
$$x^3 - 5x^2 + 6x = 0$$
an equivalent equation? Explain your reasoning.

42. Show that $x = 5$ is a solution to both the equations
$$2x - 10 = 0 \quad \text{and} \quad (x - 3)^2 - 4 = 0.$$
Does this mean these equations are equivalent? Explain your reasoning.

REVIEW EXERCISES AND PROBLEMS FOR CHAPTER ZERO

Exercises

For Exercises 1–4, write an expression that gives the result of performing the indicated operations on the number x. You do not need to simplify your answer.

1. Add 2, double the result, then subtract 4.

2. Divide by 5, subtract 2 from the result, then divide by 3.

3. Subtract 1, square the result, then add 1.

4. Divide by 2, subtract 3 from the result, then add the result to the result of first subtracting 3 from x and then dividing the result by 2.

In Exercises 5–7, evaluate the expression using the given values.

5. $\dfrac{av^2}{a - v^3}$, $a = -2, v = -3$

6. $\dfrac{a - bc^d}{e}$, $a = 64, b = -5, c = -2, d = 3, e = 6$

7. $\dfrac{\sqrt{r} - \sqrt{s}}{\sqrt{r^2 - s}}$, $r = 10, s = 36$

Give possible values for A and B that make the expression
$$\frac{1 + A + B}{AB}$$
equal to the expressions in Exercises 8–14.

8. $\dfrac{1 + r + s}{rs}$

9. $\dfrac{1 + 2r + 3s}{6rs}$

10. $\dfrac{1 + n + m + z^2}{(n + m)z^2}$

11. $\dfrac{1 + n + m + z^2}{n(m + z^2)}$

12. $\dfrac{1 + 2x}{x^2}$

13. $\dfrac{13}{35}$

14. 0

In Exercises 15–19, write in words the statement represented by the equation.

15. $1 + 3r = 15$

16. $s^3 = s^2$

17. $w + 15 = 3w$

18. $0.25t = 100$

19. $2a - 7 = 5a$

Solve each equation in Exercises 20–24.

20. $7r = 21$

21. $a + 7 = 21$

22. $J/8 = 4$

23. $b - 15 = 25$

24. $5 = 1 - d$

For Exercises 25–32, which of the given values of the variable are solutions?

25. $x^2 + 2 = 3x$ for $x = 0, 1, 2$.

26. $2x^2 + 3x^3 = 5x^5$ for $x = -1, 0, 1$.

27. $t + 1/(t^2 + 1) = (t + 1)/(t^2 + 1)$ for $t = -1, 0, 1$.

28. $2(u - 1) + 3(u - 2) = 7(u - 3)$ for $u = 1, 2, 3$, and 6.5.

29. $2(r - 6) = 5r + 12$, for $r = 8, -8$

30. $n^2 - 3n = 2n + 24$, for $n = 8, -3$

31. $n^2 - 3n = 2n - 24$, for $n = -8, 3$

32. $s^3 - 8 = -16$, for $s = -2, 2$

In Exercises 33–37, explain what operation can be used to transform the first equation into the second equation.

33. $2x + 5 = 13$ and $2x = 8$

34. $x - 11 = 26$ and $x = 37$

35. $x/2 = 40$ and $x = 80$

36. $(2y)/3 = 20$ and $y = 30$

37. $t/2 = (t + 1)/4$ and $2t = t + 1$

Verify the identities in Exercises 38–39.

38. $\frac{1}{2}(a + b) + \frac{1}{2}(a - b) = a$

39. $\frac{1}{2}(a+b) - \frac{1}{2}(a-b) = b$

In Exercises 40–45, identify each description as either an expression or equation and write it algebraically using the variables given.

40. Twice n plus three more than n.

41. Twice n plus three more than n is 21.

42. The combined salary of Jason, J, and Steve, S.

43. Twice the combined salary of Jason, J, and Steve, S, is $140,000.

44. 225 pounds is ten pounds more than Will's weight, w.

45. 50 pounds less than triple Bob's weight, w.

Problems

46. A company uses two different-sized trucks to deliver sand. The first truck can transport x cubic yards, and the second y cubic yards. The first truck makes S trips to a job site, while the second makes T trips. What do the following expressions represent in practical terms?

  (a) $S+T$
  (b) $x+y$
  (c) $xS+yT$
  (d) $(xS+yT)/(S+T)$

47. A company out-sources the manufacturing of widgets to two companies, A and B, which supply a and b widgets respectively. However, 10% of a and 5% of b are defective widgets. What do the following expressions represent in practical terms?

  (a) $a+b$
  (b) $a/(a+b)$
  (c) $10a+5b$
  (d) $0.1a$
  (e) $0.1a+0.05b$
  (f) $(0.1a+0.05b)/(a+b)$
  (g) $(0.9a+0.95b)/(a+b)$

48. A total of 5 bids are entered for a government contract. Let p_1 be the highest bid (in dollars), p_2 be the second highest, and so forth to p_5, the lowest bid. Explain what the following expression means in terms of the bids:

$$p_1 - \frac{p_1+p_2+p_3+p_4+p_5}{5}.$$

The expressions in Problems 49–52 can be put in the form

$$ax^2 + x.$$

For each expression, identify a and x.

49. $bz^2 + z$

50. $r(n+1)^2 + n + 1$

51. $\dfrac{t^2}{7} + t$

52. $12d^2 + 2d$

In Problems 53–54, rewrite the expression $2v + 3w - 7$, given that $v = 3w - 4$.

53. Without using v.

54. Without using w.

55. Group expressions (a)–(f) together so that expressions in each group are equivalent. Note that some groups may contain only one expression.

  (a) $\dfrac{2}{k} + \dfrac{3}{k}$
  (b) $\dfrac{5}{2k}$
  (c) $\dfrac{10}{2k}$

  (d) $2.5k$
  (e) $\dfrac{10k}{4}$
  (f) $\dfrac{1}{2k/5}$

56. Find the expression for the volume of a rectangular solid in terms of its width w where the length l is twice the width and the height h is five more than the width.

57. Two wells produce r_1 and r_2 barrels of oil per day. Write an expression in terms of r_1 and r_2 describing a well that produces 100 fewer barrels per day than the first two wells combined.

For Problems 58–60, assume that movie tickets cost p for adults and q for children.

58. Write expressions for the total cost of tickets for:

  (a) 2 adults and 3 kids   **(b)** No adults and 4 kids
  (c) No kids and 5 adults   **(d)** A adults and C kids

59. Rework Problem 58, this time giving expressions for the average cost per ticket.

60. A family of two adults and three children has an entertainment budget equal to the cost of 10 adult tickets. An adult ticket costs twice as much as a child ticket. How much money will they have left after seeing two movies? Can they afford to see a third movie?

61. It costs a contractor p to employ a plumber, e to employ an electrician, and c to employ a carpenter.

  (a) Write an expression for the total cost to employ 4 plumbers, 3 electricians, and 9 carpenters.
  (b) Write an expression for the fraction of the total cost in part (a) that is due to plumbers.
  (c) Suppose the contractor hires P plumbers, E electricians, and C carpenters. Write expressions for the total cost for hiring these workers and the fraction of this cost that is due to plumbers.

62. A contractor is managing three different job sites. It costs her $\$c$ to employ a carpenter, $\$p$ to employ a plumber, and $\$e$ to employ an electrician. The total cost to employ carpenters, plumbers, and electricians at each site is

$$\text{Cost at site } 1 = 12c + 2p + 4e$$
$$\text{Cost at site } 2 = 14c + 5p + 3e$$
$$\text{Cost at site } 3 = 17c + p + 5e.$$

Write expressions in terms of c, p, and e for:

(a) The total employment cost for all three sites.
(b) The difference between the employment cost at site 1 and site 3.
(c) The amount remaining in the contractor's budget after accounting for the employment cost at all three sites, given that originally the budget is $50c + 10p + 20e$.

An airline has four different types of jet in its fleet:

- p 747s with a capacity of 400 seats each,
- q 757s with a capacity of 200 seats each,
- r DC9s with a capacity of 120 seats each, and
- s Saab 340s with a capacity of 30 seats each.

Answer Problems 63–64 given this information.

63. Write an expression representing the airline's total capacity (in seats).

64. To cut costs, the airline decides to maintain only two types of airplanes in its fleet. It replaces all 747s with 757s and all DC9s with Saab 340s. It keeps all the original 757s and Saab 340s. If it retains the same total capacity (in seats), write an expression representing the total number of airplanes it now has.

65. You plan to drive 300 miles at 55 miles per hour, stopping for a two-hour rest. You want to know t, the number of hours the journey is going to take. Which of the following equations would you use?

(A) $55t = 190$ **(B)** $55 + 2t = 300$
(C) $55(t + 2) = 300$ **(D)** $55(t - 2) = 300$

66. A tank contains $20 - 2t$ cubic meters of water, where t is in days.

(a) Construct a table showing the number of m^3 of water at $t = 0, 2, 4, 6, 8, 10$.
(b) Use your table to determine when the tank
   (i) Contains 12 m^3
   (ii) Is empty.

67. A vending machine contains $40 - 8h$ bags of chips h hours after 9 am.

(a) Construct a table showing the number of bags of chips in the machine at $h = 0, 1, 2, 3, 4, 5, 6$.
(b) How many bags of chips are in the machine at 9 am?
(c) At what time does the machine run out of chips?
(d) Explain why the number of bags you found for $h = 6$ using $40 - 8h$ is not reasonable. How many bags are probably in the machine at $h = 6$?

68. (a) Does $\dfrac{x}{4} + \dfrac{1}{2} = 3x$ have the same solution as $x + 2 = 12x$?
(b) Is $\dfrac{x}{4} + \dfrac{1}{2}$ equivalent to $x + 2$?

69. (a) Does $\dfrac{5x}{3} + 2 = 1$ have the same solution as $5x + 6 = 1$?
(b) Is $\dfrac{5x}{3} + 2$ equivalent to $\dfrac{5x + 6}{3}$?

70. The equation

$$x^2 = 12 + x$$

has solutions $x = 4$ and $x = -3$. Is

$$x = \sqrt{12 + x}$$

an equivalent equation? Explain your reasoning.

71. Which of the following is *not* the result of a valid operation on the equation $3(x - 1) + 2x = -5 + x^2 - 2x$?

(a) $5x - 3 = x^2 - 2x - 5$
(b) $4x + 3(x - 1) = -5 + x^2$
(c) $3(x - 1) = -5 + x^2 - 2x + 2x$
(d) $3(x - 1) + 2x - 5 = -10 + x^2 - 2x$

72. Use the fact that $x = 2$ is a solution to the equation $x^2 - 5x + 6 = 0$ to find a solution to the equation

$$(2w - 10)^2 - 5(2w - 10) + 6 = 0.$$

73. If m is the number of males and f is the number of females in a population, which of the following expresses the fact that 47% of the population is male and 53% of the population is female?

(a) $P = 0.47m + 0.53f$
(b) $P = 0.53m + 0.47f$
(c) $m = 0.47P$ and $f = 0.53P$
(d) $P = 0.47m$ and $P = 0.53f$

74. Weight Watchers© assigns points to various foods, and limits the number of points you can accumulate in a day. If c is the number of calories, g the grams of fat, and f

the grams of dietary fiber, then the number of points for a piece of food is

$$\text{Number of points} = \frac{c}{50} + \frac{g}{12} - \frac{f}{4}.$$

How many grams of fiber can you trade for three grams of fat without changing the number of points?

75. Here are the line-by-line instructions for calculating the deduction on your federal taxes for medical expenses.

1	Enter medical expenses	
2	Enter adjusted gross income	
3	Multiply line 2 by 7.5% (0.075)	
4	Subtract line 3 from line 1. If line 3 is more than line 1, enter 0.	

Write an expression for your medical deduction (line 4) in terms of your medical expenses, E, and your adjusted gross income, I.

A government buys x fighter planes at f dollars each, and y tons of wheat at w dollars each. It spends a total of B dollars, where $B = xf + yw$. In Problems 76–78, write an equation whose solution is the given value.

76. The number of tons of wheat it can afford to buy if it spends a total of $100 million, wheat costs $300 per ton, and it must buy 5 fighter planes at $15 million each.

77. The price of fighter planes if it bought 3 of them, 10,000 tons of wheat at $500 a ton, and spent a total of $50,000,000.

78. The price of a ton of wheat if it buys 20 fighter planes and 15,000 tons of wheat for a total expenditure of $90,000,000, and a fighter plane costs 100,000 times a ton of wheat.

In Problems 79–81, decide if the statement is true or false. Justify your answers using algebraic expressions.

79. The sum of three consecutive integers is a multiple of 3.

80. The sum of three consecutive integers is even.

81. The sum of three consecutive integers is three times the middle integer.

82. An invasive species of fish is introduced into a lake. The number N (in hundreds) of fish after t months is given by the expression $\dfrac{500}{1 + 49e^{-0.04t}}$, where e is a positive constant. It can be shown that the only solution to the equation

$$\frac{500}{1 + 49e^{-0.04t}} = 16$$

is 12. What does this solution tell you about the number of fish in the lake?

Chapter One

Functions, Expressions, and Equations

Contents

1.1   WHAT IS A FUNCTION?

In everyday language, *function* expresses the notion of one thing depending on another. We might say that election results are a function of campaign spending, or that ice cream sales are a function of the weather. In mathematics, the meaning of *function* is more precise, but the idea is the same. If the value of one quantity determines the value of another, we say the second quantity is a function of the first.

> A **function** is a rule which takes numbers as inputs and assigns to each input exactly one number as output. The output is a function of the input.

Example 1   A 20% tip on a meal is a function of the cost of the meal. What is the input and what is the output to this function?

Solution   The input is the amount of the bill, and the output is the amount of the tip.

Function Notation

We use *function notation* to represent the output of a function in terms of its input. The expression $f(20)$, for example, stands for the output of the function f when the input is 20. Here the letter f stands for the function itself, not for a number. If f is the function in Example 1, we have

$$f(20) = 4,$$

since the tip for a $20 meal is $(0.2)(20) = \$4$. In words, we say "$f$ of 20 equals 4." In general,

$$f(\text{Amount of bill}) = \text{Tip}.$$

Example 2   For the function in Example 1, let B stand for the bill amount and T stand for the amount of the tip.
(a) Express the function in terms of B and T.
(b) Evaluate $f(8.95)$ and $f(23.70)$ and interpret your answer in practical terms.

Solution   (a) The input is the bill amount, B, and the output is the tip, T. Since the tip is 20% of B, or $0.2B$, we have
$$T = f(B) = 0.2B.$$
(b) For $f(8.95)$ the input is 8.95, which tells us that the meal costs $8.95. The output is

$$f(8.95) = 0.2(8.95) = 1.79,$$

which tells us the tip on a $8.95 meal is $1.79. Similarly, for $f(23.70)$ the input is 23.70, which is the dollar cost of the meal. The output is

$$f(23.70) = 0.2(23.70) = 4.74.$$

So the tip on a $23.70 meal is $4.74.

Example 3   If you buy first class stamps, the total cost in dollars is a function of the number of stamps bought.
(a) What is the input and what is the output of this function?
(b) Use function notation to express the fact that the cost of 14 stamps is $5.46.
(c) If a first class stamp costs 39 cents, find a formula for the cost, C, in dollars, in terms of the number of stamps, n.

Solution   (a) The input is the number of stamps, and the output in the total cost:
$$\text{Total cost} = f(\text{Number of stamps}).$$
(b) If the total cost is $5.46 for 14 stamps, then
$$f(14) = 5.46.$$
(c) Since 39 cents is 0.39 dollars, n stamps cost $0.39n$ dollars, so
$$C = f(n) = 0.39n.$$

Dependent and Independent Variables

The input variable, such as B in Example 2 or n in Example 3, is also called the *independent variable*, and the output variable, such as T in Example 2 or C in Example 3, is also called the *dependent variable*. Symbolically,
$$\text{Output} = f(\text{Input})$$
$$\text{Dependent} = f(\text{Independent}).$$

Example 4   The area of a circle of radius r is given by $A = q(r) = \pi r^2$. What is the independent variable? What is the dependent variable?

Solution   We use this function to compute the area when we know the radius. Thus, the independent variable is r, the radius, and the dependent variable is A, the area. Note also that q is not the dependent variable. In fact, q is simply the name of the function and is not a variable at all.

In a situation like Example 4, where the function is given by an algebraic expression in the independent variable and we have chosen a letter to represent the dependent variable, we often omit the function notation and simply write
$$A = \pi r^2.$$
We call this a *formula* for the function. However, in the next example we do not have a formula.

Example 5   The cost in dollars, T, of tuition at most colleges is a function of the number, c, of credits taken. Express the relationship in function notation, and identify the independent and dependent variables.

Solution   We have
$$\text{Tuition cost} = f(\text{Number of credits}),$$
or
$$T = f(c).$$
The independent variable is c, and the dependent variable is T.

Evaluating Functions

If $f(x)$ is given by an algebraic expression in x, then finding the value of $f(5)$, for instance, is the same as evaluating the expression at $x = 5$.

Example 6   If $f(x) = 5 - \sqrt{x}$, evaluate each of the following

(a) $f(0)$ (b) $f(16)$ (c) $f(12^2)$

Solution   (a) We have $f(0) = 5 - \sqrt{0} = 5$.
(b) We have $f(16) = 5 - \sqrt{16} = 5 - 4 = 1$.
(c) We have $f(12^2) = 5 - \sqrt{12^2} = 5 - 12 = -7$.

Example 6 (c) illustrates that the input to a function can be more complicated than a simple number. Often we want to consider inputs to functions that are numerical or algebraic expressions.

Example 7   Let $h(x) = x^2 - 3x + 5$. Evaluate the following

(a) $h(a) - 2$ (b) $h(a - 2)$ (c) $h(a - 2) - h(a)$

Solution   To evaluate a function at an expression like $a - 2$, it is helpful to remember that $a - 2$ is the input, and to rewrite the formula for f as

$$\text{Output} = h(\text{Input}) = (\text{Input})^2 - 3 \cdot (\text{Input}) + 5.$$

(a) First input a, then subtract 2:

$$h(a) - 2 = \underbrace{(a)^2 - 3(a) + 5}_{h(a)} - 2$$

$$= a^2 - 3a + 3.$$

(b) In this case, Input $= a - 2$. We substitute and multiply out

$$h(a - 2) = (a - 2)^2 - 3(a - 2) + 5$$
$$= a^2 - 4a + 4 - 3a + 6 + 5$$
$$= a^2 - 7a + 15.$$

(c) We must evaluate h at two different inputs, $a - 2$ and a. We have

$$h(a - 2) - h(a) = \underbrace{(a - 2)^2 - 3(a - 2) + 5}_{h(a-2)} - \underbrace{\left((a)^2 - 3(a) + 5\right)}_{h(a)}$$

$$= \underbrace{a^2 - 4a + 4 - 3a + 6 + 5}_{h(a-2)} - \underbrace{(a^2 - 3a + 5)}_{h(a)}$$

$$= -4a + 10.$$

Interpreting Functions

It is often useful to think about the units of measurement for the independent and dependent variables of a function.

Example 8   Let $f(A)$ be the number of gallons needed to paint a house with walls of area A ft^2. If $f(3500) = 10$, what are the units of the 3500 and the 10? What does the statement $f(3500) = 10$ tell you about painting the house?

Solution   The input, 3500, represents the area of the walls, so its units are ft^2. The output, 10, is the number of gallons of paint. The statement $f(3500) = 10$ tells us that 10 gallons are needed to paint 3500 ft^2 of walls.

Example 9   Suppose the function in Example 8 is given by $f(A) = A/350$.

(a) How many gallons do you need to paint a house whose walls measure 5000 ft^2? 10,000 ft^2?
(b) Explain in words the relationship between the number of gallons and A. What is the practical interpretation of the 350 in the expression for the function?

Solution   (a) For a house measuring 5000 ft^2 we need $f(5000) = 5000/350 = 14.3$ gallons. For a house measuring 10,000 ft^2 we need $f(10,000) = 10,000/350 = 28.6$ gallons.
(b) The expression tells us to divide the area by 350 to compute the number of gallons. This means that 350 ft^2 is the area covered by one gallon.

Example 10   Let $n = f(p)$ be the average number of days a house in a particular community stays on the market before being sold for price p (in $1000s), and let p_0 be the average sale price of houses in the community. What do the following expressions mean in terms of the housing market?

(a) $f(250)$ (b) $f(p_0 + 10)$ (c) $f(0.9p_0)$

Solution   (a) This is the average number of days a house stays on the market before being sold for $250,000.
(b) Since $p_0 + 10$ is 10 more than the average price in thousands of dollars, $f(p_0 + 10)$ is the average number of days a house stays on the market before being sold for $10,000 above the average sale price.
(c) Since $0.9p_0$ is 90% of the average sale price, or 10% less than the average sale price, $f(0.9p_0)$ is the average number of days a house stays on the market before being sold at 10% below the average sale price.

Tables And Graphs

We can often see key features of a function by making a table of output values and by drawing a graph of the function. For example, Table 1.1 and Figure 1.1 show values of $g(a) = a^2$ for $a = -2, -1, 0, 1$ and 2. We can see that the output values never seem to be negative, which is confirmed by the fact that a square is always positive or zero.

Table 1.1   *Values of* $g(a) = a^2$

a (input)	$g(a) = a^2$ (output)
-2	4
-1	1
0	0
1	1
2	4

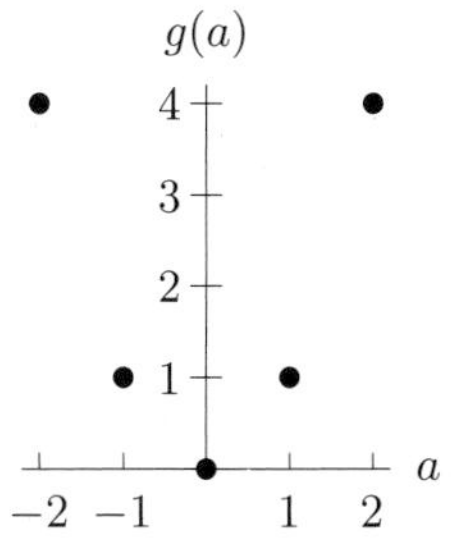

Figure 1.1: Graph of values of $g(a) = a^2$

If we plot more values, we get Figure 1.2. Notice how the points appear to be joined by a smooth curve. If we could plot all the values we would get the curve in Figure 1.3. This is the *graph* of the function. A graphing calculator or computer shows a good approximation of the graph by plotting many points on the screen.

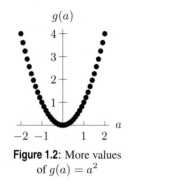

Figure 1.2: More values
of $g(a) = a^2$

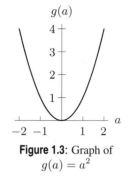

Figure 1.3: Graph of
$g(a) = a^2$

Example 11   Explain how Table 1.1 and Figure 1.3 illustrate that $g(a) = g(-a)$ for any number a.

Solution   Notice the pattern in the right column of Table 1.1: the values go from 4 to 1 to 0, then back to 1 and 4 again. This is because $(-2)^2$ and 2^2 both have the same value, namely 4, so

$$g(-2) = g(2) = 4.$$

Similarly,

$$g(-1) = g(1) = 1.$$

We can see the same thing in the symmetrical arrangement of the points in Figures 1.1 and 1.2 about the vertical axis. Since both a^2 and $(-a)^2$ have the same value, the point above a on the horizontal axis has the same height as the point above $-a$.

Example 12   For the function graphed in Figure 1.4, find $f(x)$ for $x = -3, -2, -1, 0, 1, 2, 3$.

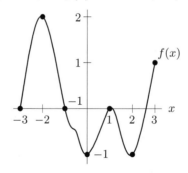

Figure 1.4

Solution   The coordinates of a point on the graph of f are $(a, f(a))$ for some number a. So, since the point $(-3, 0)$ is on the graph, we must have $f(-3) = 0$. Similarly, since the point $(-2, 2)$ is on the graph, we must have $f(-2) = 2$. Using the other marked points, we get the values in Table 1.2.

Table 1.2

x	-3	-2	-1	0	1	2	3
$f(x)$	0	2	0	-1	0	-1	1

Exercises and Problems for Section 1.1

Exercises

In Exercises 1–2, write the relationship using function notation (i.e. y is a function of x is written $y = f(x)$).

1. Weight, w, is a function of caloric intake, c.

2. Number of molecules, m, in a gas, is a function of the volume of the gas, v.

3. The number, N, of napkins used in a restaurant is $N = f(C) = 2C$, where C is the number of customers. What is the dependent variable? The independent variable?

4. A silver mine's profit, P, is $P = g(s) = -100000 + 50000s$ dollars, where s is the price per ounce of silver. What is the dependent variable? The independent variable?

In Exercises 5–6, evaluate the function for $x = -7$.

5. $f(x) = x/2 - 1$ **6.** $f(x) = x^2 - 3$

Evaluate the expressions in Exercises 7–12 for

$$f(x) = \frac{2x+1}{3-5x} \qquad g(y) = \frac{1}{\sqrt{y^2+1}}.$$

7. $f(0)$ **8.** $g(0)$ **9.** $g(-1)$

10. $f(10)$ **11.** $f(1/2)$ **12.** $g\left(\sqrt{8}\right)$

13. Let $g(x) = (12 - x)^2 - (x - 1)^3$. Evaluate

(a) $g(2)$ (b) $g(5)$
(c) $g(0)$ (d) $g(-1)$

14. Let $f(x) = 2x^2 + 7x + 5$. Evaluate

(a) $f(3)$ (b) $f(a)$
(c) $f(2a)$ (d) $f(-2)$

15. Use Table 1.3 to fill in the missing values. (There may be more than one answer.)

(a) $f(0) = ?$ (b) $f(?) = 0$
(c) $f(1) = ?$ (d) $f(?) = 1$

Table 1.3

x	0	1	2	3	4
$f(x)$	4	2	1	0	1

16. Use Figure 1.5 to fill in the missing values:

(a) $f(0) = ?$ (b) $f(?) = 0$

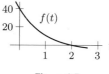

Figure 1.5

17. The sales tax on an item is 6%. Express the total cost, C, in terms of the price of the item, P.

Problems

Evaluate the expressions in Exercises 18–23 given that

$$h(t) = 10 - 3t.$$

18. $h(r)$ **19.** $h(2u)$ **20.** $h(k - 3)$

21. $h(4 - n)$ **22.** $h(5t^2)$ **23.** $h(4 - t^3)$

Evaluate the expressions in Problems 24–25 given that

$$f(n) = \frac{1}{2}n(n+1).$$

24. $f(100)$ **25.** $f(n+1) - f(n)$

The lower the price per song, the more songs are downloaded from an online music store. Let $n = r(p)$ give the average number of daily downloads as a function of the price (in cents). Let p_0 be the price currently being charged (in cents). What do the expressions in Problems 26–33 tell you about downloads from the store?

26. $r(99)$ **27.** $r(p_0)$

28. $r(p_0 - 10)$ **29.** $r(p_0 - 10) - r(p_0)$

30. $365r(p_0)$ **31.** $r(0.80p_0)$

32. $\dfrac{r(p_0)}{24}$ **33.** $p_0 \cdot r(p_0)$

Different strains of a virus survive in the air for different time periods. For a strain that survives t minutes, let $h(t)$ be the number of people infected (in thousands). The most common strain survives for t_0 minutes. What do the expressions in 34–35 tell you about the number of people infected?

34. $h(t_0 + 3)$

35. $\dfrac{h(2t_0)}{h(t_0)}$

A car with tire pressure P lbs/in^2, gets gas mileage $g(P)$ (in mpg). The recommended tire pressure is P_0. What do the expressions in Problems 36–37 tell you about the car's tire pressure and gas mileage?

36. $g(0.9P_0)$

37. $g(P_0) - g(P_0 - 5)$

38. Let $f(t)$ be the number of people, in millions, who own cell phones t years after 1990. Explain the meaning of the following statements.

   (a) $f(10) = 100.3$ **(b)** $f(a) = 20$

   (c) $f(20) = b$ **(d)** $n = f(t)$

39. (a) Using Figure 1.6, fill in Table 1.4.

Table 1.4

x	-2	-1	0	1	2	3
$h(x)$						

  (b) Evaluate $h(3) - h(1)$ **(c)** Evaluate $h(2) - h(0)$

  (d) Evaluate $2h(0)$ **(e)** Evaluate $h(1) + 3$

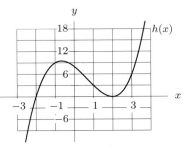

Figure 1.6

1.2 FUNCTIONS AND EXPRESSIONS

A function is often defined by an algebraic expression. We find the output for a given input by evaluating the expression. By paying attention to the form of the expression, we can learn properties of the function.

Example 1   Bernardo plans to travel 400 miles over spring break to visit his family. He can choose to fly, drive, take the train or make the journey as a bicycle road trip. If his average speed is r miles per hour, then the time taken is a function of r and is given by

$$\text{Time taken at } r \text{ miles per hour} = T(r) = \frac{400}{r}.$$

(a) Find $T(200)$ and $T(80)$ and give a practical interpretation of the answers.
(b) Use the algebraic form of the expression for $T(r)$ to explain why $T(200) < T(80)$, and explain why the inequality makes sense in practical terms.

Solution   (a) We have

$$T(200) = \frac{400}{200} = 2.$$

This means that it takes Bernardo 2 hours traveling at 200 miles per hour. We have

$$T(80) = \frac{400}{80} = 5,$$

which means it takes 5 hours traveling at 80 miles per hour.

(b) In the fraction $400/r$, the variable r occurs in the denominator. Since dividing by a larger number gives a smaller number, making r larger makes $T(r)$ smaller. This makes sense in practical terms, because if you travel at a faster speed you finish the journey in a shorter time.

It is also useful to interpret the algebraic form graphically.

Example 2   For the function $T(r)$ in Example 1,

(a) Make a table of values for $r = 25, 80, 100,$ and 200, and graph the function.
(b) Explain the shape of the graph using the form of the expression $400/r$.

Solution   (a) Table 1.5 shows values of the function, and Figure 1.7 shows the plot of these values, along with the graph of f.

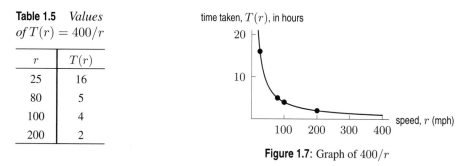

Table 1.5   *Values of $T(r) = 400/r$*

r	$T(r)$
25	16
80	5
100	4
200	2

Figure 1.7: Graph of $400/r$

(b) The graph slopes downward as you move from left to right. This is because the input variable r occurs in the denominator of the expression $400/r$, so when you input larger values of r, the output value is smaller.

What is the Difference Between a Constant and a Variable?

Sometimes an expression for a function contains letters in addition to the independent variable. We call these letters *constants* because for a given function, their value does not change.

Example 3   Einstein's famous equation $E = mc^2$ expresses energy E as a function of mass m. Which letters in this equation represent variables and which represent constants?

Solution   We are given that E is a function of m, so E is the dependent variable and m is the independent variable. The symbol c is a constant (which stands for the speed of light).

Which letters in an expression are constants and which are variables depends on the context in which the expression is being used.

Example 4   A tip of r percent on a bill of B dollars is given in dollars by

$$\text{Tip} = \frac{r}{100}B.$$

Which letters in the expression $(r/100)B$ would you call variables and which would you call constants if you were considering

(a) The tip as a function of the bill amount

(b) The tip as a function of the rate.

Solution

(a) In this situation, we regard the tip as a function of the variable B, and regard r as a constant. If we call this function f, then we can write

$$\text{Tip} = f(B) = \frac{r}{100}B.$$

(b) Here we regard the tip as a function of the variable r, and regard B as a constant. If we call this function g, then

$$\text{Tip} = g(r) = \frac{r}{100}B.$$

Functions and Equivalent Expressions

A number can be expressed in many different ways. For example, $1/4$ and 0.25 are two different ways of expressing the same number. Similarly, we can have more than one expression for the same function.

Example 5   In Example 2 on page 40 we found the expression $0.2B$ for a 20% tip on a bill. Pares says she has an easy way to figure out the tip: she moves the decimal point in the bill one place to the left then doubles the answer.

(a) Check that Pares' method gives the same answer on bill amounts of $8.95 and $23.70 as Example 2 on page 40.

(b) Does Pares' method always work? Explain your answer using algebraic expressions.

Solution

(a) For $8.95, first we move the decimal point to the left to get 0.895, then double to get a tip of $1.79. For $23.70, we move the decimal point to the left to get 2.370, then double to get a tip of $4.74. Both answers are the same as in Example 2 on page 40.

(b) Moving the decimal point to the left is the same as multiplying by 0.1. So first Pares multiplies the bill by 0.1, then multiplies the result by 2. Her calculation of the tip is

$$2(0.1B).$$

We can simplify this expression by regrouping the multiplications:

$$2(0.1B) = (2 \cdot 0.1)B = 0.2B.$$

This last expression for the tip is the same one we found in Example 2 on page 40.

For each value of B the two expressions, $2(0.1B)$ and $0.2B$, give equal values for the tip. Although the expressions appear different, they are equivalent, and therefore define the same function.

Example 6   Let $g(a) = a^2 - a$. Which of the following pairs of expressions are equivalent?

(a) $g(2a)$ and $2g(a)$
(b) $(1/2)g(a)$ and $g(a)/2$
(c) $g(a + 1)$ and $g(a) + 1$

Solution
(a) In order to find the output for g, we square the input value, a and subtract a from the square. This means that $g(2a) = (2a)^2 - (2a) = 4a^2 - 2a$. The value of $2g(a)$ is two times the value of $g(a)$. So $2g(a) = 2(a^2 - a) = 2a^2 - 2a$. The two expressions are not equivalent.
(b) Since multiplying by $1/2$ is the same as dividing by 2, the two expressions are equivalent.
(c) To evaluate $g(a + 1)$ we have $g(a + 1) = (a + 1)^2 - (a + 1) = a^2 + 2a + 1 - a - 1 = a^2 + a$. To evaluate $g(a) + 1$ we add one to $g(a)$, so $g(a) + 1 = a^2 - a + 1$. The two expressions are not equivalent.

Often we need to express a function in a standard form to recognize what type of function it is. For example, in Section 1.4 we study functions of the form $Q = kt$, where k is a constant.

Example 7   Express each of the following functions in the form $Q = kt$, and give the value of k.

(a) $Q = 5t + rt$ (b) $Q = \dfrac{-t}{10}$ (c) $Q = t(t + 1) - t(t - 1)$

Solution
(a) We have
$$Q = 5t + rt = (5 + r)t,$$
which is the form $Q = kt$, with $k = 5 + r$.
(b) Rewriting the fraction as
$$Q = \frac{-t}{10} = \left(\frac{-1}{10}\right)t = -0.1t,$$
we see that $k = -0.1$.
(c) We have
$$Q = t(t + 1) - t(t - 1) = t^2 + t - t^2 + t = 2t,$$
so $k = 2$.

The Expression for the Average Rate of Change

The average rate of change of a function measures how much the output values change relative to the difference between two input values:

The *average rate of change of f between a and b* is given by

$$\text{Average rate of change} = \frac{f(b) - f(a)}{b - a}.$$

Example 8 Let $F(t)$ be the distance in miles that a car has traveled after t hours. Interpret the following statements in terms of the car's journey.

(a) $F(5) - F(3) = 140$

(b) $\dfrac{F(5) - F(3)}{5 - 3} = 70$

(c) $\dfrac{F(a + 5) - F(a)}{5} = 63$

Solution

(a) Since $F(5)$ is the distance the car has traveled after 5 hours and $F(3)$ is the distance it has traveled after 3 hours, $F(5) - F(3) = 140$ means that after 5 hours the car has traveled 140 miles beyond the point it had reached at 3 hours.

(b) Notice that $5 - 3 = 2$ is the two-hour period during which the car traveled the 140 miles. This statement tells us that the average rate of change of F during the 2 hour period is 70 mph. In this case the average rate of change of F is the average speed of the car.

(c) We have

$$\frac{F(a + 5) - F(a)}{5} = \frac{F(a + 5) - F(a)}{(a + 5) - a} = \text{Average rate of change between } t = a \text{ and } t = a + 5.$$

This statement tells us that during the 5 hour period from $t = a$ to $t = a + 5$ the car travels at an average speed of 63 mph.

Example 9 Table 1.6 shows the population $P(t)$ of a colony of termites t months after it was started. What is the average rate of change of the population

(a) during the first 6 months? (b) during the second 6 months?

Table 1.6 *Population of a colony of termites*

t (months)	0	3	6	9	12
$P(t)$	0	1000	3000	7500	1800

Solution

(a) The average rate of change during the first 6 months is

$$\frac{P(6) - P(0)}{6 - 0} = \frac{3000}{6} = 500 \text{ termites per month.}$$

Thus, on average, the colony adds 500 termites per month during the first 6 months. Note that this is only an average. The actual change during any given month could be more or less than 500.

(b) The average rate of change during the second 6 months is

$$\frac{P(12) - P(6)}{12 - 6} = \frac{-1200}{6} = -200 \text{ termites per month.}$$

Notice that the average rate of change is negative, because there is a net decrease in the termite population. In general, if $f(b) < f(a)$ then the numerator in the expression for the average rate of change is negative, so the rate of change is negative (provided $b > a$).

Example 10   A ball thrown in the air has height $h(t) = 90t - 16t^2$ feet after t seconds.

(a) What are the units of measurement for the average rate of change of h? What does your answer tell you about how to interpret the rate of change in this case?

(b) Find the average rate of change of h between

    (i) $t = 0$ and $t = 2$ (ii) $t = 1$ and $t = 2$ (iii) $t = 2$ and $t = 4$

(c) Use a graph of h to interpret your answers in (b) in terms of the motion of the ball.

Solution   (a) In the expression

$$\frac{h(b) - h(a)}{b - a}$$

the numerator is measured in feet and the denominator is measured in seconds, so the ratio is measured in ft/sec. Thus it measures a velocity.

(b) We have

    (i) Average rate of change between $t = 0$ and $t = 2$

$$\frac{h(2) - h(0)}{2 - 0} = \frac{(90 \cdot 2 - 16 \cdot 4) - (0 - 0)}{2 - 0} = \frac{116}{2} = 58 \text{ ft/sec}$$

    (ii) Average rate of change between $t = 1$ and $t = 2$

$$\frac{h(2) - h(1)}{2 - 1} = \frac{(90 \cdot 2 - 16 \cdot 4) - (90 - 16)}{2 - 1} = \frac{42}{1} = 42 \text{ ft/sec}$$

    (iii) Average rate of change between $t = 2$ and $t = 4$

$$\frac{h(4) - h(2)}{4 - 2} = \frac{(90 \cdot 4 - 16 \cdot 16) - (90 \cdot 2 - 16 \cdot 4)}{4 - 2} = \frac{-12}{2} = -6 \text{ ft/sec}$$

(c) Figure 1.8 shows the ball rising to a peak somewhere between 2 and 3 seconds, and then starting to fall. The average velocity is positive during the first two seconds because the height is increasing during that time period. The height is also increasing between $t = 1$ and $t = 2$, but since the ball is rising more slowly the velocity is less. Between $t = 2$ and $t = 4$ the ball rises and falls, experiencing a net loss in height, so its average velocity is negative for that time period.

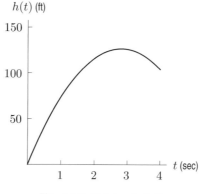

Figure 1.8: Height of a ball

Exercises and Problems for Section 1.2

Exercises

In Exercises 1–4

(a) Evaluate the function at the given input values. Which gives the greater output value?

(b) Explain the answer to part (a) in terms of the algebraic expression for the function.

1. $f(x) = 9 - x$, $x = 1, 3$

2. $g(a) = a - 2$, $a = -5, -2$

3. $C(p) = \dfrac{-p}{5}$, $p = 100, 200$

4. $h(t) = \dfrac{t}{5}$, $t = 4, 6$

5. Let $f(T)$ be the volume in liters of a balloon at temperature $T°$C. If $f(40) = 3$,

(a) What are the units of the 40 and the 3?

(b) What is the volume of the balloon at $40°$C?

6. The price of apartments near a subway station is given by

$$\text{Price} = \frac{\$1000 \cdot A}{10d},$$

where A is the area of the apartment in square feet and d is the distance in miles from the subway station. Which letters would you call constants and which would you call variables if

(a) You want an apartment of 1000 square feet?

(b) You want an apartment 1 mile from the subway station?

(c) You want an apartment that costs $200,000?

In Exercises 7–10, $f(t) = t/2 + 7$. Determine whether the two expressions are equivalent.

7. $\dfrac{f(t)}{3}, \dfrac{1}{3}f(t)$

8. $f(t^2), (f(t))^2$

9. $2f(t), f(2t)$

10. $f(4t^2), f((2t)^2)$

In Exercises 11–16, put the functions in the form $Q = kt$ and state the value of k.

11. $Q = \dfrac{t}{4}$

12. $Q = t(n+1)$

13. $Q = bt + rt$

14. $Q = \dfrac{1}{2}t\sqrt{3}$

15. $Q = \dfrac{\alpha t - \beta t}{\gamma}$

16. $Q = (t-3)(t+3) - (t+9)(t-1)$

17. Sea levels were most recently at a low point about 22,000 years ago. [1] Since then they have risen approximately 130 meters. Find the average rate of change of the sea level during this time period.

18. Global temperatures may increase by up to $10°$F between 1990 and 2100. [2] Find the average rate of change of global temperatures between 1990 and 2100.

Find the average rate of change of $f(x) = x^3 - x^2$ on the intervals indicated in Exercises 19–21.

19. Between 2 and 4.

20. Between -2 and 4.

21. Between -4 and -2.

Problems

In Problems 22–28, both a and x are positive. What is the effect of increasing a on the value of the expression? Does the value increase? Decrease? Remain unchanged?

22. $ax + 1$

23. $x + a$

24. $x - a$

25. $\dfrac{x}{a} + 1$

26. $x + \dfrac{1}{a}$

27. $ax - \dfrac{1}{a}$

28. $a + x - (2 + a)$

29. The children in a family contribute equally for a gift to their mother. Each child contributes $\dfrac{c}{5}$, where c is the cost of the gift.

(a) How many children are there?

(b) If the gift costs $200, how much would each child contribute?

(c) Find the value of c if each child contributes $50.

(d) Write an expression for the amount contributed by each child if there are 3 children.

---

[1] See http://en.wikipedia.org/wiki/Sea_level_rise, http://en.wikipedia.org/wiki/Last_glacial_maximum, and related links. Pages last accessed September 13, 2006.

[2] See http://en.wikipedia.org/wiki/Global_warming. Page last accessed September 13, 2006.

30. The number of people who attend a concert is $160 - p$ when the price of a ticket is $\$p$.

   (a) What is the practical interpretation of the 160?
   (b) Why is it reasonable that the p term has a negative sign?
   (c) The number of people who attend a movie at ticket price $\$p$ is $175 - p$. If tickets are the same price, does the concert or the movie draw the larger audience?
   (d) The number of people who attend a dance performance at ticket price $\$p$ is $160 - 2p$. If tickets are the same price, does the concert or the dance performance draw the larger audience?

For Problems 31–34, put the function in the required form and state the values of all constants.

31. $y = 3x + 8$ in the form $y = 5 + m(x - x_0)$.

32. $y = 3\left(x\sqrt{7}\right)^3$ in the form $y = kx^p$.

33. $y = 3x^3 - 2x^2 + 4x + 5$ in the form

$$y = a + x\left(b + x(c + dx)\right)$$

34. $y = \dfrac{\left(\sqrt{3}\right)^{4t}}{5}$ in the form $y = ab^t$.

Let $g(t)$ give the market value (in $\$1000$s) of a house in year t. Say what the following statements tell you about the house.

35. $g(5) - g(0) = 30$

36. $\dfrac{g(10) - g(4)}{6} = 3$

37. $\dfrac{g(20) - g(12)}{20 - 12} = -1$

Let $f(t)$ be the population of a town that is growing over time. Say what must be true about a in order for the expressions in Problems 38–40 to be positive.

38. $f(a) - f(3)$

39. $\dfrac{f(3) - f(a)}{3 - a}$

40. $f(t + a) - f(t + b)$

1.3 FUNCTIONS AND EQUATIONS

In the last section we saw how to evaluate an expression to find the output of a function, given an input. In this section we see how to solve an equation to find the input (or inputs) that gives a certain output.

Example 1   For $f(x) = x^2$, give the solutions to the equation $f(x) = 16$.

Solution   Replacing $f(x)$ in the equation by the expression for it, we obtain the equation $x^2 = 16$. There are two numbers whose square is 16, $x = 4$, and $x = -4$.

Example 2   In Example 1 on page 46 we saw that the function

$$T(r) = \frac{400}{r}$$

gives the time taken for Bernardo to travel 400 miles at a speed of r miles per hour.

   (a) Write an equation whose solution is the speed Bernardo would have to maintain to make the trip in 10 hours.
   (b) Solve the equation and represent your solution on a graph.

Solution   (a) We want the time taken to be 10 hours, so we want $T(r) = 10$. Since $T(r) = 400/r$, we want to solve the equation

$$10 = \frac{400}{r}.$$

     (b) We multiply both sides of the equation by r to get

$$10r = 400,$$

then divide both sides by 10 to get

$$r = \frac{400}{10} = 40.$$

Thus, the solution to the equation is $r = 40$. Using function notation, we write

$$T(40) = 10.$$

See Figure 1.9. We also know the shape of the graph from Figure 1.7 on page 47.

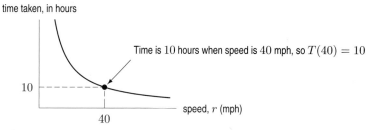

Figure 1.9: Solution to $400/r = 10$ is $r = 40$

In Example 2 we visualized the solution on a graph. Sometimes it is possible to see the solution directly from a table or a graph.

Example 3   A town's population t years after it was incorporated is given by the function $f(t) = 30{,}000 + 2000t$.

     (a) Make a table of values for the population at five-year intervals over a 20-year period starting at $t = 0$, and plot the results.

     (b) Using the table, find the solution to the equation

$$f(t) = 50{,}000$$

and indicate the solution on your plot.

Solution   (a) The initial population in year $t = 0$ is given by

$$f(0) = 30{,}000 + 2000(0) = 30{,}000.$$

In year $t = 5$ the population is given by

$$f(5) = 30{,}000 + 2000(5) = 30{,}000 + 10{,}000 = 40{,}000.$$

In year $t = 10$ the population is given by

$$f(10) = 30{,}000 + 2000(10) = 30{,}000 + 20{,}000 = 50{,}000.$$

Similar calculations for year $t = 15$ and year $t = 20$ give the values in Table 1.7 and Figure 1.10.

Table 1.7   *Population over 20 years*

t, years	Population
0	30,000
5	40,000
10	50,000
15	60,000
20	70,000

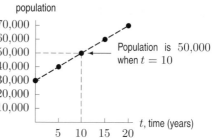

Figure 1.10: Town's population over 20 years

(b) Looking down the right-hand column of the table we see that the population reaches 50,000 when $t = 10$, so the solution to the equation

$$f(t) = 50{,}000$$

is $t = 10$. The practical interpretation of the solution $t = 10$ is that the population reaches 50,000 in 10 years. See Figure 1.10.

Example 4   For the function graphed in Figure 1.11, give

(a) $f(0)$ (b)   The value of x such that $f(x) = 0$.

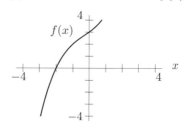

Figure 1.11

Solution   (a) Since the graph crosses the y-axis at the point $(0, 3)$, we have $f(0) = 3$.
(b) Since the graph crosses the x-axis at the point $(-2, 0)$, we have $f(-2) = 0$, so $x = -2$.

The value where the graph crosses the vertical axis is called the *vertical intercept* or *y-intercept*, and the values where it crosses the horizontal axis are called the *horizontal intercepts*, or *x-intercepts*. In Example 4, the vertical intercept is $y = 3$, and the horizontal intercept is $x = -2$.

How do We Find when Two Functions are Equal?

Often, we want to know when two functions are equal to each other. That is, we want to find the input value that produces the same output value for both functions. To do this, we set the two outputs equal to each other and solve for the input value.

Example 5   The populations, in year t, of two towns are given by the functions

$$\text{Town A}: \quad P(t) = 600 + 100(t - 2000)$$
$$\text{Town B}: \quad Q(t) = 200 + 300(t - 2000).$$

(a) Write an equation whose solution is the year in which the two towns have the same population.
(b) Make a table of values of the populations for the years 2000-2004 and find the solution to the equation in part (a).

Solution   (a) We want to find the value of t that makes $P(t) = Q(t)$, so we must solve the equation

$$600 + 100(t - 2000) = 200 + 300(t - 2000).$$

(b) From Table 1.8, we see that the two populations are equal in the year $t = 2002$.

Table 1.8

t	2000	2001	2002	2003	2004
$P(t)$	600	700	800	900	1000
$Q(t)$	200	500	800	1100	1400

Checking the populations in that year, we see

$$P(2002) = 600 + 100(2002 - 2000) = 800$$
$$Q(2002) = 200 + 300(2002 - 2000) = 800.$$

Functions and Inequalities

Sometimes, rather than wanting to know where two functions are equal, we want to know when one is bigger than the other.

Example 6   (a) Write an inequality whose solution is the years for which the population of Town A is greater than the population of Town B in Example 5.
(b) Solve the inequality by graphing the populations.

Solution   (a) We want to find the values of t that make $P(t) > Q(t)$, so we must solve the inequality

$$600 + 100(t - 2000) > 200 + 300(t - 2000).$$

(b) In Figure 1.12, the point where the two graphs intersect is $(2002, 800)$, because the populations are both equal to 800 in the year 2002. To the left of this point, the graph of P (the population of town A) is higher than the graph of Q (the population of town B), so town A has the larger population when $t < 2002$.

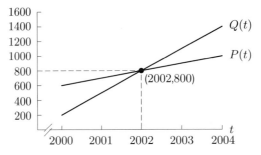

Figure 1.12: When is the population of Town A greater than Town B?

Exercises and Problems for Section 1.3

Exercises

In Exercises 1–4, solve $f(x) = 8$ for x.

1. $f(x) = 5x + 3$

2. $f(x) = x - 7$

3. $f(x) = x^2 - 8$

4. $f(x) = \sqrt{x} + 1$

5. Chicago's average monthly rainfall, $R = f(t)$ inches, is given as a function of month, t, in Table 1.9. (January is $t = 1$.) Solve and interpret:

   (a) $f(t) = 3.7$ **(b)** $f(t) = f(2)$

Table 1.9

t	1	2	3	4	5	6	7	8
R	1.8	1.8	2.7	3.1	3.5	3.7	3.5	3.4

In Exercises 6–7, give

(a) $f(0)$ **(b)** The values of x such that $f(x) = 0$.
Your answers may be approximate.

6.   7.

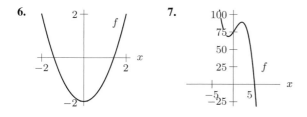

Answer Exercises 8–9 based on the graph of $y = v(x)$ in Figure 1.13.

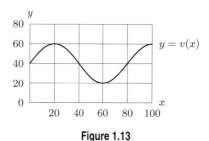

Figure 1.13

8. Solve $v(x) = 60$.   9. Evaluate $v(60)$.

Problems

10. The volume of a cone of height 2 and radius r is $V = \frac{2}{3}\pi r^2$. What is the radius of such a cone whose volume is 3π?

11. Let $V = s^3$ give the volume of a cube of side length s centimeters. For what side length is the cube's volume 27 cm^3?

12. The balance in a checking account set up to pay rent, m months after its establishment, is given by $\$4800 - 400m$.

    (a) Write an equation whose solution is the number of months it takes for the account balance to reach $2000.
    (b) Make a plot of the balance for $m = 1, 3, 5, 7, 9, 11$, and indicate the solution $m = 7$ to the equation in part (a).

13. The number of gallons of gas in a car's tank, d miles after stopping for gas, is given by $15 - d/20$.

    (a) Write an equation whose solution is the number of miles it takes for the amount of gas in the tank to reach 10 gallons.
    (b) Make a plot of the balance for $d = 40, 60, 80, 100, 120, 140$, and indicate the solution $m = 100$ to the equation in part (a).

14. Table 1.10 shows values of x and the expression $3x + 2$. For which values of x in the table is

    (a) $3x + 2 < 8$ **(b)** $3x + 2 > 8$
    (c) $3x + 2 = 8$

Table 1.10

x	0	1	2	3	4
$3x + 2$	2	5	8	11	14

15. The height (in meters) of a diver s seconds after beginning his dive is given by the expression $10 + 2s - 9.8s^2$. For which values of s in Table 1.11 is

    (a) $10 + 2s - 9.8s^2 < 9.89$?
    (b) His height greater than 9.89 meters?
    (c) Height $= 9.89$ meters?

Table 1.11

s	0	0.25	0.5	0.75	1
$10 + 2s - 9.8s^2$	10	9.89	8.55	5.99	2.2

16. Table 1.12 shows values of v and the expressions $12 - 3v$ and $-3 + 2v$. For which values of v in the table is

(a) $12 - 3v < -3 + 2v$?
(b) $12 - 3v > -3 + 2v$?
(c) $12 - 3v = -3 + 2v$?

Table 1.12

v	0	1	2	3	4	5	6
$12 - 3v$	12	9	6	3	0	-3	-6
$-3 + 2v$	-3	-1	1	3	5	7	9

17. If a company sells p software packages, its profit is given by $\$10{,}000 - \dfrac{100{,}000}{p}$, as shown in Figure 1.14.

(a) From the graph, estimate the number of packages sold when profits are $8000.
(b) Check your answer to part (a) by substituting it into the equation
$$10{,}000 - \frac{100{,}000}{p} = 8000.$$

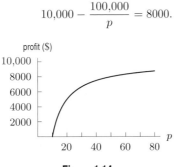

Figure 1.14

18. The tuition for a semester at a small public university t years from now is given by $\$3000 + 100t$, as shown in Figure 1.15.

(a) From the graph, estimate how many years it will take for tuition to reach $3700.
(b) Check your answer to part (a) by substituting it into the equation
$$3000 + 100t = 3700.$$

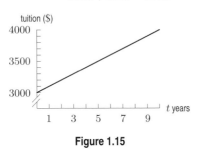

Figure 1.15

19. Figure 1.16 shows the number of pigeons living in a town in various years t.

(a) In what year is the population 500?
(b) The number of pigeons is $300 + 100t$. From the graph, when does $300 + 100t$ equal 600?

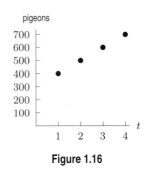

Figure 1.16

20. Antonio and Lucia are both driving through the desert from Tucson to San Diego, which takes each of them 7 hours of driving time. Antonio's car starts out full with 14 gallons of gas and uses 2 gallons per hour. Lucia's SUV starts out full with 30 gallons of gas and uses 6 gallons per hour.

(a) Construct a table showing how much gas is in each of their tanks at the end of each hour into the trip. Assume each stops for gas just as the tank is empty, and then the tank is filled instantaneously.
(b) Use your table to determine when they have the same amount of gas.
(c) If they drive at the same speed while driving and only stop for gas, which of them gets to San Diego first?
(d) Now suppose that between 1 hour and 6.5 hours outside of Tucson, all of the gas stations are closed unexpectedly. Does Antonio arrive in San Diego? Does Lucia?
(e) The amount of gas in Antonio's tank after t hours is $14 - 2t$ gallons, and the amount in Lucia's tank is $30 - 6t$ gallons. When does
(i) $14 - 2t = 30 - 6t$?
(ii) $14 - 2t = 0$?
(iii) $30 - 6t = 0$?

1.4 FUNCTIONS AND MODELING

Often when we want to apply mathematics to real-world situations, we are not given a function, but have to construct one. This process is called *modeling*. In this section we consider situations that can be modeled by a proportional relationship between the variables.

Direct Proportionality

Suppose a state sales tax rate is 6%. Then the tax, T, on a purchase of price P, is given by the function

$$\text{Tax} = 6\% \times \text{Price} \quad \text{or} \quad T = 0.06P.$$

Rewriting this equation as a ratio, we see that T/P is constant, $T/P = 0.06$, so the tax is proportional to the purchase price. In general

---

A quantity Q is **directly proportional** to a quantity t if

$$Q = f(t) = k \cdot t,$$

where k is the *constant of proportionality*. We often omit the word "directly" and simply say Q is proportional to t.

In the tax example, the constant of proportionality is $k = 0.06$, because $T = 0.06P$. Similarly, in Example 2 on page 40, the constant of proportionality is $k = 0.2$, and in Example 3 on page 3, the constant of proportionality is 0.39.

Example 1   The distance d, in miles, that a car travels is proportional to the number of gallons, g, of gas that it uses, with constant of proportionality 18.

(a) Express d as a function of g.
(b) How far can the car travel on 1 gallon? 2 gallons? 10 gallons? 20 gallons?

Solution   (a) Since d is proportional to g, we have $d = kg$ for some constant k. We are told that $k = 18$, so

$$d = 18g.$$

(b) We have

$$
\begin{aligned}
\text{Distance on 1 gallon} &= 18 \cdot 1 \text{ gallon} &&= 18 \text{ miles} \\
\text{Distance on 2 gallons} &= 18 \cdot 2 \text{ gallons} &&= 36 \text{ miles} \\
\text{Distance on 10 gallons} &= 18 \cdot 10 \text{ gallons} &&= 180 \text{ miles} \\
\text{Distance on 20 gallons} &= 18 \cdot 20 \text{ gallons} &&= 360 \text{ miles}.
\end{aligned}
$$

Notice that in Example 1, when the number of gallons doubles from 1 to 2, the number of miles doubles from 18 to 36, and when the number of gallons doubles from 10 to 20, the number of miles doubles from 180 to 360. In general:

The Behavior of Proportional Quantities

If $Q = kt$, then doubling the value of t will double to value of Q, tripling the value of t will triple the value of Q, and so on. Likewise, halving the value of t will halve the value of Q, and so on.

Example 2   Vincent pays five times as much for a car as Dominic. Dominic pays $300 sales tax. How much sales tax does Vincent pay (assuming they pay the same rate)?

Solution   Since Vincent's car costs five times as much as Dominic's car, Vincent's sales tax should be five times as large as Dominic's, or $5 \cdot 300 = \$1500$.

Sometimes we have to rewrite an expression to see that it is a direct proportion.

Example 3   Do the following functions represent a direct proportionality? If so, give the constant of proportionality, k.

(a) $f(x) = 19x$
(b) $g(x) = x/53$
(c) $F(a) = 2a + 5a$
(d) $u(t) = \sqrt{5}t$
(e) $A(n) = n\pi^2$
(f) $P(t) = 2 + 5t$

Solution
(a) Here $f(x)$ is proportional to x with constant of proportionality $k = 19$.
(b) We rewrite this as $g(x) = (1/53)x$, so $g(x)$ is proportional to x with constant of proportionality $k = 1/53$.
(c) Simplifying the right-hand side we get $F(a) = 2a + 5a = 7a$, so $F(a)$ is proportional to a with constant $k = 7$.
(d) Here $u(t)$ is proportional to t with constant $k = \sqrt{5}$.
(e) Rewriting this as $A(n) = \pi^2 n$, we see that it represents a direct proportionality with constant $k = \pi^2$.
(f) Here $P(t)$ is not proportional to t, because of the constant 2 on the right-hand side. However, $P(t) - 2$ is directly proportional to t, because $P(t) - 2 = 5t$.

Solving for the Constant of Proportionality

If we do not know the constant of proportionality, we can find it using one pair of known values for the quantities that are proportional.

Example 4   A graduate assistant at a college earned $80 for 10 hours of work. Express her earnings as a function of hours worked, assuming that her pay rate is constant. What is the constant of proportionality and what is its practical interpretation?

Solution   We have
$$E = f(t) = kt,$$
where E is the amount earned and t is the number of hours worked. We are given
$$80 = f(10) = 10k,$$
and solving for k we have
$$k = \frac{80 \text{ dollars}}{10 \text{ hours}} = \$8/\text{hour}.$$
Thus, $k = \$8/\text{hour}$, which is her hourly wage.

Example 5   A person's heart mass is proportional to his or her body mass.[3]

(a) A person with a body mass of 70 kilograms has a heart mass of 0.42 kilograms. Find the constant of proportionality, k.

(b) Estimate the heart mass of a person with a body mass of 60 kilograms.

Solution   (a) We have
$$H = k \cdot B.$$

for some constant k. We substitute $B = 70$ and $H = 0.42$ and solve for k:

$$0.42 = k \cdot 70$$
$$k = \frac{0.42}{70} = 0.006.$$

So the formula for heart mass as a function of body mass is

$$H = 0.006B,$$

where H and B are measured in kilograms.

(b) With $B = 60$, we have
$$H = 0.006 \cdot 60 = 0.36 \text{ kg}.$$

Rates of Change

In Example 4 on page 60, we saw that the constant of proportionality could be interpreted as a pay rate in dollars per hour. Often, thinking about the units of the constant helps us interpret it.

Example 6   Suppose that the distance you travel, in miles, is proportional to the time spent traveling, in hours:

$$\text{Distance} = k \times \text{Time}.$$

What is the practical interpretation of the constant k?

Solution   To see the practical interpretation of k, we write

$$\text{miles} = \text{Units of } k \times \text{hours}.$$

Therefore,
$$\text{Units of } k = \frac{\text{miles}}{\text{hours}}.$$

The units of k are miles per hour, and it represents speed.

---

[3]K. Schmidt-Nielsen: *Scaling, Why is Animal Size So Important?* (Cambridge: CUP, 1984).

Example 7   The *data rate* of an Internet connection[4] is the rate in bytes per second that data, such as a web page, image, or music file, can be transmitted across the connection. Suppose the data rate is 300 bytes per second. How long does it take to download a file of 42,000 bytes?

Solution   Let T be the time in seconds and N be the number of bytes. Then

$$N = 300T$$
$$42,000 \text{ bytes} = 300 \text{ bytes/sec} \cdot T$$
$$T = \frac{42,000 \text{ bytes}}{300 \text{ bytes/sec}} = 140 \text{ sec},$$

so it takes 140 seconds to download the file.

Example 8   Give the units of the constant 18 in Example 1 on page 59 and a give a practical interpretation of it.

Solution   In the equation $d = 18g$, the units of g are gallons and the units of distance, d, are miles. Thus

$$\text{miles} = \text{Units of } k \cdot \text{gallons}$$
$$\text{Units of } k = \frac{\text{miles}}{\text{gallons}}$$
$$= \text{miles per gallon, or mpg.}$$

The constant tells us that the car gets 18 miles to a gallon of gas.

Families of Functions

Because the functions $Q = f(t) = kt$ all have the same algebraic form, we think of them as a family of functions, one function for each value of the constant k. Figure 1.17 shows graphs of various functions in this family. All the functions in the family share some common features: their graphs are all straight lines, and they all pass through the origin. There are also differences within the family, corresponding to different values of the constant k. The constant k is called a *parameter* for the family. Throughout this book we consider various different families of functions. Modeling with these families is a matter of finding and interpreting the parameters in the algebraic expressions for them.

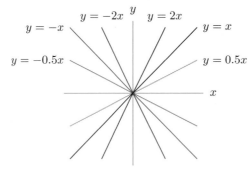

Figure 1.17: The family of functions $y = kx$

---

[4]Sometimes mistakenly called "bandwidth." See the Free Online Dictionary of Computing, http://foldoc.doc.ic.ac.uk/foldoc/contents.html.

The Rate of Change of the Functions $Q = f(t) = kt$

We have seen that the constant k can be interpreted as a rate. In Section 1.1 we considered the average rate of change of a function between two input values.

Example 9   Show algebraically that the average rate of change of the function $f(t) = kt$ between any two different values of t is equal to the constant k.

Solution   We have

$$\frac{f(b) - f(a)}{b - a} = \frac{kb - ka}{b - a} = \frac{k(b - a)}{b - a} = k.$$

Exercises and Problems for Section 1.4

Exercises

For each of the formulas in Exercises 1–9, is y directly proportional to x? If so, give the constant of proportionality.

1. $y = 5x$

2. $y = x \cdot 7$

3. $y = x \cdot x$

4. $y = \sqrt{5} \cdot x$

5. $y = x/9$

6. $y = 9/x$

7. $y = x + 2$

8. $y = 3(x + 2)$

9. $y = 6z$ where $z = 7x$

10. If r is directly proportional to s, and $r = 36$ when $s = 4$, find r when s is 5.

11. If p is directly proportional to q, and $p = 24$ when $q = 6$, find q when p is 32.

12. If y is directly proportional to x, and $y = 16$ when $x = 12$, find y when x is 9.

13. If s is directly proportional to t, and $s = 35$ when $t = 25$, find t when s is 14.

14. Suppose y is directly proportional to x. If $y = 6$ when $x = 4$, find the constant of proportionality and write the formula for y as a function of x. Use your formula to find x when $y = 8$.

Problems

15. Model trains are proportional in size to real trains.

   (a) Write an equation that models this information, using m to represent the size of the model train and r to represent the size of the real train.
   (b) The HO scale is the most popular size model train.[5] An HO train is $1/87^{\text{th}}$ the size of a real train. What is the length, in feet, of a real locomotive if the HO locomotive is 10.5 inches long?
   (c) A Z scale train is $1/220^{\text{th}}$ the size of a real train. What is the length, in inches, of a Z scale locomotive if the real locomotive is 75 feet long?

16. A store is having a 50% off sale. Liza buys a dress that costs twice as much as Diana's dress. If Diana saves $42, how much did Liza's dress originally cost?

17. During the holiday season, a store advertises "Spend $50, save $5. Spend $100, save $10." Assuming that the savings are directly proportional to the amount spent, what is the constant of proportionality? Interpret this in terms of the sale.

18. The number of Mathematics 101 sections is directly proportional to the enrollment of students in the course. During the fall semester, 570 students are placed into 19 sections. What is the constant of proportionality and what is its practical interpretation?

19. In the US, workers under the age of 20 can be paid less than minimum wage for their first 90 consecutive calendar days of employment with an employer. After that, the

---

[5]www.internettrains.com, accessed December 11, 2004.

employee must receive the federal minimum wage.[6]

(a) A worker's salary is directly proportional to the number of hours worked. If a 19-year-old worker is paid $106.25 for working 25 hours at minimum wage, what is the constant of proportionality? What does this tell you about the minimum wage for a worker under the age of 20?

(b) After working 90 consecutive calendar days, the same worker is paid $128.75 for working 25 hours. What is the constant of proportionality? What does this tell you about the federal minimum wage?

(c) Full-time students employed in certain jobs can be paid 85% of the federal minimum wage. What is the constant of proportionality? Round your answer to two decimal places so that there are no fractions of cents.

(d) What is the least a full-time student can be paid to work 25 hours?

20. New York State increased the minimum wage to an amount above the federal minimum wage law, effective January 1, 2005, with additional increases for the next two consecutive years.[7]

(a) If a worker in New York State was paid $180.25 for working 35 hours during the week of December 13, 2004, what is the constant of proportionality? What does this tell you about the minimum wage?

(b) After January 1, 2005, the same worker is paid $210 for working 35 hours. What is the constant of proportionality? What does this tell you about New York's new minimum wage?

21. A 30-second commercial during Super Bowl XL in 2006 cost advertisers $2.5 million. For the first Super Bowl in 1967, an advertiser could have purchased approximately 28.699 minutes of advertising time for the same amount of money.[8]

(a) Assuming that cost is proportional to time, find the cost of advertising, in dollars/second, during the 1967 and 2006 Super Bowls.

(b) How many times more expensive was Super Bowl advertising in 2006 than in 1967?

22. Throughout its long history, the annual number of crimes in a city has been directly proportional to the population. In 1900, the population was 23,000 and there were 415 crimes.

(a) By 1925, the population had doubled. How many crimes were committed in the city that year?

(b) In 1850, the population was only one-fourth as large as it was in 1900. How many crimes were committed in 1850?

(c) By 1970, the population was about 160,000. How many crimes were there in 1970?

(d) How large will the city be when there are 5000 crimes?

23. If a car travels at a constant speed of 30 mph, the distance the car travels is proportional to the time spent traveling. Write a formula for the distance, D, in miles, in terms of the time, t, in hours. What is the constant of proportionality? Show that the units on each side of the proportionality equation agree. If the car travels for 5 hours, what is the distance traveled?

24. The amount of interest a savings account pays in one year is proportional to the starting balance, with constant of proportionality 0.06 if the account pays an annual interest rate of 6%. Write a formula for I, the amount of interest earned, in terms of B, the starting balance. Find the interest earned if the starting balance is

(a) $500   **(b)** $1000   **(c)** $5000.

25. The cost of denim fabric is directly proportional to the amount that you buy. Let C be the cost, in dollars, of x yards of denim fabric.

(a) Write a formula for the cost, C, in terms of x.

(b) One type of denim costs $28.50 for 3 yards. Find the value of the constant of proportionality and rewrite the formula for C using it.

(c) What are the units of the constant of proportionality?

(d) How much will it cost to buy 5.5 yards of this denim?

26. On a map, 1/2 inch represents 5 miles. Use the fact that map distance is proportional to the actual distance between two locations to write a formula for map distance, M, in terms of actual distance d. Give units for the constant of proportionality. How far apart are two towns if the distance between them on the map is 3.25 inches?

27. The blood mass of a mammal is proportional to its body mass. A rhinoceros with body mass 3000 kilograms has blood mass of 150 kilograms. Find a formula for the blood mass of a mammal in terms of the body mass and estimate the blood mass of a human with body mass 70 kilograms.

[6]www.dol.gov/esa/minwage/q-a.htm, accessed December 10, 2004. Note that the minimum wage law does not apply to all enterprises.

[7]www.latimes.com, accessed December 10, 2004.

[8]money.cnn.com/2006/01/03/news/companies/superbowlads, accessed January 15, 2006.

28. The distance a car travels on the highway is proportional to the quantity of gas consumed. A car travels 225 miles on 5 gallons of gas. Find the constant of proportionality, give units for it, and explain its meaning.

29. When a businessman drives to a meeting, he is reimbursed for his mileage expenses at a rate of $0.36 per mile. How much is he reimbursed for driving 82 miles?

30. The number of batteries needed for a set of calculators is directly proportional to the number of calculators. If the constant of proportionality is $k = 4$, how many batteries are needed for 30 calculators?

31. If z is proportional to y and y is proportional to x, is z proportional to x?

32. If z is proportional to x and y is proportional to x, is $z+y$ proportional to x?

33. If z is proportional to x and y is proportional to x, is zy proportional to x?

REVIEW EXERCISES AND PROBLEMS FOR CHAPTER ONE

Exercises

1. When there are c cars on campus, the number of cars without a parking space is $w(c) = -2000 + c$. Express in function notation the number of cars without parking spaces when there are 8000 cars on campus, and evaluate your expression.

2. Let $g(s) = \dfrac{5s + 3}{2s - 1}$. Evaluate

  (a) $g(4)$ **(b)** $g(a)$

  (c) $g(a) + 4$ **(d)** $g(a + 4)$

3. Let $g(t) = \dfrac{t^2 + 1}{5 + t}$. Evaluate

  (a) $g(3)$ **(b)** $g(-1)$ **(c)** $g(a)$

Evaluate the expressions in Exercises 4–9 given that

$$f(x) = \frac{x + 1}{2x + 1}.$$

4. $f(0)$ **5.** $f(-1)$ **6.** $f(0.5)$

7. $f(-0.5)$ **8.** $f\left(\dfrac{1}{3}\right)$ **9.** $f(\pi)$

10. If $f(x) = 2x+1$, (a) Find $f(0)$ (b) Solve $f(x) = 0$.

11. If $f(t) = t^2 - 4$, (a) Find $f(0)$ (b) Solve $f(t) = 0$.

12. If $g(t) = \dfrac{1}{t + 2} - 1$, (a) Find $g(0)$ (b) Solve $g(t) = 0$.

13. (a) You are going to graph $p = f(w)$. Which variable goes on the horizontal axis?
  (b) If $10 = f(-4)$, give the coordinates of a point on the graph of f.
  (c) If 6 is a solution of the equation $f(w) = 1$, give a point on the graph of f.

In Exercises 14–17, label the axes for a sketch to illustrate the given statement.

14. "Over the past century we have seen changes in the population, P (in millions), of the city. . ."

15. "Sketch a graph of the cost of manufacturing q items. . ."

16. "Graph the pressure, p, of a gas as a function of its volume, v, where p is in pounds per square inch and v is in cubic inches."

17. "Graph D in terms of y. . ."

18. Let $f(N) = 3N$ give the number of glasses a cafe should have if it has an average of N clients per hour.

  (a) How many glasses should the cafe have if it expects an average of 50 clients per hour?
  (b) What is the relationship between the number of glasses and N? What is the practical meaning of the 3 in the expression $f(N) = 3N$?

In Exercises 19–22, $g(z) = 4z^2 - 3$. Determine whether the two expressions are equivalent.

19. $2^2 g(z), 4g(z)$ **20.** $4g(z), g(4z)$

21. $g(\sqrt{z}), \sqrt{g(z)}$ **22.** $g(zz), g(z^2)$

In Problems 23–26, find the x value that results in $f(x) = 3$.

23. $f(x) = 5x - 2$

24. $f(x) = 5x - 5$

25. $f(x) = \dfrac{2x}{5}$

26. $f(x) = \dfrac{5}{2x}$

In Exercises 27–32, is y directly proportional to x? If so, give the constant of proportionality.

27. $x = 5y$

28. $y = 2x - x$

29. $2y = -3x$

30. $y - 2 = 3x$

31. $y/3 = \sqrt{5}x$

32. $y = ax - 2x$

Problems

Let $n = f(p)$ be the function giving the average number of days a house in a particular community stays on the market before being sold for price p (in $1000s), and let p_0 be the average sale price of houses in the community. What do the expressions in Problems 33–35 mean in terms of the housing market?

33. $f(p_0)$

34. $f(p_0) + 10$

35. $f(p_0) - f(0.9p_0)$

The investment portfolio in Problems 36–39 includes stocks and bonds. Let $v(t)$ be the dollar value after t years of the portion held in stocks, and let $w(t)$ be the value held in bonds.

36. Explain what the following expression tells you about the investment:
$$\frac{w(t)}{v(t) + w(t)}.$$

37. The equation $w(t) = 2v(t - 1)$ has a solution at $t = 5$. What does this solution tell you about the investment?

38. Write an expression that gives the difference in value of the stock portion of the investment in year t and the bond portion of the investment the preceding year.

39. Write an equation whose solutions are the years in which the value of the bond portion of the investment exceed the value of the stock portion by exactly $3000.

40. The population of a town, t years after it was founded, is given by $5000 + 350t$.

    (a) Write an equation whose solution is the number of years it takes for the population to reach $12{,}000$.

    (b) Make a plot of the population for $t = 14, 16, 18, 20, 22, 24$ and indicate the solution $t = 20$ to the equation in part (a).

41. The number of stamps in a person's passport, t years after the person gets a new job which involves overseas travel, is given by $8 + 4t$.

    (a) Write an equation whose solution is the number of years it takes for the passport to have 24 stamps.

    (b) Make a plot of the number of stamps for $t = 1, 2, 3, 4, 5, 6$, and indicate the solution $t = 4$ to the equation in part (a).

42. **(a)** Plot the values of $7 + 5q$ for $q = 0, 1, 2, 3, 4$.

    (b) Use your graph to determine at which q-value $7 + 5q$ is equal to 17.

43. **(a)** Plot the values of $2b^2 - b^3$ for $b = 0, 1, 2, 3, 4$.

    (b) Use your graph to determine at which b-values $2b^2 - b^3 = 0$.

44. The number of dirty socks on your roommate's floor, t days after the start of exams, is given by $10 + 2t$.

    (a) Write an equation whose solution is the number of days it takes for the number of socks to reach 26.

    (b) Make a plot of the number of socks for $t = 2, 4, 6, 8, 10, 12$, and indicate the solution $t = 8$ to the equation in part (a).

45. For accounting purposes, the value of a machine, t years after it is purchased, is given by $\$100{,}000 - 10{,}000t$.

    (a) Write an equation whose solution is the number of years it takes for the machine's value to reach $\$70{,}000$.

    (b) Make a plot of the value of the machine at $t = 1, 2, 3, 4, 5, 6$, and indicate, on the graph, the solution $t = 3$ to the equation in part (a).

46. The amount of glass used in a door is $4500 + w^2$ cm^2, where w cm is the width of a window in the door. Figure 1.18 shows the graph of this expression.

    (a) From the graph, estimate the width of the window if the total glass used is 7000 cm^2.

    (b) Check your answer to part (a) by substituting it into the equation $4500 + w^2 = 7000$.

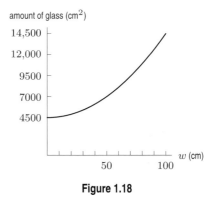

Figure 1.18

47. The population, in millions, of bacteria in some rotting vegetables after t days is given by $2000(2)^t$, as shown in Figure 1.19.

  (a) From the graph, estimate how many days it takes for the population to reach 32 billion (32,000 million).
  (b) Check your answer to part (a) by substituting it into the equation

$$2000(2)^t = 32,000.$$

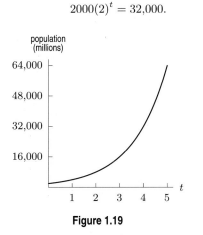

Figure 1.19

48. Table 1.13 shows values of z and the expression $4 - 2z$. For which values of z in the table is

  (a) $4 - 2z < 2$ **(b)** $4 - 2z > 2$
  (c) $4 - 2z = 2$

Table 1.13

z	0	1	2	3	4
$4 - 2z$	4	2	0	-2	-4

49. Table 1.14 shows values of a and the expressions $2 + a^2$ and $10 - 2a$. For which values of a in the table is

  (a) $2 + a^2 < 10 - 2a$?
  (b) $2 + a^2 > 10 - 2a$?
  (c) $2 + a^2 = 10 - 2a$?

Table 1.14

a	0	1	2	3	4	5	6
$2 + a^2$	2	3	6	11	18	27	38
$10 - 2a$	10	8	6	4	2	0	-2

50. Tuition cost T (in dollars) for part-time students at Stonewall College is given by $T = 300 + 200C$, where C represents the number of credits taken.

  (a) Make a table showing costs for taking from one to twelve credits. For each value of C, give both the tuition cost, T, and the cost per credit, T/C. Round to the nearest dollar.
  (b) Which of these values of C has the smallest cost per credit?

51. Observations show that the heart mass H of a mammal is 0.6% of the body mass M, and that the blood mass B is 5% of the body mass.[9]

  (a) Construct a formula for M in terms of H.
  (b) Construct a formula for M in terms of B.
  (c) Construct a formula for B in terms of H. Is this consistent with the statement that the mass of blood in a mammal is about 8 times the mass of the heart?

52. When you convert British pounds (£) into US dollars ($), the number of dollars you receive is proportional to the number of pounds you exchange. A traveler receives $400 in exchange for £250. Find the constant of proportionality, give units for it, and explain its meaning.

53. Three ounces of broiled ground beef contains 245 calories.[10] The number of calories, C, is proportional to the number of ounces of ground beef, b. Write a formula for C in terms of b. How many calories are there in 4 ounces of ground beef?

Table 1.15 gives values of $D = f(t)$, the total US debt (in $ billions) t years after 2000.[11] Answer Problems 54–57 based on this information.

Table 1.15

t	D ($ billions)
0	5674.2
1	5807.5
2	6228.2
3	6783.2
4	7379.1
5	7932.7
6	8494.1

---

[9]K. Schmidt-Nielsen, *Scaling, Why is Animal Size so Important?* (Cambridge: CUP, 1984).
[10]The World Almanac Book of Facts, 1999, p. 718.
[11]See http://www.publicdebt.treas.gov/opd/opdpdodt.htm. Page last accessed September 21, 2006; figure for 2006 is based on the September 20 value. Other values are at the end of fiscal year (on or about September 30).

54. Evaluate the following expression, and say what it tells you about the US debt.

$$\frac{f(5) - f(1)}{5 - 1}.$$

55. Which expression has the larger value,

$$\frac{f(5) - f(3)}{5 - 3} \quad \text{or} \quad \frac{f(3) - f(0)}{3 - 0}?$$

Say what this tells you about the US debt.

56. Show that

$$\begin{array}{ccc} \text{Average rate of change} & < & \text{Average rate of change} \\ \text{from 2005 to 2006} & & \text{from 2004 to 2005} \end{array}.$$

Does this mean the US debt is starting to go down? Discuss.

57. Project the value of $f(10)$ by assuming

$$\frac{f(10) - f(6)}{10 - 6} = \frac{f(6) - f(0)}{6 - 0}.$$

Explain the assumption that goes into making your projection, and what your answer tells you about the US debt.

Answer Problems 58–60 given that a family uses $q(t)$ gallons of gas in week t at an average price per gallon of $p(t)$ dollars.

58. Explain what this expression tells you about the family's use of gasoline:

$$p(t)q(t) - p(t - 1)q(t - 1).$$

59. Suppose $p(t_2) > p(t_1)$ and $p(t_2)q(t_2) < p(t_1)q(t_1)$. What does this tell you about the family's gasoline usage?

60. Write an expression for the family's average weekly expenditure on gasoline during the four-week period $2 \leq t \leq 5$.

Chapter Two

Linear Functions, Expressions, and Equations

Contents

2.1   LINEAR FUNCTIONS

Linear functions describe quantities that have a constant rate of increase (or decrease).

Example 1   A video store's policy for overnight rentals is

Overnight rentals cost $2.50 plus a late fee of $2.99 per day.

Express C, the total cost, as a function of t, the number of days late.

Solution   We have

$$f(0) = \text{Rental plus 0 days late fee} = 2.50 \qquad\qquad = 2.50 + 0(2.99) = 2.50$$
$$f(1) = \text{Rental plus 1 day late fee } = 2.50 + 2.99 \qquad = 2.50 + 1(2.99) = 5.49$$
$$f(2) = \text{Rental plus 2 days late fee} = 2.50 + \underbrace{2.99 + 2.99}_{2} \qquad = 2.50 + 2(2.99) = 8.48$$
$$f(3) = \text{Rental plus 3 days late fee} = 2.50 + \underbrace{2.99 + 2.99 + 2.99}_{3} = 2.50 + 3(2.99) = 11.47$$

$$\vdots \qquad\qquad\qquad \vdots$$

$$f(t) = \text{Rental plus } t \text{ days late fee} = 2.50 + \underbrace{2.99 + \cdots + 2.99}_{t} \quad = 2.50 + 2.99t.$$

So

$$C = f(t) = 2.50 + 2.99t.$$

See Table 2.1 and Figure 2.1. Notice that the values of C go up by 2.99 each time t increases by 1. This has the effect that the points in Figure 2.1 lie in a straight line.

Table 2.1

t	C
0	2.50
1	5.49
2	8.48
3	11.47
4	14.46

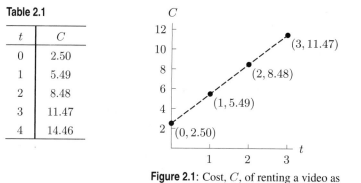

Figure 2.1: Cost, C, of renting a video as a function of t, the number of days late

In Section 1.4, we discussed the family of functions having the algebraic form $Q = kt$. The function $f(t) = 2.50 + 2.99t$ cannot be written in this form, so it is not a member of this family. Instead, it is a member of the family of *linear functions*: functions that can be written in the form

$$f(t) = \text{A constant} + \text{another constant} \times t.$$

This function family has two parameters, not one. For linear functions, it is customary to label these parameters b and m. For instance, the function $f(t) = 2.50 + 2.99t$ has $b = 2.50$ and $m = 2.99$. In general:

A **linear function** of t is a function that can be written

$$f(t) = b + mt, \quad \text{for constants } b \text{ and } m.$$

What is the Meaning of b and m?

Example 2   The population of a town t years after it is incorporated is given by the linear function

$$P(t) = 30{,}000 + 2000t.$$

(a) What is the town's population when it is incorporated?
(b) What is the population of the town one year after it is incorporated? By how much does the population increase every year?
(c) Sketch a graph of the population.

Solution   (a) The town is incorporated in year $t = 0$,

$$\text{Initial population} = P(0) = 30{,}000 + 2000(0) = 30{,}000.$$

Thus the 30,000 in the formula for $P(t)$ represents the starting population of the town.
(b) After one year $t = 1$, and

$$P(1) = 30{,}000 + 2000(1) = 32{,}000,$$

an increase of 2000 over the starting population. After two years we have $t = 2$, so

$$P(2) = 30{,}000 + 2000(2) = 34{,}000,$$

an increase of 2000 over the year one population. In fact, the 2000 in the formula for $P(t)$ represents the amount by which the population increases every year.
(c) The population is 30,000 when $t = 0$, so the graph passes through the point $(0, 30{,}000)$. It slopes upward since the population is increasing at a rate of 2000 people per year. See Figure 2.2.

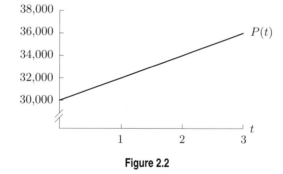

Figure 2.2

The next example gives a decreasing linear function—one whose output value goes down when the input value goes up.

Example 3   The value of a car in dollars t years after it is purchased is given by the linear function $V(t) = 18{,}000 - 1700t$.

(a) What is the value of the car when it is new?
(b) What is the value of the car after one year? How much does the value decrease each year?
(c) Sketch a graph of the value.

Solution   (a) As in Example 2, the 18,000 represents the starting value of the car, when it is new:

$$\text{Initial value} = V(0) = 18{,}000 - 1700(0) = \$18{,}000.$$

(b) The value after one year is

$$V(1) = 18{,}000 - 1700(1) = \$16{,}300,$$

which is $1700 less than the value when new. The car decreases in value by $1700 every year until it is worthless. Another way of saying this is that the rate of change in the value of the car is −$1700 per year.

(c) Since the value of the car is 18,000 dollars when $t = 0$, the graph passes through the point $(0, 18{,}000)$. It slopes downward, since the value is decreasing at the rate of 1700 dollars per year. See Figure 2.3.

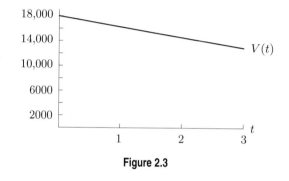

Figure 2.3

Example 2 involves a population given by $P(t) = 30{,}000 + 2000t$, so

$$\text{Current population} = \underbrace{\text{Initial population}}_{30{,}000 \text{ people}} + \underbrace{\text{Growth rate}}_{2000 \text{ people per year}} \times \underbrace{\text{Number of years}}_{t}.$$

In Example 3, the value is given by $V(t) = 18{,}000 + (-1700)t$, so

$$\text{Total cost} = \underbrace{\text{Initial value}}_{\$18{,}000} + \underbrace{\text{Change per year}}_{-\$1700 \text{ per year}} \times \underbrace{\text{Number of years}}_{t}.$$

In each case the coefficient of t is a rate of change, and the resulting function is linear, $Q = b + mt$. Notice the pattern:

$$\underbrace{\text{Output}}_{y} = \underbrace{\text{Initial value}}_{b} + \underbrace{\text{Rate of change}}_{m} \times \underbrace{\text{Input}}_{t}.$$

Graphical and Numerical Interpretation

We can also interpret the constants b and m in terms of the graph of $y = b + mx$.

Example 4   (a) Make a table of values for the function $y = 5 + 2x$, and sketch its graph.
(b) Interpret the constants 5 and 2 in terms of the table and the graph.

Solution   (a) See Table 2.2 and Figure 2.4.

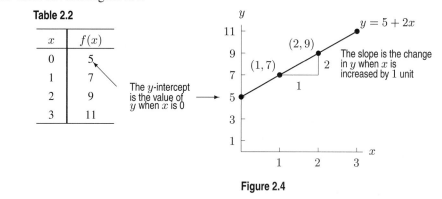

Table 2.2

x	$f(x)$
0	5
1	7
2	9
3	11

The y-intercept is the value of y when x is 0

The slope is the change in y when x is increased by 1 unit

Figure 2.4

(b) The constant 5 gives the y-value where the graph crosses the y-axis. This is called *y-intercept* or *vertical intercept* of the graph. From the table and the graph we see that 2 represents the amount by which y increases when x is increased by 1 unit. This is called the *slope*.

In Example 4, m is positive, so the y-value increases when x is increased by 1 unit, and consequently the graph rises from left to right. If the value of m is negative, the y-value decreases when x is increased by 1 unit, resulting in a graph that falls from left to right.

Example 5   For the function $g(x) = 16 - 3x$,

(a) What does the coefficient of x tell you about the graph?
(b) What is the y-intercept of the graph?

Solution   (a) Since the coefficient of x is negative, the graph falls from left to right. The value of the coefficient, -3, tells us that each time the value of x is increased by 1 unit, the value of y goes down by 3 units. The slope of the graph is -3.
(b) Although we cannot tell from Figure 2.5 where the graph crosses the y-axis, we know that $x = 0$ is on the y-axis, so the y-intercept is $y = 16 - 3 \cdot 0 = 16$.

Table 2.3

x	y
3	7
4	4
5	1

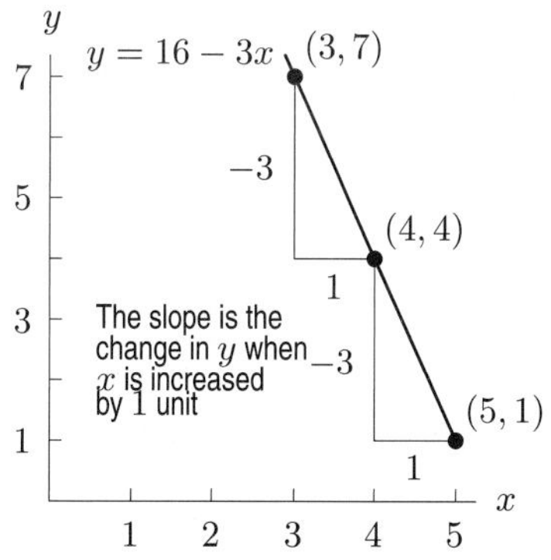

$y = 16 - 3x$

The slope is the change in y when x is increased by 1 unit

Figure 2.5

In general, we have

> For the linear function of x
> $$f(x) = b + mx,$$
> - $b = f(0)$ is the initial value, and gives the vertical intercept of the graph.
> - m is the rate of change and gives the slope of the graph.
>
> If the slope is positive, then the graph rises from left to right. If the slope is negative, then the graph falls from left to right.

Example 6   Find the vertical intercept and slope, and use this information to graph the functions:

(a) $f(x) = 100 + 25x$ (b) $g(x) = 6 - 0.5x$

Solution   (a) The vertical intercept is 100, and $f(x)$ increases by 25 each time x increases by 1. See Figure 2.6.

(b) The vertical intercept is 6, and $g(x)$ decreases by 0.5 each time x goes up by 1. See Figure 2.7.

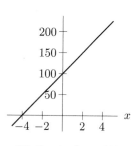

Figure 2.6: Graph of $y = 100 + 25x$

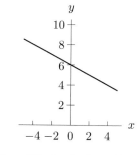

Figure 2.7: Graph of $y = 6 - 0.5x$

Units

It is often useful to consider units of measurement when interpreting a linear function $Q = f(t) = b + mt$. Since b is the initial value of Q, the units of b are the same as the units of Q. Since m is the rate of change of Q with respect to t, the units of m are the units of Q divided by the units of t.

Example 7   The height in inches, h, of a plant d days after it was planted is given by

$$h = 10 + 2d.$$

What do the constants 10 and 2 mean in terms of the height of the plant?

Solution   The constant 10, is the height in inches at $d = 0$, the day the plant was planted. The constant 2 has the units of h/d (inches/day) and is the rate of change of the plant's height, 2 inches/day. It tells us that the plant is growing at a rate of 2 inches every day.

Exercises and Problems for Section 2.1

Exercises

1. The cost of producing x blank CDs is $0.05x + 50,000$ dollars.

   (a) Find the cost of producing
      (i)   100 CDs   (ii) 100,000 CDs
      (iii) No CDs
   (b) Give the practical meaning of your answer to (iii).

2. A homing pigeon starts 1000 miles from home and flies 50 miles toward home each day. Express distance from home in miles, D, as a function of the number of days, d.

3. You buy a saguaro cactus 5 ft high and it grows at a rate of 0.2 inches each year. Express its height in inches, h, as a function of time t in years since the purchase.

4. You have 150 books and you join a club that sends you 3 books a month. Write an expression for the number, N, of books in your collection as a function of the number of years, t, after you join the club.

5. The temperature of the soil is $30°$ C at the surface and decreases by $0.04°$ C each centimeter. Express temperature T as a function of depth d, in centimeters, below the surface.

Give the values for b and m for the linear functions in Exercises 6–11.

6. $f(x) = 3x + 12$
7. $g(t) = 250t - 5300$
8. $h(n) = 0.01n + 100$
9. $u(k) = 0.007 - 0.003k$
10. $v(z) = 30$
11. $w(c) = 0.5c$

12. The cost, C, of hiring a repairman for h hours is given by $C = 50 + 25h$.

   (a) What does the repairman charge to walk in the door?
   (b) What is his hourly rate?

13. The cost, C, of renting a limousine for h hours above the 4 hour minimum is given by $C = 300 + 100h$.

   (a) What does the 300 represent?
   (b) What is the hourly rate?

14. The population of a town, t years after it is founded, is given by $P(t) = 5000 + 350t$.

   (a) What is the population when it is founded?
   (b) What is the population of the town one year after it is founded? How much does it increase by during the first year? During the second year?

15. An orbiting spaceship releases a probe that travels directly away from Earth. The probe's distance s (in km) from Earth after t seconds is given by $s = 600 + 5t$. Identify the initial value and rate of change of this function. What do they tell you about the probe?

16. After a rain storm, the water in a trough begins to evaporate. The amount in gallons remaining after t days is given by $V = 50 - 1.2t$. Identify the initial value and rate of change of this function. What do they tell you about the water in the trough?

In Exercises 17–25, identify the initial value and the rate of change, and explain their meanings in practical terms.

17. The monthly charge of a cell phone is $25+0.06n$ dollars, where n is the number of minutes used.

18. The number of people enrolled in Mathematics 101 is $200 - 5y$, where y is the number of years since 2004.

19. On a spring day the temperature in degrees Fahrenheit is $50 + 1.2h$, where h is the number of hours since noon.

20. The value of an antique is $2500 + 80n$ dollars, where n is the number of years since the antique is purchased.

21. A professor calculates a homework grade of $100 - 3n$ for n missing homework assignments.

22. The cost, C, in dollars of a high school dance attended by n students is given by $C = 500 + 20n$.

23. The total amount, C, in dollars, spent by a company on a piece of heavy machinery after t years in service is given by $C = 20,000 + 1500t$ dollars.

24. The population, P, of a city is predicted to be $P = 9000 + 500t$ in t years from now.

25. The distance, d, in meters from the shore, of a surfer riding a wave is given by $d = 120 - 5t$, where t is the number of seconds since she caught the wave.

In Exercises 26–29, sketch a graph for each of the following equations by making a table of values and plotting the points.

26. $y = -3 + 2x$.
27. $y = 5 - x$.
28. $y = -1 - 3x$.
29. $y = -2 + 5x$.

In Exercises 30–35, identify the slope and y-intercept and graph the function.

30. $y = 2x + 3$
31. $y = 4 - x$
32. $y = -2 + 0.5x$
33. $y = 3x - 2$
34. $y = -2x + 5$
35. $y = -0.5x - 0.2$

Problems

36. Long Island Power Authority charges its residential customers a monthly service charge plus an energy charge based on the amount of electricity used.[1] The monthly cost of electricity is approximated by the function: $C = f(h) = 36.60 + 0.14h$, where h represents the number of kilowatt hours (kWh) of electricity used in excess of 250 kWh.

   (a) What does the coefficient 0.14 mean in terms of the cost of electricity?

   (b) Find $f(50)$ and interpret its meaning.

37. The following functions describe four different collections of baseball cards. The collections begin with different numbers of cards and cards are bought and sold at different rates. The number, B, of cards in each collection is a function of the number of years, t, that the collection has been held. Describe each of these collections in words.

   (a) $B = 200 + 100t$ **(b)** $B = 100 + 200t$

   (c) $B = 2000 - 100t$ **(d)** $B = 100 - 200t$

38. If the tickets for a concert cost $\$p$ each, the number of people who will attend is $2500 - 80p$. Which of the following best describes the meaning of the 80 in this expression?

   (i) The price of an individual ticket.

   (ii) The slope of the graph of attendance against ticket price.

   (iii) The price at which no one will go to the concert.

   (iv) The number of people who will decide not to go if the price is raised by one dollar.

39. Match the equations in (a)–(e) with the graphs in Figure 2.8. The constants s and k are the same in each equation.

   (a) $y = s$ **(b)** $y = kx$ **(c)** $y = kx - s$

   (d) $y = 2s - kx$ **(e)** $y = 2s - 2kx$

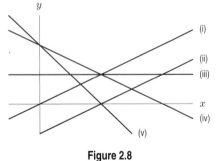

Figure 2.8

40. The velocity of an object tossed up in the air is modeled by the function $v(t) = 48 - 32t$, where t is measured in seconds, and $v(t)$ is measured in feet per second.

   (a) Create a table of values for the function.

   (b) Graph the function.

   (c) Explain what the constants 48 and -32 tell you about the velocity.

   (d) What does a positive velocity indicate? a negative velocity?

41. If a is a constant, does the equation $y = ax + 5a$ define y as a linear function of x? If so, identify the slope and vertical intercept.

42. Two linear functions have the same slope, but different x-intercepts. Can they have the same y-intercept?

43. Two linear functions have the same x-intercept, but different slopes. Can they have the same y-intercept?

Give the slope and y-intercept for the functions in Problems 44–49.

44. $y = 220 - 12x$ **45.** $y = \dfrac{1}{3}x - 11$

46. $y = \dfrac{x}{7} - 12$ **47.** $y = \dfrac{20 - 2x}{3}$

48. $y = 15 - 2(3 - 2x)$ **49.** $y = \pi x$

How do the graphs of the equations in Problems 50–57 compare to the graph of $y = b + mx$ where b, m are both positive.

50. $y = b + 1 + mx$ **51.** $y = b + (1 + m)x$

52. $y = b + m(x + 1)$ **53.** $y = b + mx + x + 1$

54. $y = b + 2mx$ **55.** $y = 0.5b + mx$

56. $y = mx - b$ **57.** $y = mx$

In Problems 58–62, use units to interpret the practical meaning of the expression, where the variables refer to a truck driven at a constant speed of 70 mph for d miles, and[2]

$$m = \text{mileage in miles per gallon}$$
$$f = \text{fuel cost per gallon in dollars}$$
$$w = \text{driver's wages per hour in dollars}$$
$$c = \frac{\text{operating cost per mile in dollars,}}{\text{excluding fuel cost and driver's wages.}}$$

58. dc **59.** $d/70$ **60.** $dw/70$ **61.** df/m

62. $df/m + dw/70 + dc$

[1] www.lipower.org, accessed December 10, 2004.

[2] Adapted from Cliff Sloyer, *Fantastiks of Mathematiks* (Rhode Island: Janson, 1986).

63. If n birds eating continuously consume V in^3 of seed in T hours, how much does one bird consume per hour?

64. If n birds eating continuously consume W ounces of seed in T hours, what are the units of $W/(nT)$? What does $W/(nT)$ represent in practical terms?

2.2   EXPRESSING LINEAR FUNCTIONS IN DIFFERENT FORMS

So far we have considered linear functions that are given in the form $f(t) = b + mt$. However, sometimes functions are given a different form which is equivalent to $b + mt$. Since equivalent expressions define the same function, we say that any expression equivalent to $b + mt$ is *a linear expression in* t. When we are talking about the expression, rather than the function it defines, we call b the *constant term* and m the *coefficient*. In this section we consider different forms of linear expressions.

How do we Recognize a Linear Expression?

Sometimes a simple rearrangement of the terms is enough to recognize an expression as linear.

Example 1   Identify the constant term and the coefficient in the expression for the following linear functions.

(a) $u(t) = 20 + 4t$ (b) $v(t) = 8 - 0.3t$ (c) $w(t) = t/7 + 5$

Solution   (a) We have constant term 20 and coefficient 4.

(b) Writing $v(t) = 8 + (-0.3)t$, we see that the constant term is 8 and the coefficient is -0.3.

(c) Writing $w(t) = 5 + (1/7)t$, we see that the constant term is 5 and the coefficient is 1/7.

Example 2   Is the expression linear in x? If it is, give the constant term and the coefficient.

(a) $5 + 0.2x^2$ (b) $3 + 5\sqrt{x}$ (c) $3 + x\sqrt{5}$

(d) $2(x + 1) + 3x - 5$ (e) $(1 + x)/2$ (f) $ax + x + b$

Solution   (a) This expression is not linear because of the x^2 term.

(b) This expression is not linear because of the $\sqrt{x}$ term.

(c) This expression is linear. Writing this as $y = 3 + \left(\sqrt{5}\right)x$, we see that the constant term is 3 and the coefficient is $\sqrt{5}$.

(d) We distribute the 2 and combine like terms:

$$2(x + 1) + 3x - 5 = 2x + 2 + 3x - 5 = -3 + 5x,$$

so we see that the expression is linear. The constant term is -3, and the coefficient is 5.

(e) Distributing the 1/2 we get

$$\frac{1 + x}{2} = \frac{1}{2} + \frac{1}{2}x,$$

so we see that the expression is linear. The constant term is 1/2, and the coefficient is 1/2.

(f) We collect the x terms to get $ax + x + b = b + (a + 1)x$, so we see that the expression is linear. The constant term is b, and the coefficient is $a + 1$.

Expressions Involving More Than One Letter

In Example 2(f) there was a constant a in the expression. An expression involving more than one letter can be viewed in more than one way, depending on which letter we choose as the variable.

Example 3   (a) Is the expression $xy^2 + 5xy + 2y - 8$ linear in x? In y?

(b) Is the expression $\pi r^2 h$ linear in r? In h?

Solution   (a) To see if the expression is linear in x, we try to match it with the form $b + mx$. We have

$$xy^2 + 5xy + 2y - 8 = (2y - 8) + (y^2 + 5y)x$$

which is linear in x with $b = 2y - 8$ and $m = y^2 + 5y$. Thus the expression is linear in x. It is not linear in y because of the y^2 term.

(b) The expression is not linear in r because of the r^2 term. It is linear in h, with constant term 0 and coefficient πr^2.

Modeling with Linear Functions

Sometimes a linear function can be used to model a real-world phenomenon.

Example 4   Is each function linear? If so, give the coefficient and the constant term.

(a) The share of a community garden plot with area A square feet divided between 5 families is $f(A) = A/5$ square feet per family.
(b) The gasoline remaining in an electric generator running for h hours is $G = 0.75 - 0.3h$.
(c) The circumference of a circle of radius r is $C = 2\pi r$.
(d) The time it takes to drive 300 miles at v mph is $T = 300/v$ hours.

Solution   (a) This function is linear, since we can rewrite it in the form $f(A) = 0 + (1/5)A$. The constant term is 0 and the coefficient is $1/5$.

(b) Writing $G = 0.75 + (-0.3)h$, we see this is linear with constant term 0.75 and coefficient -0.3.
(c) This function is linear, with constant term 0 and coefficient 2π.
(d) This function is not linear, since it involves dividing by v, not multiplying v by a constant.

When is it Useful to Express a Function in a Different Form?

Slope-Intercept Form

When we express a linear function in the form

$$f(x) = b + mx$$

we say it is in *slope-intercept form*, because the coefficient m gives the slope of the graph (or rate of change of the function), and the constant term b give its vertical intercept (or initial value). Putting a function in slope-intercept form allows us to see these numbers clearly.

Example 5   The cost C of a holiday lasting d days consists of the air fare, \$350, plus accommodation expenses of \$55 times the number of days, plus food expenses of \$40 times the number of days.

(a) Give an expression for C as a function of d that shows air fare, accommodation, and food expenses separately.
(b) Express the function in slope-intercept form. What is the significance of the constant term and the coefficient?

Solution   (a) The cost is obtained by adding together the air fare of \$350, the accommodation, and the food. The accommodation for d days costs $55d$ and the food costs $40d$. Thus

$$C = 350 + 55d + 40d.$$

(b) Collecting like terms, we get

$$C = 350 + 95d$$

dollars, which is linear in d with constant term 350 and coefficient 95. The constant term represents the initial cost (the air fare) and the coefficient represents the total daily cost of the holiday, \$95 per day.

Point-Slope Form

Example 6   The population of a town t years after it is founded is given by

$$P(t) = 16{,}000 + 400(t - 5).$$

(a) What is the practical interpretation of the constants 5 and 16,000 in the expression for P?
(b) Express $P(t)$ in slope-intercept form and interpret the slope and intercept.

Solution   (a) The difference between the form given and slope-intercept form is that the coefficient 400 multiplies the expression $t - 5$, rather than just t. Whereas the slope-intercept form tells us the value of the function when $t = 0$, this form tells us the value when $t = 5$:

$$P(5) = 16{,}000 + 400(5 - 5) = 16{,}000 + 0 = 16{,}000.$$

Thus, the population is 16,000 after 5 years.
(b) We have

$$
\begin{aligned}
P(t) &= 16{,}000 + 400(t - 5) &&\text{Population in year 5 is 16,000.}\\
&= 16{,}000 + 400t - 400 \cdot 5 &&\text{Five years ago, it was 2000 fewer.}\\
&= 14{,}000 + 400t &&\text{Thus, the starting value is 14,000.}
\end{aligned}
$$

From the slope-intercept form we see that 400 represents the growth rate of the population per year. When we subtract $400 \cdot 5$ from 16,000 in the above calculation, we are deducting five years of growth in order to obtain the initial population, 14,000.

In Example 6 we see that the form

$$P(t) = 16{,}000 + 400(t - 5)$$

tells us that $P(5) = 16{,}000$. Another way of saying this is that the graph of P passes through the point $(5, 16{,}000)$. In general, a linear function

$$f(x) = y_0 + m(x - x_0)$$

is said to be in *point-slope form*, because
- $f(x_0) = y_0$, so the graph passes through the point (x_0, y_0)
- m is the same as in slope-intercept form, giving the rate of change and the slope of the graph.

Example 7   Graph the function
$$f(x) = 5 + 3(x - 2)$$
by (a)   Putting it in slope-intercept form   (b)   Finding $f(2)$.

Solution   (a) Slope-intercept form is $f(x) = -1 + 3x$, so the graph has slope 3 and intercepts the vertical axis at $y = -1$. See Figure 2.9.
(b) The point-slope form makes it easy to see that

$$f(2) = 5 + 3(2 - 2) = 5 + 0 = 5.$$

So the graph passes through point $(2, 5)$. Thus we can sketch it by drawing a line with slope 3 though that point.

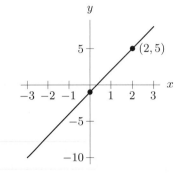

Figure 2.9: Graph of $y = 5 + 3(x - 2)$

Notice that the coefficient 3 in $f(x) = 5 + 3(x - 2)$ is also the coefficient of x when we express the function in slope-intercept form $f(x) = -1 + 3x$, so we can read the slope from either form.

Exercises and Problems for Section 2.2

Exercises

In Exercises 1–8, is the expression linear?

1. $5t - 3$

2. $5^x + 1$

3. $6r + r - 1$

4. $(3a + 1)/4$

5. $(3a + 1)/a$

6. $5r^2 + 2$

7. $4^2 + (1/3)x$

8. $6A - 3(1 - 3A)$

In Exercises 9–19, is the given expression linear in the indicated variable? Assume all constants are non-zero.

9. $\dfrac{a + b}{2}, a$

10. $2\pi r^2 + \pi rh, r$

11. $2\pi r^2 + \pi rh, h$

12. $ax^2 + bx + c^3, a$

13. $ax^2 + bx + c^3, x$

14. $2ax + bx + c, a$

15. $2ax + bx + c, x$

16. $3xy + 5x + 2 - 10y, x$

17. $3xy + 5x + 2 - 10y, y$

18. $P(P - c), P$

19. $P(P - c), c$

In Exercises 20–23, does the description lead to a linear expression? If so, give the expression.

20. The distance traveled is the speed, 45 mph, times the number of hours, t.

21. The area of a circle of radius r is πr^2.

22. The area of a rectangular plot of land w ft wide and 20 ft long is $20w$ ft^2.

23. The area of a square plot of land x ft on a side is x^2 ft^2.

For Exercises 24–26, write an expression in x representing the result of the given operations on x. Is the expression linear in x?

24. Add 5, multiply by 2, subtract x.

25. Add 5, multiply by x, subtract 2.

26. Add x, multiply by 5, subtract 2.

For each of the linear expressions in x in Exercises 27–32, give the constant term and the coefficient of x.

27. $3x + 4$

28. $5x - x + 5$

29. $w + wx + 1$

30. $x + rx$

31. $mx + mn + 5x + m + 7$

32. $5 - 2(x + 4) + 6(2x + 1)$

In Exercises 33–36, rewrite the function in slope-intercept form.

33. $y = 12 + 3(x - 1)$

34. $f(x) = 1800 + 500(x + 3)$

35. $g(n) = 14 - 2/3(n - 12)$

36. $j(t) = 1.2 + 0.4(t - 5)$

In Exercises 37–40, the form of the expression for the function tells you a point on the graph and the slope of the graph. What are they? Sketch the graph.

37. $f(x) = 3(x - 1) + 5$ **38.** $f(t) = 4 - 2(t + 2)$

39. $g(s) = (s - 1)/2 + 3$ **40.** $h(x) = -5 - (x - 1)$

41. For working n hours a week, where $n \geq 40$, a personal trainer is paid, in dollars,

$$P(n) = 500 + 18.75(n - 40)$$

What is the practical interpretation of the 500 and the 18.75?

42. When n guests are staying in a room, where $n \geq 2$, the Happy Place Hotel charges, in dollars,

$$C(n) = 79 + 10(n - 2).$$

What is the practical interpretation of the 79 and the 10?

43. A salesperson receives a weekly salary plus a commission when the weekly sales exceed \$1000. The person's total income in dollars for weekly sales of s dollars is given, in dollars, by

$$T(s) = 600 + 0.15(s - 1000).$$

What is the practical interpretation of the 600 and the 0.15?

Problems

44. A gas company charges residential customers \$8 per month even if they use no gas, plus 82¢ per therm used. (A therm is a quantity of gas.) In addition, the company is authorized to add a rate adjustment, or surcharge, per therm. The total cost of g therms of gas is given by the expression

$$\text{Total cost} = 8 + 0.82g + 0.109g.$$

(a) Which term represents the rate adjustment? What is the rate adjustment in cents per therm?
(b) Is the expression for the total cost linear?

45. A car trip costs \$1.50 per fifteen miles for gas and 30¢ per mile for other expenses, plus \$20 for car rental. The total cost for a trip of d miles is given by the expression

$$\text{Total cost} = 1.5\left(\frac{d}{15}\right) + 0.3d + 20.$$

(a) Explain what each of the three terms in the expression represents in terms of the trip.
(b) What units for cost and distance are being used?
(c) Is the expression for cost linear?

46. A boy's height h, in feet, t years after his 10th birthday, is given by $h = 4 + 0.2t$. Which of the following equivalent expressions for this function shows most clearly his height at age 20? What is that height?

(i) $h = 4 + 0.2t$ (ii) $h = 6 + 0.2(t - 10)$
(iii) $h = 10 + 0.2(t - 60)$

47. The number of books you can afford to buy, b, is a function of the number of CDs, c, you buy and is given by $b = 10 - 0.5c$. Which of the following equivalent expressions for this function most clearly shows the number of books you can afford if you buy 6 CDs?

(i) $b = 10 - 0.5c$ (ii) $b = 6 - 0.5(c - 8)$
(iii) $b = 7 - 0.5(c - 6)$

48. A cyclist's distance, d, from the finish line, 12 minutes after reaching the flat, is 45 km. If t is in minutes since reaching the flat, and the cyclist is riding at 1/2 km/min, in point-slope form we have $d = 45 - (1/2)(t - 12)$. Rewrite this equation in a form which clearly shows the distance from the start of the flat to the finish line.

49. The number of butterflies, B, in a collection in 1980 is 50. If x is in years since 1960 and the collection grows by two butterflies per year, we have $B = 50 + 2(x - 20)$. Rewrite this formula in a form which clearly shows the size of the collection when it started in 1960.

50. A company's profit after t months of operation is given by $P(t) = 1000 + 500(t - 4)$.

(a) What is the practical interpretation of the constants 4 and 1000?
(b) Rewrite the function in slope-intercept form and give a practical interpretation of the constants.

51. After t hours, Liza's distance from home, in miles, is given by $D(t) = 138 + 40(t - 3)$.

(a) What is the practical interpretation of the constants 3 and 138?
(b) Rewrite the function in slope-intercept form and give a practical interpretation of the constants.

52. The cost, $C(w)$, of mailing a letter weighing w ounces, $0 < w \leq 13$ can be modeled by the equation, $C(w) = 0.39 + 0.24(w - 1)$. (Note that all fractional ounces are rounded up to the next integer.)

(a) What is the practical interpretation of the constants 0.39 and 1?
(b) What does the 0.24 represent?
(c) How much would it cost to mail a letter weighing 9.1 ounces?

53. The cost of a holiday lasting d days consists of the travel expenses, \$350, plus accommodation expenses of \$55 times the number of days, plus food expenses of \$40 times the number of days. Why would you expect the cost to be given by a linear expression in d? Give the expression.

54. The total cost of making q widgets is the cost, \$1000, of setting up a production line, plus the cost per widget, \$15, times the number of widgets. The profit is the selling price, \$27, times the number of widgets, minus the total cost. Explain why you would expect the profit to be given by a linear expression in q, and give the expression.

55. The construction costs of a toll road are \$5000 times the length, d, in miles plus \$10,000 for each toll booth. Why do you expect the cost of a toll road with 4 toll booths to be given by a linear expression in d? Give the expression.

56. A farmer builds a fence with two gates, each 4 meters wide, around a square field x meters on a side. The cost of the project is \$10 times the length of the fence in meters plus \$300 for each gate. Why do you expect the cost

of the project to be given by a linear expression in x? Give the expression.

57. Take any three successive integers, n, $n + 1$, and $n + 2$, and add them together. Is the resulting sum linear in n?

58. Take any two successive integers, n and $n + 1$, and multiply them together. Is the resulting product linear in n?

59. Put the expression $2 + 5x$ in the form $\frac{1}{\Gamma}(1 - \sigma x)$.

60. Put $30 - x/3$ in the form $\frac{1}{\delta - 1}(\rho - x)$ and give the values of all the constants.

61. A movie theater is filled to capacity with 550 people. After the movie ends, people start leaving at the rate of 100 each minute.

    (a) Write an expression for N, the number of people in the theater, as a function of t, the number of minutes after the movie ends.

    (b) For what values of t does the expression make sense in practical terms?

2.3   LINEAR EQUATIONS IN ONE VARIABLE

An equation in which the expressions on each side of the equal sign are linear is a linear equation. We are often interested in knowing which input values to a linear function give a particular output value. This corresponds to the algebraic problem of solving a linear equation of the form

$$\text{Linear expression} = \text{Constant}.$$

Example 1   For the town in Example 2 on page 71, the population t years after incorporation is given by $P(t) = 30{,}000 + 2000t$. How many years does it take for the population to reach 50,000?

Solution   We want to know the value of t that makes $P(t)$ equal to 50,000, so we solve the equation

$$30{,}000 + 2000t = 50{,}000.$$

The expression on the left is built by first multiplying t by 2000 and then adding 30,000, and we solve the equation by performing the opposite operations in the reverse order.

$$30{,}000 + 2000t = 50{,}000$$
$$2000t = 20{,}000 \qquad \text{subtract 30,000 from both sides}$$
$$t = \frac{20{,}000}{2000} = 10 \quad \text{divide both sides by 2000.}$$

Thus, it takes 10 years for the population to reach 50,000.

Example 2   In Example 3 on page 71, the car's value t years after it was purchased is given by $V(t) = 18{,}000 - 1700t$. How long does it take for the car's value to drop to \$2000?

Solution   We want to know the value of t that makes $V(t)$ equal to \$2000, so we solve

$$18{,}000 - 1700t = 2000$$

$$-1700t = -16{,}000 \qquad \text{subtract 18,000 from both sides}$$

$$t = \frac{-16{,}000}{-1700} = 9.412. \qquad \text{divide both sides by } -1700.$$

Thus the car is worth \$2000 about halfway through its 10th year—after about nine and a half years.

Finding Where Linear Functions are Equal

Sometimes we want to know what input makes two linear functions $f(x)$ and $g(x)$ have the same output. The corresponding algebraic problem is solving an equation of the form

$$\text{Linear expression} = \text{Linear expression}.$$

We solve such equations by collecting all the terms involving the variable to be solved for on one side and all the other terms on the other side.

Example 3   Solve the equation $7 + 3p = 1 + 5p$.

Solution   For this equation we collect the p terms on the left and the constant terms on the right.

$$7 + 3p = 1 + 5p$$

$$3p = 5p - 6 \qquad \text{subtract 7 from both sides}$$

$$-2p = -6 \qquad \text{subtract } 5p \text{ from both sides}$$

$$p = \frac{-6}{-2} = 3 \qquad \text{divide both sides by } -2.$$

We check that $p = 3$ satisfies the original equation:

$$\text{Left-hand side} = 7 + 3(3) = 7 + 9 = 16$$

$$\text{Right-hand side} = 1 + 5(3) = 1 + 15 = 16.$$

Note that we could also have collected the p terms on the right-hand side and the constant terms on the left, and we would have arrived at the same result.

Example 4   Solve $1 - 2(3 - x) = 10 + 5x$.

Solution   First we distribute the -2 and collect like terms:

$$1 - 2(3 - x) = 10 + 5x$$

$$1 - 6 + 2x = 10 + 5x \qquad \text{distribute } -2$$

$$-5 + 2x = 10 + 5x \qquad \text{simplify left side}$$

$$-5 - 3x = 10 \qquad \text{subtract } 5x \text{ from both sides}$$

$$-3x = 15 \qquad \text{add 5 to both sides}$$

$$x = \frac{15}{-3} = -5 \qquad \text{divide both sides by } -3.$$

Checking $x = -5$:

$$\text{Left-hand side} = 1 - 2(3 - (-5)) = 1 - 2(8) = -15$$
$$\text{Right-hand side} = 10 + 5(-5) = 10 - 25 = -15.$$

Example 5   Incandescent light bulbs are cheaper to buy but more expensive to operate than fluorescent bulbs. The total cost in dollars to purchase a bulb and operate it for t hours is given by

$$f(t) = 0.50 + 0.004t \quad \text{(for incandescent bulbs)}$$
$$g(t) = 5.00 + 0.001t \quad \text{(for fluorescent bulbs)}.$$

How many hours of operation gives the same cost with either choice?

Solution   We need to find a value of t that makes $f(t) = g(t)$, so

$$0.50 + 0.004t = 5.00 + 0.001t \quad \text{set expressions for } f(t) \text{ and } g(t) \text{ equal}$$
$$0.50 + 0.003t = 5.00 \quad \text{subtract } 0.001t$$
$$0.003t = 4.50 \quad \text{subtract } 0.50$$
$$t = \frac{4.50}{0.003} = 1500 \quad \text{divide by } 0.003.$$

After 1500 hours of use, the cost to buy and operate an incandescent bulb equals the cost to buy and operate a fluorescent bulb.[3] Let's verify our solution:

$$\text{Cost for incandescent bulb:} \quad f(1500) = 0.50 + 0.004(1500) = 6.50$$
$$\text{Cost for fluorescent bulb:} \quad g(1500) = 5.00 + 0.001(1500) = 6.50.$$

We see that the cost of buying and operating either type of bulb for 1500 hours is the same: $6.50.

Solving Equations with Constants Represented by Letters

In the previous examples the coefficients and the constant terms were specific numbers. Often we have to solve an equation where there are unspecified constants represented by letters. The general method for solving such equations is the same.

Example 6   Solve each of the equations $2x + 12 = 20$ and $2x + z = N$ for the variable x.

Solution

$$2x + 12 = 20$$
$$2x = 20 - 12 \quad \text{subtract 12}$$
$$x = \frac{20 - 12}{2} \quad \text{divide by 2.}$$

$$2x + z = N$$
$$2x = N - z \quad \text{subtract } z$$
$$x = \frac{N - z}{2} \quad \text{divide by 2.}$$

We use the same operations to solve both equations: subtraction in the first step and division in the second. We can further simplify the solution on the left by writing $(20 - 12)/2$ as 4, but there is no corresponding simplification of the solution for the equation on the right.

---

[3]In fact incandescent bulbs typically last less than 1000 hours, which this calculation does not take into account.

Example 7   Solve the equation $2q^2p + 5p + 10 = 0$ for p.

Solution   We collect terms involving p on the left of the equal sign and the remaining terms on the right.

$$2q^2p + 5p + 10 = 0$$
$$2q^2p + 5p = -10 \qquad \text{subtract 10 from both sides}$$
$$p(2q^2 + 5) = -10 \qquad \text{factor out } p$$
$$p = \frac{-10}{2q^2 + 5} \qquad \text{divide by } 2q^2 + 5.$$

Example 8   Suppose you pay a total of \$16,368 for a car whose list price (the price before taxes) is \$$p$. Find the list price of the car if you buy it in:

(a) Arizona, where the sales tax is 5.6%.
(b) New York, where the sales tax is 8.25%.
(c) A state where the sales tax is r.

Solution   (a) If p is the list price in dollars then the tax on the purchase is $0.056p$. The total amount paid is $p + 0.056p$, so

$$p + 0.056p = 16{,}368$$
$$(1 + 0.056)p = 16{,}368$$
$$p = \frac{16{,}368}{1 + 0.056} = \$15{,}500.$$

(b) The total amount paid is $p + 0.0825p$, so

$$p + 0.0825p = 16{,}368$$
$$(1 + 0.0825)p = 16{,}368$$
$$p = \frac{16{,}368}{1 + 0.0825} = \$15{,}120.55.$$

(c) The total amount paid is $p + rp$, so

$$p + rp = 16{,}368$$
$$(1 + r)p = 16{,}368$$
$$p = \frac{16{,}368}{1 + r} \text{ dollars.}$$

Example 9   If P dollars is invested at an annual interest rate r compounded monthly, then its value in dollars after T years is

$$P\left(1 + \frac{r}{12}\right)^{12T}.$$

What amount must be invested to produce a balance of \$10,000 after T years?

Solution   We must solve the equation

$$P\left(1 + \frac{r}{12}\right)^{12T} = 10{,}000$$

for P. Although the expression on the left is complicated, it is linear in P, because it is of the form $P \times \text{Constant}$. Thus, we can solve for P by dividing through by the constant:

$$P = \frac{10{,}000}{\left(1 + \frac{r}{12}\right)^{12T}}.$$

Equations That Can Be Transformed into Linear Equations

An equation involving a fractional expression, or fractional coefficients, can sometimes be solved by first transforming it into a linear equation. To do this we clear the equation of fractions by multiplying by the product of the denominators. The resulting equation is equivalent to the original, except at values of the variable that make one of the denominators zero.

Example 10 Solve for x (a) $\dfrac{2}{3+x} = \dfrac{5}{6-x}$ (b) $s = \dfrac{a}{1-x}$, if $s \neq 0$ and $a \neq 0$

Solution (a) Multiply both sides of the equation by $(3+x)(6-x)$:

$$\frac{2}{3+x} \cdot (3+x)(6-x) = \frac{5}{6-x} \cdot (3+x)(6-x).$$

Canceling $(3+x)$ on the left and $(6-x)$ on the right, we have

$$2(6-x) = 5(3+x).$$

This is now a linear equation, with the solution given by

$$12 - 2x = 15 + 5x$$
$$-7x = 3$$
$$x = -\frac{3}{7}.$$

Since we cleared our fractions by multiplying by a quantity which can be zero, we need to check the solution by substitution in the original equation.

$$\frac{2}{3+x} = \frac{2}{3-3/7} = \frac{2}{(21-3)/7} = 2 \cdot \frac{7}{18} = \frac{7}{9}$$
$$\frac{5}{6-x} = \frac{5}{6+3/7} = \frac{5}{(42+3)/7} = 5 \cdot \frac{7}{45} = \frac{7}{9}.$$

Thus the solution is valid.

(b) Multiplying both sides by $1 - x$ gives

$$s(1-x) = a$$
$$s - sx = a$$
$$s - a = sx$$
$$x = \frac{s-a}{s}.$$

We can divide by s because it cannot be zero. Again, we need to check our solution. Substituting it into the right side of the equation, we get

$$\text{Right side} = \frac{a}{1 - \frac{s-a}{s}} = \frac{a}{\frac{s-(s-a)}{s}} = \frac{sa}{a} = s = \text{Left side}.$$

Using a Graph to Visualize Solutions

If we graph two functions f and g on the same axes, then the values of t where $f(t) = g(t)$ correspond to points where the two graphs intersect.

Example 11   In Example 5 we saw that the cost to buy and operate an incandescent bulb for 1500 hours is the same as the cost for a fluorescent bulb.

(a) Graph the cost for each bulb and indicate the solution to the equation in Example 5.
(b) Which bulb is cheaper if you use it for less than 1500 hours? More than 1500 hours?

Solution   See Figure 2.10. The t-coordinate of the point where the two graphs intersect is 1500. At this point, both functions have the same value. In other words, $t = 1500$ is a solution to the equation $f(t) = g(t)$. Figure 2.10 shows that show that the incandescent bulb is cheaper if you use it for less than 1500 hours. For example, it costs \$4.50 to operate the incandescent bulb for 1000 hours, whereas it costs \$6.00 to operate the fluorescent bulb for the same time. On the other hand, if you operate the bulb for more than 1500 hours, the fluorescent bulb is cheaper.

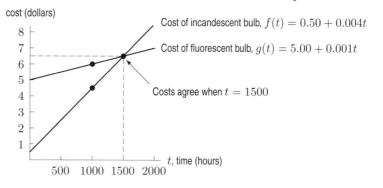

Figure 2.10: Costs of operating different types of bulb

Using the Algebraic Form of the Equation to Predict Properties of Solutions

In Example 11 the graph makes it easy to see that the equation has a solution which is a positive number. We can also predict this from the algebraic form of the equation

$$0.50 + 0.004t = 5.00 + 0.001t.$$

Notice that the coefficient on the left is larger than the coefficient on the right ($0.004 > 0.001$), and that the constant term on the right is larger than the constant term on the left ($5.00 > 0.50$). So, without doing the actual calculation, we know that when like terms are gathered on each side, the equation has the form

$$(\text{Positive number})t = \text{Positive number}.$$

So the solution t is a ratio of two positive numbers and therefore is also positive.

Example 12   Without solving, say whether the equations have a solution which is

(i) Positive   (ii) Negative   (iii) Zero.

(a) $53x = -29$ (b) $29x + 53 = 53x + 29$
(c) $29x + 13 = 53x + 29$ (d) $29x + 53 = 53x + 53$

Solution   (a) The solution is negative, since it is the ratio of a negative number and a positive number.
         (b) When we collect all the x terms on the right and the constant term on the left, we get

$$\text{Positive number} = (\text{Positive number})x,$$

because $53 > 29$, so the solution is positive.
         (c) This time the equation is equivalent to

$$\text{Negative number} = (\text{Positive number})x,$$

because $13 < 29$, so the solution is negative.
         (d) Since the constant terms are the same on both sides, the equation is equivalent to

$$0 = (\text{Positive number})x,$$

so the solution is 0.

How Many Solutions Does a Linear Equation Have?

A linear equation may have one solution, no solutions, or infinitely many solutions.

One Solution

So far we have seen linear equations that have one solution. For example, $3x + 5 = 14$ has one solution, $x = 3$.

No Solution

In contrast, the equation $3x + 5 = 3(x + 5)$ has *no* solutions:

$$3x + 5 = 3(x + 5)$$
$$3x + 5 = 3x + 15$$
$$5 = 15.$$

The last equation is false no matter what the value of x. If we look at the second line, we can see that no value of x satisfies the equation. This is because $3x + 5$ can never equal $3x + 15$, not matter what value $3x$ takes on. Figure 2.11 shows why the equation has no solution. The graphs of $y = 3x + 5$ and $y = 3(x + 5)$ have the same slope, so never meet.

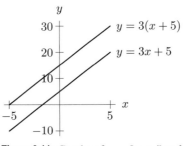

Figure 2.11: Graphs of $y = 3x + 5$ and $y = 3(x + 5)$ do not intersect.

Many Solutions

Something equally unexpected happens when we try to solve the following equation:

$$3x + 5 = 3(x + 1) + 2$$
$$3x + 5 = 3x + 5$$
$$0 = 0$$

The last equation is true no matter what the value of x. In fact, the expressions $3x+5$ and $3(x+1)+2$ are equivalent, so the graphs of $y = 3x + 5$ and $y = 3(x + 1) + 2$ are the same.

Exercises and Problems for Section 2.3

Exercises

1. The tuition cost for part-time students taking C credits at Stonewall College is given by $300 + 200C$ dollars.

   (a) Find the tuition cost for eight credits.
   (b) How many credits were taken if the tuition cost was $1700?

In Exercises 2–6, a company offers three formulas for the weekly salary of its sales people, depending on the number of sales, s, made each week:

(a) $100 + 0.10s$ dollars   (b) $150 + 0.05s$ dollars
(c) 175 dollars

2. How many sales must be made under option (a) to receive $200 a week?

3. How many sales must be made under option (c) to receive $200 a week?

4. At what sales level do options (a) and (b) produce the same salary?

5. At what sales level do options (b) and (c) produce the same salary?

6. At what sales level do options (a) and (c) produce the same salary?

Solve the equations in Exercises 7–30.

7. $3x = 18$

8. $-2y = 14$

9. $3z = 22$

10. $x + 3 = 13$

11. $y - 7 = 21$

12. $w + 23 = -34$

13. $2x + 5 = 13$

14. $2x + 5 = 4x - 9$

15. $0.5x - 3 = 11$

16. $\frac{5}{3}(y + 4) = \frac{1}{2} - y$

17. $2(a + 3) = 10$

18. $-9 + 10r = -3r$

19. $4p - 1.3 = -6p - 16.7$

20. $6n - 3 = -2n + 37$

21. $\frac{1}{2}r - 2 = 3r + 5$

22. $0.2(g - 6) = 0.6(g - 4)$

23. $-4(2m - 5) = 5$

24. $5 = \frac{1}{3}(t - 6)$

25. $-6(2k - 1) = 5(3 - 2k)$

26. $4 - (r - 3) = 6(1 - r)$

27. $\frac{2}{3}(3n - 12) = \frac{3}{4}(4n - 3)$

28. $3d - \frac{1}{2}(2d - 4) = -\frac{5}{4}(d + 4)$

29. $B - 4(B - 3(1 - B)) = 57$

30. $1.06s - 0.01(240 - s) = 22.67s$

Without solving them, say whether the equations in Exercises 31–42 have a positive solution, a negative solution, a zero solution, or no solution. Give a reason for your answer.

31. $3x = 5$

32. $3a + 7 = 5$

33. $5z + 7 = 3$

34. $3u - 7 = 5$

35. $7 - 5w = 3$

36. $4y = 9y$

37. $4b = 9b + 6$

38. $6p = 9p - 4$

39. $8r + 3 = 2r + 11$

40. $8 + 3t = 2 + 11t$

41. $2 - 11c = 8 - 3c$

42. $8d + 3 = 11d + 3$

In Exercises 43–49, does the equation have no solution, one solution, or an infinite number of solutions?

43. $4x + 3 = 7$

44. $4x + 3 = -7$

45. $4x + 3 = 4(x + 1) - 1$

46. $4x + 3 = 4(x + 1) + 1$

47. $4x + 3 = 3$

48. $4x + 3 = 4(x - 1) + 5$

49. $4x + 3 = 4(x - 1) + 7$

In Exercises 50–64, solve for the indicated variable. Assume all constants are non-zero.

50. $A = l \cdot w$, for w.

51. $y = 3\pi t$, for t.

52. $t = t_0 + \dfrac{k}{2}w$, for w.

53. $s = v_0 t + \dfrac{1}{2}at^2$, for a.

54. $bx - d = ax + c$, for x, if $a - b \neq 0$.

55. $ab + aw = c - aw$, for w

56. $3xt + 1 = 2t - 5x$, for t, if $x \neq 2/3$.

57. $u(m + 2) + w(m - 3) = z(m - 1)$, for m, if $u + w - z \neq 0$.

58. $S = \dfrac{tL - a}{t - 1}$, for t, if $S \neq L$.

59. $x + y = z$, for y.

60. $ab = c$, for b

61. $2m + n = p$, for m.

62. $2r - t = r + 2t$, for r.

63. $6w - 4x = 3w + 5x$, for w.

64. $3(3g - h) = 6(g - 2h)$, for g.

65. Solve

$$t(t + 3) - t(t - 5) = 4(t - 5) - 7(t - 3).$$

66. Solve

$$\frac{3}{z - 2} = \frac{2}{z - 3}$$

Problems

67. A car rental company charges $37 per day and $0.25 per mile.

(a) Compute the cost of renting the car for one day, assuming the car was driven 100 miles.

(b) Compute the cost of renting the car for three days, assuming the car was driven 400 miles.

(c) Andy rented a car for five days, but he did not keep track of how many miles he drove. He got a bill for $385. How many miles did he drive?

68. You have a coupon worth $20 off the purchase of a scientific calculator. At the same time the calculator is offered with a discount of 20%, and no further discounts apply. For what tag price on the calculator do you pay the same amount for each discount?

69. Apples are 99 cents a pound, and pears are $1.25 a pound. If I spend $4 and the weight of the apples I buy is twice the weight of the pears, how many pounds of pears do I buy?

70. Using Figure 2.12, determine the value of s.

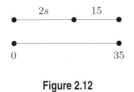

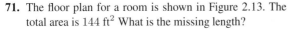

Figure 2.12

71. The floor plan for a room is shown in Figure 2.13. The total area is 144 ft^2 What is the missing length?

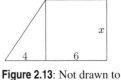

Figure 2.13: Not drawn to scale

72. You drive 100 miles. Over the first 50 miles you drive 50 mph, and over the second 50 miles you drive V mph.

(a) Calculate the time spent on the first 50 miles and on the second 50 miles.

(b) Calculate the average speed for the entire 100 mile journey (the average speed is the total distance traveled divided by the total time taken).

(c) If you want to average 75 mph for the entire journey, what is V?

(d) If you want to average 100 mph for the entire journey, what is V?

In Problems 73–80, decide for what value(s) of the constant A (if any) the equation has

(a) The solution $x = 0$ (b) A positive solution in x
(c) No solution in x

73. $3x = A$

74. $Ax = 3$

75. $3x + 5 = A$

76. $3x + A = 5$

77. $3x + A = 5x + A$

78. $Ax + 3 = Ax + 5$

79. $\dfrac{7}{x} = A$

80. $\dfrac{A}{x} = 5$

In Exercises 81–87, $f(t) = 2t + 7$. Does the equation have no solution, one solution, or an infinite number of solutions?

81. $f(t) = 7$ **82.** $2f(t) = f(2t)$

83. $f(t) = f(t+1) - 2$ **84.** $f(t) = f(-t)$

85. $f(t) = -f(t)$ **86.** $f(t) + 1 = f(t+1)$

87. $f(t) + f(3t) - 2f(2t) = 0$

88. For the function $f(t) = \dfrac{2t+3}{5}$,

     (a) Evaluate $f(11)$ **(b)** Solve $f(t) = 2$.

Solve the equations in Exercises 89–92.

89. $\dfrac{2}{2-x} - \dfrac{3}{x-5} = 0$ **90.** $\dfrac{3}{2x-1} + \dfrac{5}{3-2x} = 0$

91. $\dfrac{-3}{x-2} - \dfrac{2}{x-3} = 0.$

92. $\dfrac{1}{1 + \dfrac{1}{2-x}} = \dfrac{2}{3 + \dfrac{1}{2-x}}.$

93. Solve $g(x) = 7$ given that $g(x) = \dfrac{5x}{2x-3}$.

94. If $b_1 + m_1 x = b_2 + m_2 x$, what can be said about the constants $b_1, m_1, b_2,$ and m_2 if the equation has

     (a) One solution?
     (b) No solutions?
     (c) An infinite number of solutions?

The equation $3 - 2x = 8$ is solved below using two different approaches:

$3 - 2x = 8$	$3 - 2x = 8$ (i)
$-2x = 5$	$3 = 8 + 2x$ (ii)
$x = -\dfrac{5}{2}.$	$2x = -5$ (iii)
	$x = -\dfrac{5}{2}.$ (iv)

Show how to solve the equations in Exercises 95–96 using these same two approaches, and number your steps. (You may need to take additional steps to obtain a solution.) For Example, we can solve $7 - 3(1 - z) = 22$ using the first approach as follows:

$$7 - 3(1 - z) = 22 \qquad \text{(i)}$$
$$-3(1 - z) = 15 \qquad \text{(ii)}$$
$$1 - z = -5. \qquad \text{(iii)}$$

Solving this equation the second way gives:

$$7 - 3(1 - z) = 22 \qquad \text{(i)}$$
$$7 = 22 + 3(1 - z) \qquad \text{(ii)}$$
$$3(1 - z) = -15 \qquad \text{(iii)}$$
$$1 - z = -5. \qquad \text{(iv)}$$

In both cases, we must take additional steps to show that $z = 6$.

95. $5 - 4x = 25$ **96.** $\sqrt{5} - \sqrt{7}\alpha = 8$

2.4 LINEAR EQUATIONS IN TWO VARIABLES

We have seen that the graph of a linear function is a line. For example, the graph of $f(x) = 5 + 3x$ is the line with slope 3 and vertical intercept 5. Another way of looking at this is that every point (x, y) on the line is a solution to the equation

$$y = 5 + 3x.$$

Instead of thinking of this as a formula for a function, we think of it as an equation in two variables, x and y. A solution to the equation is a pair (x, y). Unlike a typical linear equation in one variable, a linear equation in two variables has many solutions that form a line in the xy-plane. To find the line, we put the equation in slope-intercept form.

Example 1   For each of the following equations in two variables, find the vertical intercept and slope of its graph by putting the equation in slope-intercept form. Use this information to match the equations with the graphs.

(a) $y - 5 = 8(x + 1)$ (b) $3x + 4y = 20$ (c) $6x - 15 = 2y - 3$

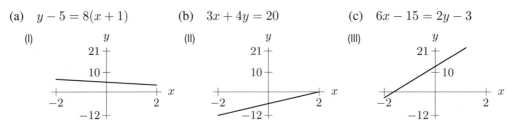

Figure 2.14: Which graph goes with which equation?

Solution   We solve for y in each case:

(a)
$$y - 5 = 8(x + 1)$$
$$y - 5 = 8x + 8$$
$$y = 8x + 8 + 5 = 13 + 8x.$$

The y-intercept is 13 and the slope is 8. This matches graph (III).

(b)
$$3x + 4y = 20$$
$$4y = 20 - 3x$$
$$y = \frac{20 - 3x}{4} = \frac{20}{4} - \frac{3}{4}x = 5 - \frac{3}{4}x.$$

The y-intercept is 5 and the slope is $-3/4$. Since the slope is negative, the y-values decrease as the x-values increase, and the graph falls. This matches graph (I).

(c)
$$6x - 15 = 2y - 3$$
$$2y = 6x - 15 + 3 = 6x - 12$$
$$y = \frac{6x - 12}{2} = \frac{6}{2}x - \frac{12}{2} = -6 + 3x.$$

The y-intercept is -6 and the slope is 3. This matches graph (II). Notice that this graph rises more gently than graph (III) because its slope is 3, which is less than 8, the slope of graph (III).

Forms of Linear Equations in Two Variables

The slope-intercept and point-slope forms for linear functions give rise to corresponding forms for equations of lines. Each of these forms has the dependent variable isolated on one side of the equation. In addition, linear equations can be written in a form like $3x + 4y = 20$ from Example 1. This is called *standard form*.

Forms for Linear Equations in Two Variables

- **Slope-Intercept Form**. A line with slope m and vertical intercept b has the equation

$$y = b + mx.$$

- **Point-Slope Form**. A line passing through (x_0, y_0) with slope m has equation

$$y = y_0 + m(x - x_0).$$

- **Standard Form**. Any line, including horizontal and vertical lines, has an equation of the form

$$Ax + By = C, \qquad \text{where } A, B, \text{ and } C \text{ are constants.}$$

How Do We Find the Slope?

To write an equation for a line in either slope-intercept or point-slope form, we need to find the slope. The slope is the change in y per unit increase in x.

Example 2   Find the slope of each of the lines in Figure 2.15.

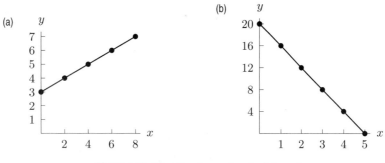

Figure 2.15: Find the slope of each of these lines

Solution   (a) The y-value increases by 1 unit for each 2-unit increase in x, so

$$\text{Slope} = \frac{\text{Change in } y}{\text{Change in } x} = \frac{1}{2}.$$

(b) Here the y-values decrease by 4 units for each 1-unit increase in x, so

$$\text{Slope} = \frac{\text{Change in } y}{\text{Change in } x} = \frac{-4}{1} = -4.$$

In general, using the Greek letter Δ (pronounced delta) to indicate change,

---

Slope between two points

Given any two points (x_1, y_1) and (x_2, y_2) on a line $y = mx + b$, we can find the slope m using the formula

$$m = \frac{\text{Change in } y}{\text{Change in } x} = \frac{\Delta y}{\Delta x} = \frac{y_2 - y_1}{x_2 - x_1}.$$

Using Slope-Intercept Form

Example 3   Find an equation for each of the following lines.

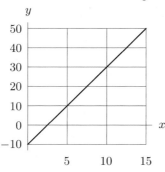

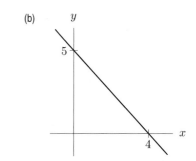

Solution

(a) We use any two points to find the slope. If we use the points $(0, -10)$ and $(5, 10)$, we have

$$m = \frac{\Delta y}{\Delta x} = \frac{10 - (-10)}{5 - 0} = \frac{20}{5} = 4.$$

The slope is $m = 4$. We see in the graph that the y-intercept is -10, so the equation of the line is

$$y = -10 + 4x.$$

(b) We use the two points $(0, 5)$ and $(4, 0)$ to find the slope:

$$m = \frac{\Delta y}{\Delta x} = \frac{0 - 5}{4 - 0} = \frac{-5}{4} = -\frac{5}{4}.$$

The slope is $m = -5/4$. (We can also see this quickly by noticing that to get from one intercept to the other, the vertical change is -5 while the horizontal change is 4, so the slope must be $-5/4$.) The slope is negative since the graph falls from left to right. We see in the graph that the y-intercept is 5, so the equation of the line is

$$y = 5 - \frac{5}{4}x.$$

Using Point-Slope Form

Example 4

Find an equation for the line that

(a) Passes through the point $(5, -8)$ and has slope $m = -3$
(b) Contains the points $(5, 20)$ and $(8, 32)$.

Solution

(a) We have $m = -3$ and $(x_0, y_0) = (5, -8)$, so

$$y = y_0 + m(x - x_0)$$
$$y = (-8) + (-3)(x - 5)$$
$$y = -8 - 3x + 15$$
$$y = 7 - 3x.$$

(b) We first find the slope:

$$m = \frac{\Delta y}{\Delta x} = \frac{32 - 20}{8 - 5} = \frac{12}{3} = 4.$$

Using the point slope form and letting $(x_0, y_0) = (5, 20)$, we have

$$y = y_0 + m(x - x_0)$$
$$y = 20 + 4(x - 5)$$
$$y = 20 + 4x - 20$$
$$y = 4x.$$

Notice that in either case, we could have left the equation of the line in point-slope form instead of putting it in slope-intercept form.

Equations of Horizontal and Vertical Lines

A line with positive slope rises and one with negative slope falls as we move from left to right. What about a line with slope $m = 0$? Such a line neither rises nor falls, but is horizontal.

Example 5   Explain why the equation $y = 4$ represents a horizontal line and the equation $x = 4$ represents a vertical line.

Solution   We can think of $y = 4$ as an equation in two variables by rewriting it

$$y = 4 + 0 \cdot x.$$

The value of y is 4 for all values of x, so all points with y-coordinate 4 lie on the graph. As we see in Figure 2.16, these points lie on a horizontal line. Similarly, the equation $x = 4$ means that x is 4 no matter what the value of y is. Every point on the vertical line in Figure 2.17 has x equal to 4.

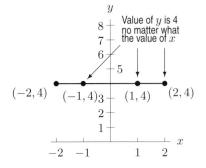

Figure 2.16: The horizontal line $y = 4$ has slope 0

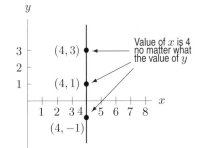

Figure 2.17: The vertical line $x = 4$ has an undefined slope

In Example 5 the equation $x = 4$ cannot be put into slope-intercept form. This is because the slope is defined as $\Delta y / \Delta x$ and Δx is zero. A vertical line does not have an equation of the form $y = mx + b$, and its slope is undefined.

In summary,

For any constant k:
- The graph of the equation $y = k$ is a horizontal line, and its slope is zero.
- The graph of the equation $x = k$ is a vertical line, and its slope is undefined.

Slopes of Parallel and Perpendicular Lines

Figure 2.18 shows two parallel lines. These lines are parallel because they have equal slopes.

Figure 2.18: Parallel lines: l_1 and l_2 have equal slopes

Figure 2.19: Perpendicular lines: l_1 has a positive slope and l_2 has a negative slope

What about perpendicular lines? Two perpendicular lines are graphed in Figure 2.19. We can see that if one line has a positive slope, then any line perpendicular to it must have a negative slope. Perpendicular lines have slopes with opposite signs.

We show (on page 97) that if l_1 and l_2 are two perpendicular lines with slopes, m_1 and m_2, then m_1 is the negative reciprocal of m_2. If m_1 and m_2 are not zero, we have the following result:

Let l_1 and l_2 be two lines having slopes m_1 and m_2, respectively. Then:
- These lines are parallel if and only if $m_1 = m_2$.
- These lines are perpendicular if and only if $m_1 = -\dfrac{1}{m_2}$.

In addition, any two horizontal lines are parallel and $m_1 = m_2 = 0$. Any two vertical lines are parallel and m_1 and m_2 are undefined. A horizontal line is perpendicular to a vertical line. See Figures 2.20–2.22.

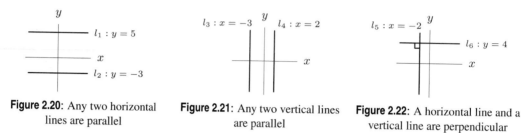

Figure 2.20: Any two horizontal lines are parallel

Figure 2.21: Any two vertical lines are parallel

Figure 2.22: A horizontal line and a vertical line are perpendicular

Example 6   Show that the lines $3x + 2y = 20$ and $12x + 8y = 5$ are parallel.

Solution   Lines are parallel if they have the same slope. We find the slopes of these lines by putting them in slope-intercept form. We first find the slope of the line $3x + 2y = 20$.

$$3x + 2y = 20$$
$$2y = 20 - 3x$$
$$y = \frac{20 - 3x}{2}$$
$$y = 10 - \left(\frac{3}{2}\right)x.$$

The slope of the first line is $-3/2$.

For the line $12x + 8y = 5$, we have:

$$12x + 8y = 5$$
$$8y = 5 - 12x$$
$$y = \frac{5 - 12x}{8}$$
$$y = \left(\frac{5}{8}\right) - \left(\frac{3}{2}\right)x.$$

The slope of the second line is also $-3/2$, so the lines are parallel.

Example 7   Find an equation for

(a) The line parallel to the graph of $y = 12 - 3x$ with a y-intercept of 7.
(b) The line parallel to the graph of $5x + 3y = -6$ containing the point $(9, 4)$.
(c) The line perpendicular to the graph of $y = 5x - 20$ that intersects the graph at $x = 6$.

Solution (a) Since the lines are parallel, the slope is $m = -3$. We know the y-intercept is $b = 7$, and so a possible equation is $y = 7 - 3x$.

(b) To find the slope of $5x + 3y = -6$, we put it in slope-intercept form $y = -2 - (5/3)x$. So the slope is $m = -5/3$. Since the parallel line has the same slope and contains the point $(9, 4)$, we can use point-slope form to get

$$y = y_0 + m(x - x_0)$$
$$y = 4 + \left(-\frac{5}{3}\right)(x - 9)$$
$$y = 4 + \left(-\frac{5}{3}\right)x + 15$$
$$y = 19 - \frac{5}{3}x.$$

(c) The slope of the original line is $m_1 = 5$, so the slope of the perpendicular line is

$$m_2 = -\frac{1}{5} = -0.2.$$

The lines intersect at $x = 6$. From the original equation, this means

$$y = 5x - 20 = 5 \cdot 6 - 20 = 10,$$

so the lines intersect at the point $(6, 10)$. Using the point-slope formula, we have

$$y = 10 + (-0.2)(x - 6)$$
$$= 10 - 0.2x + 1.2$$
$$= 11.2 - 0.2x.$$

Justification of the Formula for Slopes of Perpendicular Lines

Figure 2.23 shows l_1 and l_2, two perpendicular lines through the origin with slope m_1 and m_2. Neither line is horizontal or vertical, so m_1 and m_2 are both defined and nonzero.

We measure a distance 1 along the x-axis and mark the point on l_1 above it. This point has coordinates $(1, m_1)$, since $x = 1$ and since the equation for the line is $y = m_1 x$:

$$y = m_1 x = m_1 \cdot 1 = m_1.$$

Likewise, we measure a distance 1 down along the y-axis and mark the point on l_2 to the right of it. This point has coordinates $(-1/m_2, -1)$, since $y = -1$ and the equation for the line is $y = m_2 x$:

$$-1 = m_2 x \qquad \text{so} \qquad x = -\frac{1}{m_2}.$$

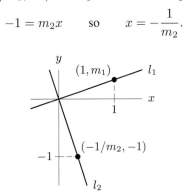

Figure 2.23: Perpendicular lines

By drawing in the dashed lines shown in Figure 2.23, we form two triangles. These triangles have the same shape and size, meaning they have three equal angles and three equal sides. (Such triangles are called *congruent* triangles.) We know this is true because both triangles have a side of length 1 along one of the axes and the angle between that side and and the longest side (the *hypotenuse*) is the same in both triangles, since the axes are perpendicular to each other, and l_1 and l_2 are also perpendicular to each other. Therefore the sides drawn with dashed lines are also equal. From the coordinates of the two labeled points, we see that these two dashed lines measure m_1 and $-1/m_2$, respectively, so

$$m_1 = \frac{-1}{m_2}.$$

A similar argument works for lines that do not intersect at the origin.

Constraint Equations

The standard form of a linear equation is useful for describing *constraints*, or situations involving limited resources.

Example 8 A newly designed motel has S small rooms measuring 250 ft² and L large rooms measuring 400 ft². The designers have 10,000 ft² of available space. Write an equation relating S and L.

Solution Altogether, we have

$$\text{Total room space} = \underbrace{\text{space for small rooms}}_{250S} + \underbrace{\text{space for large rooms}}_{400L}$$
$$= 250S + 400L.$$

We know that the total available space is 10,000 ft², so

$$250S + 400L = 10,000.$$

We graph this by determining the axis intercepts:

$$\text{If } S = 0 \text{ then } L = \frac{10,000}{400} = 25.$$

$$\text{If } L = 0 \text{ then } S = \frac{10,000}{250} = 40.$$

See Figure 2.24.

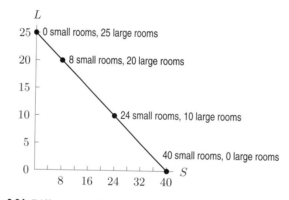

Figure 2.24: Different combinations of small and large rooms for the motel

The equation $250S + 400L = 10,000$ in Example 8 is called a *constraint equation* because it describes the constraint that floor space places on the number of rooms built. Constraint equations are usually written in standard form.

Example 9   Revised plans for the motel in Example 8 on the preceding page provide for a total floor space of 16,000 ft². Find the new constraint equation. Sketch its graph together with the graph of the original constraint. How do the two graphs compare?

Solution   The constraint in this case is the total floor area of 16,000 ft². Since small rooms have an area of 250 ft², the total area of S small rooms is $250S$. Similarly, the total area of L large rooms is $400L$. Since

$$\text{Area of small rooms} + \text{Area of large rooms} = \text{Total area,}$$

we have a constraint equation

$$250S + 400L = 16{,}000.$$

From Figure 2.25, we see that the second line is parallel to the first line, but that at every value of $S \leq 40$, the value of L is 15 units larger than before. Since one large room uses 400 ft², we see that 6000 ft² provides for the additional $6000/400 = 15$ additional large rooms.

Likewise, for every value of $L \leq 25$, the value of S is 24 units larger than before. Again, this is a consequence of the extra space: at 250 ft² each, there is space for $6000/250 = 24$ additional small rooms.

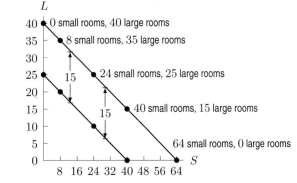

Figure 2.25: Graph of the new motel room constraint equation $250S + 400L = 16{,}000$ together with the old constraint equation $250S + 400L = 10{,}000$

Exercises and Problems for Section 2.4

Exercises

Write the linear equations in Exercises 1–4 in slope-intercept form $y = b + mx$. What are the values of m and b?

1. $y = 100 - 3(x - 20)$ **2.** $80x + 90y = 100$

3. $\dfrac{x}{100} + \dfrac{y}{300} = 1$ **4.** $x = 30 - \dfrac{2}{3}y$

5. Without a calculator, match the equations (a)–(g) to the graphs (I)–(VII).

  (a) $y = x - 3$ **(b)** $-3x + 2 = y$

  (c) $2 = y$ **(d)** $y = -4x - 3$

  (e) $y = x + 2$ **(f)** $y = x/3$

  (g) $4 = x$

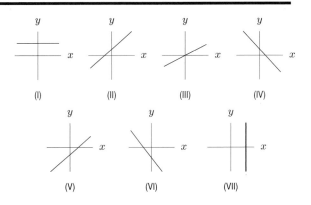

In Exercises 6–11, graph the equation.

6. $y = 3x - 6$

7. $y = 5$

8. $2x - 3y = 24$

9. $x = 7$

10. $y = -\dfrac{2}{3}x - 4$

11. $y = 200 - 4x$

In Exercises 12–22, write an equation in point-slope form for the line.

12. Through $(2, 3)$ with slope $m = 5$

13. Through $(-1, 7)$ with slope $m = 6$

14. Through $(8, 10)$ with slope $m = -3$

15. Through $(2, -9)$ with slope $m = -2/3$

16. Passes through $(4, 7)$ and $(1, 1)$

17. Passes through $(6, 5)$ and $(7, 1)$

18. Passes through $(-2, -8)$ and $(2, 4)$

19. Passes through $(6, -7)$ and $(-6, -1)$

20. Passes through $(-1, -8)$ and is parallel to the line $y = 5x - 2$

21. Passes through $(3, -6)$ and is parallel to the line $y = 5/4(x + 10)$

22. Passes through $(12, 20)$ and is perpendicular to the line $y = -4x - 3$.

23. Find a possible formula for $y = h(x)$, a linear function whose graph contains the points $(-30, 80)$ and $(40, -60)$.

24. Find a possible formula for $y = f(x)$, a linear function whose graph contains the points $(20, 70)$ and $(70, 10)$.

25. Find a possible formula for $y = f(x)$, a linear function whose graph contains the points $(-12, 60)$ and $(24, 42)$.

In Exercises 26–28, find an equation for the line and give a practical interpretation of the slope and vertical intercept.

26.

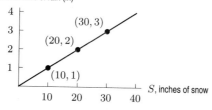

27.

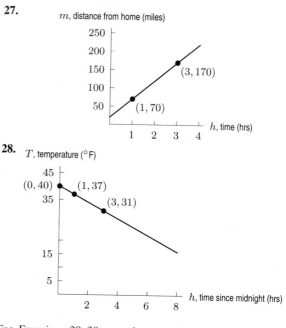

28.

For Exercises 29–38, put the equation in standard form. (Many answers are possible.)

29. $x = 3y - 2$

30. $y = 2 + 4(x - 3)$

31. $5x = 7 - 2y$

32. $y - 6 = 5(x + 2)$

33. $x + 4 = 3(y - 1)$

34. $6(x + 4) = 3(y - x)$

35. $9(y + x) = 5$

36. $3(2y + 4x - 7) = 5(3y + x - 4)$

37. $y = 5x + 2a$, with a constant

38. $5b(y + bx + 2) = 4b(4 - x + 2b)$, with b constant

For Exercises 39–42,

(a) Write an equation for the given relation

(b) Choose two solutions

(c) Graph the equation and mark your solutions.

39. The relation between quantity of chicken and quantity of steak if chicken costs \$1.29/lb and steak costs \$3.49/lb, and you have \$100 to spend on a barbecue.

40. The relation between the time spent walking and driving if you walk at 3 mph then hitch a ride in a car at 75 mph, covering a total distance of 60 miles.

41. The relation between the volume of titanium and iron in a bicycle weighing 5kg, if titanium has a density of 4.5g/cm^3 and iron has a density of 7.87 g/cm^3 (ignore other materials).

42. The relation between the time spent walking and the time spent canoeing on a 30 mile trip if you walk 4 mph and canoe 7 mph.

43. Find an equation for the line parallel to the graph of

    (a) $y = 3 + 5x$ with a y-intercept of 10.
    (b) $4x + 2y = 6$ with a y-intercept of 12.
    (c) $y = 7x + 2$ and containing the point $(3, 22)$.
    (d) $9x + y = 5$ and containing the point $(5, 15)$.

In Exercises 44–51, are the lines parallel?

44. $y = 12 + ax; y = 20 + ax$, where a is a constant

45. $y = 1 + x; y = 1 + 2x$

46. $y = 5 + 4(x - 2); y = 2 + 4x$

47. $y = 2 + 3(x + 5); y = 2 + 4(x + 5)$

48. $2x + 3y = 5; 4x + 6y = 7$

49. $qx + ry = 3; qx + ry = 4$, where q and r are non-zero constants

50. $y = 7 + 4(x - 2); y = 8 + 2(2x + 3)$

51. $y = 5 + 6(x + 2) \ y = 5 + 6(3x - 1)$

Find possible formulas for the linear functions described in Exercises 52–54.

52. The graph of $y = g(x)$ is perpendicular to the graph of the equation $5x - 3y = 6$, and the two lines intersect at $x = 15$.

53. The graph of $y = u(x)$ is perpendicular to the graph of $y = 0.7 - 0.2x$ and intersects it at $x = 1.5$.

54. The graphs of $y = v(x)$ and $y = 400 + 25x$ are perpendicular and intersect at $x = 12$.

Problems

Match the statements in Problems 55–56 with the lines I–VI.

I. $y = 2(x - 4) + 9$ II. $y - 9 = -3(x - 4)$
III. $y + 9 = -2(x - 4)$ IV. $y = 4x + 9$
V. $y = 9 - 2(4 - x)$ VI. $y = 9 - \dfrac{4 - 8x}{4}$

55. These three lines pass through the same point.

56. These three lines have the same slope.

Match the statements in Problems 57–59 with equations I–VI.

I. $y = 20 + 2(x - 8)$ II. $y = 20 - 2(x - 8)$
III. $y = 5x + 30$ IV. $y = -5(6 - x)$
V. $y = \dfrac{2x + 90}{3}$ VI. $y = -\dfrac{2}{3}(x - 8) + 20$

57. These three lines pass through the same point.

58. These two lines have the same y-intercept.

59. These two lines have the same slope.

60. Find the equation of the line intersecting the graph of $y = x^3 - x + 3$ at $x = -2$ and $x = 2$.

61. Find a possible formula for the linear function $y = g(x)$ given that:

    • the value of the expression $g(100)$ is 30, and
    • the solution to the equation $g(x) = 15$ is -50.

In Problems 62–66, is the point-slope form or slope-intercept form the easier form to use when writing an equation for the line?

62. Slope $= 3$, Intercept $= -6$

63. Passes through $(2, 3)$ and $(-6, 7)$

64. Passes through $(-5, 10)$ and has slope 6

65. Is parallel to the line $y = 0.4x - 5.5$ and has the same y-intercept as the line $y = -2x - 3.4$

66. Is parallel to the line $y = 4x - 6$ and contains the point $(2, -3)$

If two populations have the same constant rate of change, then the graphs that describe them are parallel lines. Are the graphs of the populations in Problems 67–70 parallel lines?

67. Towns A and B are each growing by 1000 people per decade.

68. Bacteria populations C and D are each growing at 30% each hour.

69. Country E is growing by 1 million people per decade. Country F is growing by 100 thousand people per year.

70. Village G has 200 people and is growing by 2 people per year. Village H has 100 people and is growing by 1 person per year.

In Problems 71–74, write the equation in the form $y = b + mx$, and identify the values of b and m.

71. $y - y_0 = r(x - x_0)$ **72.** $y = \beta - \dfrac{x}{\alpha}$

73. $Ax + By = C$, if $B \neq 0$

74. $y = b_1 + m_1 x + b_2 + m_2 x$

75. Put the equation $y = 3xt + 2xt^2 + 5$ in the form $y = b + mx$. What are the values of b and m? [Note: Your answers could include t.]

76. Show that the points $(0, 12), (3, 0)$, and $(17/3, 2/3)$ form the corners of a right triangle (that is, a triangle with a right angle).

In Problems 77–82, which line has the greater

(a) Slope? **(b)** y-intercept?

77. $y = 3 + 6x$, $y = 5 - 3x$

78. $y = \frac{1}{5}x$, $y = 1 - 6x$

79. $2x = 4y + 3$, $y = -x - 2$

80. $3y = 5x - 2$, $y = 2x + 1$

81. $y + 2 = 3(x - 1)$, $y = 6 - 50x$

82. $y - 3 = -4(x + 2)$, $-2x + 5y = -3$

83. Which equation, (a)–(d), has the graph that crosses the y-axis at the highest point?

(a) $y = 3(x - 1) + 5$ **(b)** $x = 3y + 2$
(c) $y = 1 - 6x$ **(d)** $2y = 3x + 1$

84. Which of the following equations has a graph that slopes down the most steeply as you move from left to right?

(a) $y + 4x = 5$ **(b)** $y = 5x + 3$
(c) $y = 10 - 2x$ **(d)** $y = -3x + 2$

85. Explain the differences between the graphs of the equations $y = 14x - 18$ and $y = -14x + 18$.

86. Using the window $-10 \le x \le 10, -10 \le y \le 10$, graph $y = x, y = 10x, y = 100x$, and $y = 1000x$.

(a) Explain what happens to the graphs of the lines as the slopes become larger.
(b) Write an equation of a line that passes through the origin and is horizontal.

87. Graph $y = x + 1, y = x + 10$, and $y = x + 100$ in the window $-10 \le x \le 10, -10 \le y \le 10$.

(a) Explain what happens to the graph of a line, $y = b + mx$, as b becomes large.
(b) Write a linear equation whose graph cannot be seen in the window $-10 \le x \le 10, -10 \le y \le 10$ because all its y-values are less than the y-values shown.

88. (a) Find the equation of the line with intercepts
  (i) $(2, 0)$ and $(0, 5)$
  (ii) Double those in part (i)

(b) Are the two lines in part (a) parallel? Justify your answer.
(c) In words, generalize your conclusion to part (b). (There are many ways to do this; pick one. No justification is necessary.)

Italian coffee costs $10/lb, and Kenyan costs $15/lb. A workplace has a $60/week budget for coffee. If it buys I lbs of Italian coffee and K pounds of Kenyan coffee, which of the following equations best matches the statements in Problems 89–92?

(a) The total amount spent each week on coffee equals $60.

(b) The amount of Kenyan coffee purchased can be found by subtracting the amount spent on Italian coffee from the total and then dividing by the price per pound of Kenyan.

(c) The amount of Italian coffee that can be purchased is proportional to the amount of Kenyan coffee *not* purchased (as compared to the largest possible amount of Kenyan coffee that can be purchased).

(d) For every two pounds of Kenyan purchased, the amount of Italian purchased goes down by 3 lbs.

89. $I = 6 - \frac{3}{2}K$ **90.** $K = \dfrac{60 - 10I}{15}$

91. $I = 1.5(4 - K)$ **92.** $10I + 15K = 60$

93. A gram of fat contains 9 dietary calories, whereas a gram of carbohydrates contains only 4.[4]

(a) Write an equation relating the amount f, in grams, of fat and the amount c, in grams, of carbohydrates that one can eat if limited to a total of 2000 calories/day.
(b) The USDA recommends that calories from fat should not exceed 30% of all calories. What does this tell you about f?

94. Put the equation $250S + 400L = 10{,}000$ from Example 8 into slope-intercept form in two different ways by solving for **(a)** L **(b)** S.

Which form fits best with Figure 2.24?

95. The graph in Figure 2.26 shows the relationship between hearing ability score h and age a.

(a) What is the expected hearing ability score for a 40 year old?
(b) What age is predicted to have a hearing ability score of 40?
(c) What is the equation that relates hearing ability and age?

[4]Food and Nutrition Information Center, USDA, http://www.nalusda.gov/fnic/Dietary/9dietgui.htm

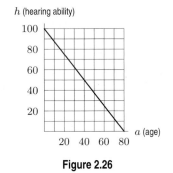

Figure 2.26

96. The final plans for the motel in Example 8 call for a total floor space of 16,000 ft^2, and for less spacious small rooms measuring 200 ft^2 instead of 250 ft^2. The large rooms are to remain 400 ft^2. Sketch a graph of the resulting constraint equation together with the constraint from Example 9. How do the two graphs compare?

The coffee variety *Arabica* yields about 750 kg of coffee beans per hectare, while *Robusta* yields about 1200 kg/hectare. In Problems 97–98, suppose that a plantation has a hectares of *Arabica* and r hectares of *Robusta*. [5]

97. Write an equation relating a and r if the plantation yields 1,000,000 kg of coffee.

98. On August 14, 2003, the world market price of coffee was $1.42/kg for *Arabica* and $0.73/kg for *Robusta*.[6] Write an equation relating a and r if the plantation produces coffee worth $1,000,000.

2.5 MODELING WITH LINEAR FUNCTIONS

Linear functions model many situations in the real world. In this section, we explore general linear models. If we are given the starting value and rate of change, we can write an expression for the function directly using the slope-intercept form.

Example 1 The amount C charged by a band for playing h hours includes a fixed fee of $100 plus an additional $40 per hour played. Express C as a linear function of h.

Solution The total cost of hiring the band for h hours is given by

$$\text{Total cost} = \text{Starting cost} + \text{Hourly fee} \times \text{Number of hours}$$
$$C = 100 + 40h.$$

In the next two examples the independent variable is not a measure of time. However, the coefficient of the variable can still be interpreted as a rate.

Example 2 The total cost of ownership of an inkjet printer, C, is a function of n, the number of 8×10 color photos printed. Express C as a linear function of n for the following printers:[7]

(a) The Hewlett-Packard Deskjet 6540 costs $130 and the cost per photo is $1.15.
(b) The Canon Pixma iP8500 costs $345 and the cost per photo is $0.95.

Solution In both cases,

$$\text{Total cost} = \text{Cost of printer} + \text{Cost per photo} \times \text{Number of photos},$$

so the cost functions are (a) $C = 130 + 1.15n$ (b) $C = 345 + 0.95n$.

---

[5]http://www.da.gov.ph/tips/coffee.html
[6]http://www.cafedirect.co.uk/about/gold_prices.php
[7]http://www.consumerreports.org. Figures are quoted from the May 2005 review of inkjet printers.

Example 3   A borehole is a hole dug deep in the earth, usually during exploration for oil or other minerals. The temperature-depth profile of a borehole is a description of how the temperature in the hole changes at different depths. It is often the case that the deeper one drills, the warmer the temperature in the borehole becomes. Consider a borehole whose average temperature at the surface is 4°C, and whose temperature rises by 0.02°C with each additional meter of depth. Use a linear function to describe the temperature-depth profile of this borehole.

Solution   Here, the starting value is $b = 4°C$, and the rate of change is $m = 0.02°C/m$. Since the rate of change is a constant, we can model this borehole's temperature-depth profile using the linear function $T = 4 + 0.02d$ for the temperature T in °C at depth d meters.

If we are not given the rate of change, we can use the formula for slope to find it.

Example 4   Worldwide, soda is the third most popular commercial beverage, after tea and milk. The global consumption of soda[8] rose at an approximately constant rate from 150 billion liters in 1995 to 179 billion liters in 2000.

(a) Find a linear function for the quantity of soda consumed, S, in billions of liters, t years after 1995.

(b) Give the units and practical interpretation of the slope and the vertical intercept.

Solution   (a) To find the linear function, we first find the rate of change, or slope. Since consumption increased from $S = 150$ to $S = 179$ over a period of 5 years, we have

$$\text{Rate of change} = \frac{\Delta S}{\Delta t} = \frac{179 - 150}{5} = \frac{29}{5} = 5.8.$$

When $t = 0$ (the year 1995), we have $S = 150$, so the vertical intercept is 150. Therefore

$$S = 150 + 5.8t.$$

(b) Since the rate of change is equal to $\Delta S/\Delta t$, its units are S-units over t-units, or billion liters per year. The slope tells us that world soda production has been increasing at a constant rate of 5.8 billion liters per year.

The vertical intercept 150 is the value of S when t is zero. Since it is a value of S, the units are S-units, or billion liters of soda. The vertical intercept tells us that the global consumption of soda in 1995 was 150 billion liters.

Deciding When to Use a Linear Model

Table 2.4 gives the temperature-depth profile measured in a borehole in 1988 in Belleterre, Quebec.[9] How can we decide whether a linear function models the data in Table 2.4? There are two ways to answer this question. We can plot the data to see if the points fall on a straight line, or we can calculate the slope to see if it is constant.

Table 2.4   *Temperature in a borehole at different depths*

d, depth (m)	150	175	200	225	250	275	300
H, temp (°C)	5.50	5.75	6.00	6.25	6.50	6.75	7.00

[8]The Worldwatch Institute, *Vital Signs 2002*, p. 140, (New York: W.W. Norton, 2002)

[9]Hugo Beltrami of St. Francis Xavier University and David Chapman of the University of Utah posted this data at http://geophysics.stfx.ca/public/borehole/borehole.html. Page last accessed on March 3, 2003.

Example 5   Is the temperature data in Table 2.4 linear with respect to depth? If so, find a formula for temperature, H, as a function of depth, d, for depths ranging from 150 m to 300 m.

Solution   Since the temperature rises by 0.25°C for every 25 additional meters of depth, the rate of change of temperature with respect to depth is constant. Thus, we use a linear function to model this relationship. The slope is the constant rate of change:

$$\text{Slope} = \frac{\Delta H}{\Delta d} = \frac{0.25}{25} = 0.01 \text{ °C/m}.$$

To find the linear function, we substitute this value for the slope and any point from Table 2.4 into the point-slope formula. For example, using the first entry in the table, $(d, H) = (150, 5.50)$, we get

$$H = 5.50 + 0.01(d - 150)$$
$$H = 5.50 + 0.01d - 1.5$$
$$H = 4.0 + 0.01d.$$

The linear function $H = 4.0 + 0.01d$ gives temperature as a function of depth.

Recognizing Values of a Linear Function

Values of x and y in a table could be values of a linear function $f(x) = b + mx$ if the same change in x-values always produces the same change in the y-values.

Example 6   Which of the following tables could represent values of a linear function?

(a)

x	20	25	30	35
y	17	14	11	8

(b)

x	2	4	6	8
y	10	20	28	34

Solution   (a) The x-values go up in steps of 5, and the corresponding y-values go down in steps of 3:

$$14 - 17 = -3 \quad \text{and} \quad 11 - 14 = -3 \quad \text{and} \quad 8 - 11 = -3.$$

Since the y-values change by the same amount each time, the table satisfies a linear equation.
(b) The x-values go up in steps of 2. The corresponding y-values do not go up in steps of constant size, since

$$20 - 10 = 10 \quad \text{and} \quad 28 - 20 = 8 \quad \text{and} \quad 34 - 28 = 6.$$

Thus, even though the value of Δx is the same for consecutive entries in the table, the value of Δy is not. This means the slope changes, so the table does not satisfy a linear equation.

Using Models to Make Predictions

In Example 5 we found a function for the temperature in a borehole based on temperature data for depths ranging from 150 meters to 300 meters. We can use our function to make predictions.

Example 7   Use the function from Example 5 to predict the temperature at the following depths. Do you think these predictions are reasonable?

(a)   260 m   (b)   350 m   (c)   25 m

Solution   (a) We have $H = 4 + 0.01(260) = 6.6$, so the function predicts a temperature of 6.6°C at a depth of 260 m. This is a reasonable prediction, because it is a little warmer than the temperature at 250 m and a little cooler than the temperature at 275 m:

	(shallower)	(in between)	(deeper)
Depth (m)	250	**260**	275
Temp (°C)	6.50	**6.60**	6.75
	(cooler)	(in between)	(warmer)

Keep in mind that even though this is a reasonable prediction, without additional data we have no real way of knowing what the temperature is at $d = 260$. There could be a hot spot at this depth or an underground stream that lowers the surrounding temperature.

(b) We have $H = 4 + 0.01(350) = 7.5$, so the function predicts a temperature of 7.5°C at a depth of 350 m. This seems plausible, since the temperature at 300 m is 7°C. However, it may be that the temperature at this depth is considerably warmer or colder than 7.5°C; without additional data, there is no way to be sure.

(c) We have $H = 4 + 0.01(25) = 4.25$, so the function predicts a temperature of 4.25°C at a depth of 25 m. Since this prediction is using a value of d that is well outside the range of our original data set, we should treat it with some caution. (In fact it turns out that at depths less than 100 m, the temperature actually rises as you get closer to the surface, instead of falling as our function predicts, so this is not a good prediction.)

Exercises and Problems for Section 2.5

Exercises

1. In 2002, the tuition cost for a part-time student at Nassau Community College was $112 per credit plus $35 in fees. Find the tuition cost for

   (a) 6 credits   **(b)** c credits.

2. The monthly charge for cell phone usage is $39.99 plus $0.40 for each daytime minute used over 300 minutes. Find the monthly cost for

   (a) 450 daytime minutes
   (b) m daytime minutes, if $m > 300$.

In Exercises 3–12, express the quantity as a linear function of the indicated variable.

3. The cost for h hours if a handyman charges $50 to come to the house and $45 for each hour that he works.

4. The population of a town in year y if the initial population is 23,400 and decreases by 200 people per year.

5. The value of a computer in year y, if it costs $2400 when new and depreciates $500 each year.

6. A collector starts with 5 figurines, and each year, y, she adds 2 more to her collection.

7. A population of p people increases by 3%.

8. A company's cumulative operating costs after m months if it needs $7600 for equipment, furniture, etc. and if the rent is $3500 for each month.

9. A student's test score if he answers p extra credit problems for 2 points each and has a base score of 80.

10. A driver's distance from home after h hours if she starts 200 miles from home and drives 50 miles per hour away from home.

11. A driver's distance from home after h hours if she starts 200 miles from home and drives 50 miles per hour toward home.

12. The cumulative cost of a gym membership after m months if the membership fee is $350 and the monthly rate is $30.

13. The value of a Honda Civic car is $11,835 one year after it was bought and $9765 three years after it was bought.[10] Assuming the value of the car decreases linearly, find

   (a) A formula for the value, V, of the car t years after it was bought.
   (b) The value 8 years after it was bought.
   (c) The value 13 years after it was bought.

[10]Data from Kelly Blue Book, www.kbb.com, January 18, 2004.

14. Snow begins falling steadily at time $t = 0$. The ground already has some snow on it, and at time $t = 2$, it is covered to a depth of 8 inches. Four hours later (at time $t = 6$), the depth is 14 inches. Find a possible formula for d, the depth of snow coverage (in inches), as a function of t, the amount of time elapsed (in hours). What does your formula tell you about the snow cover?

For Exercises 15–18, describe the journey of the person whose distance from home is shown in the graph, including their starting position and speed.

15. **16.**

17. **18.**

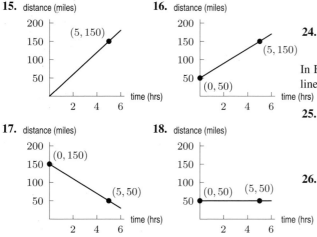

19. Logging and other activities have reduced the number of forested acres in a wilderness preserve, as shown by Figure 2.27. Find a formula for N, the number of forested acres remaining, as a function of t, the year.

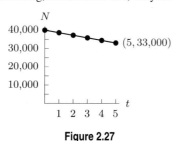

Figure 2.27

Find formulas for the linear functions in Exercises 20–24.

20. The graph of f contains the points $(-3, -8)$ and $(5, -20)$.

21. $g(100) = 2000$, $g(400) = 3800$

22. $P = h(t)$ gives the size of a population at time t in years, that begins with 12,000 members and grows by 225 members each year.

23. In year $t = 0$, a previously uncultivated region is planted with crops. The depth d of topsoil is initially 77 cm and begins to decline by 3.2 cm each year.

24. The graph of h intersects the graph of $y = x^2$ at $x = -2$ and $x = 3$.

In Exercises 25–29, could the table represent the values of a linear function?

25.

x	-6	-3	0	3	6
y	12	8	4	0	-4

26.

x	7	9	11	13	15
y	43	46	49	52	55

27.

x	2	4	8	16
y	5	10	15	20

28.

x	2	4	8	16	32
y	5	7	11	19	35

29.

x	-2	0	4	10
y	3	2	0	-3

Problems

30. A full printer ink cartridge holds 21 ml, enough to print 300 pages. Find a possible formula for V, the amount of remaining ink (in ml), as a function of n, the number of pages printed using an initially full cartridge. Assume each page uses the same amount of ink.

31. A line of people waiting for a concert is 80 people long twelve hours before the show and 200 people long ten hours before the show. If the line continues to grow at a steady rate, find a formula for N, the number of people waiting t hours before the show.

Match equations I–VI, which describe the value of investments after t years, to the statements in Problems 32–34.

I. $V = 5000 + 200t$.   II. $V = 500t$
III. $V = 8000$ IV. $V = 10,000 + 200(t - 10)$
V. $V = 200t + 7000$ VI. $V = 8500 - 550t$

32. These investments begin with the same amount of money.

33. This investment's value never changes.

34. This investment begins with the most money.

35. A car rental company charges $37.00 per day and $0.25 per mile. Find the cost of renting the car for

(a) One day and 100 miles
(b) One day and m miles
(c) d days and 100 miles.

36. Water is released from a dam holding 7,500,000,000 ft^3 of water at a rate of 45,000 ft^3/sec. Water flows into the dam at a rate of 20,000 ft^3/sec. Write an expression for the quantity of water in the dam, Q, in ft^3, t days after the release starts.

37. The number, D, of dollars per capita in circulation in the US between 1996 and 1998 can be modeled by $D = 1477 + 94.5t$, where $t = 1$ corresponds to the year 1996.[11] What do the numbers 1477 and 94.5 mean in practical terms? Find the value of t where $D = 0$, and explain its significance in terms of this model. Is this model realistic?

38. A family pays $117.50 in December for 100 therms of gas to heat their home (a therm is a unit of gas) and $172.50 for 150 therms in January. Assuming that the cost of gas depends linearly on the quantity used, find

(a) A formula for cost, C, in dollars in terms of g, the number of therms used.
(b) The cost of using 200 therms.
(c) The service fee charged even if no gas is used.

39. A study showed the average income of radiologists increased from about $145,000 in 1984 to $220,000 in 1990.[12] Assume that average incomes climbed linearly during the period 1982–1992.

(a) Find a formula for I, the average income in thousands of dollars t years after 1982.
(b) Estimate the average income in
  (i) 1987   (ii) 2050   (iii) 1980
(c) Which of your answers in part (b) do you think are likely to be reliable? Why?

40. As an airplane climbs, for several kilometers the outside air temperature drops approximately linearly with altitude. If the temperature is 8°C at 1500 meters and −12°C at 4 km, find the temperature at ground level. [Note: a kilometer is 1000 meters.]

41. A 200 gallon container contains 100 gallons of water. At time $t = 0$ minutes, water is added to the container at the rate of 5 gallons per minute, but drains out at the rate of 1 gallon per minute. How much water is in the container after t minutes? Is the container emptying or filling? When is the container empty/full? What happens after that?

42. The ice cap on Mount Kilimanjaro is melting at a rate of 1.6 feet per year, and, if current climate conditions continue, will vanish by the year 2020.[13] Express the height x of the ice cap in feet as a function of the year y.

43. In the 1990s the population of Phoenix was growing at approximately 100,000 people per year; by 2000, it had reached 3.25 million. Assuming that it continues to grow at the same rate, write a formula for the population, P, in millions, in terms of time, t, in years

(a) Since 2000   (b) Since 1990

In Problems 44–46, could the table represent the values of a linear function? Give a formula if it could.

44.

t	0	1	2	3	4	5
Q	100.00	95.01	90.05	85.11	80.20	75.31

45.

t	3	7	19	21	26	42
Q	5.79	6.67	9.31	9.75	10.85	14.37

46.

x	0	2	10	20
y	50	58	90	130

47. Table 2.5 shows the readings given by an instrument for measuring weight when various weights are placed on it.[14]

Table 2.5

w, weight (lbs)	0.0	0.5	1.0	1.5	2.0	2.5
I, reading	46.0	272.8	499.6	726.4	953.2	1180.0

(a) Is the reading, I, a linear function of the weight, w? If so, find a formula for it.
(b) Give possible interpretations for the slope and vertical intercept in your formula from part (a). [Hint: 1 lb = 453.6 g]

48. Table 2.6 shows the air temperature T as a function of the height h above the earth's surface.[15] Is T a linear function of h? Give a formula if it is.

Table 2.6

h, height (m)	0	2000	4000	6000	8000	10,000
T, temperature (°C)	15	2	−11	−24	−37	−50

---

[11] *The World Almanac 2002*, p. 112 (New York: World Almanac Books 2002).

[12] Adapted from Martin Gonzalez, ed. "Socioeconomic Indicators of Medical Practice", AMA, 1994. Reported by Ann Watkins et al. in *Statistics in Action*.

[13] Thompson, Lonnie G., et al. *Kilimanjaro Ice Core Records: Evidence of Holocene Climate Change in Tropical Africa* Science 2002 298: pp. 589-593

[14] Adapted from J.G. Greeno, *Elementary Theoretical Psychology*, (Massachusetts: Addison-Wesley, 1968).

[15] Adapted from H. Tennekes, *The Simple Science of Flight*, (Cambridge: MIT Press, 1996).

2.6 SYSTEMS OF LINEAR EQUATIONS

In Section 2.4 we considered linear equations in two variables, whose solutions are pairs of numbers. For any given x-value, we can solve for y to find a corresponding y-value, and vice versa.

Example 1

Find solutions to the equation $3x + y = 14$ given the following values of x and y.

(a) $x = 1, 2, 3$ (b) $y = 1, 2, 3$.

Solution

(a) We substitute these values for x then solve for y:

$$
\begin{array}{llll}
x = 1: & 3(1) + y = 14 & \rightarrow & y = 11 \\
x = 2: & 3(2) + y = 14 & \rightarrow & y = 8 \\
x = 3: & 3(3) + y = 14 & \rightarrow & y = 5.
\end{array}
$$

The solutions are the (x, y)-pairs $(1, 11), (2, 8), (3, 5)$.

(b) We substitute these values for y then solve for x:

$$
\begin{array}{llll}
y = 1: & 3x + 1 = 14 & \rightarrow & x = 13/3 \\
y = 2: & 3x + 2 = 14 & \rightarrow & x = 4 \\
y = 3: & 3x + 3 = 14 & \rightarrow & x = 11/3.
\end{array}
$$

The solutions are the (x, y)-pairs $(13/3, 1), (4, 2), (11/3, 3)$.

Sometimes we are interested in finding pairs that satisfy more than one equation at the same time.

Example 2

Admission to a play for one adult and two children costs $11. Admission for two adults and three children costs $19. What are the admission prices for each?

Solution

If A is the admission price for an adult and C the admission cost for a child, then

$$A + 2C = 11 \quad \text{and} \quad 2A + 3C = 19.$$

We want a pair of values for A and C that satisfies both of these equations simultaneously. Because adults are often charged more than children, we might first guess that $C = 1$ and $A = 9$. While these values satisfy the first equation, substituting them into the second gives

$$2 \cdot 9 + 3 \cdot 1 = 19,$$

which is not true, since the left-hand side adds up to 21. However, the values $C = 3$ and $A = 5$ satisfy both equations. Thus, the price of admission is $3 for a child and $5 for an adult.

What is a System of Equations?

In Example 2, we saw that the (A, C)-pair $(5, 3)$ is a solution to the equation $A + 2C = 11$ and also to the equation $2A + 3C = 19$. Since this pair of values makes both equations true, we say that it is a solution to the *system of equations*

$$
\begin{cases}
A + 2C = 11 \\
2A + 3C = 19.
\end{cases}
$$

> ## Systems of Equations
>
> A **system of equations** is a set of two or more equations. A solution to a system of equations is a set of values for the variables that makes all of the equations true.

We write the system of equations with a brace to indicate that a solution must satisfy both equations.

Example 3   Solve the system of equations

$$\begin{cases} 3x + y = 14 & \text{(equation 1)} \\ 2x + y = 11. & \text{(equation 2)} \end{cases}$$

Solution   In Example 1 we saw that solutions to equation 1 include the pairs $(1, 11), (2, 8)$, and $(3, 5)$. We can also find pairs of solutions to equation 2:

$$\begin{aligned} x = 1: &\quad 2(1) + y = 11 &\rightarrow&\quad y = 9 \\ x = 2: &\quad 2(2) + y = 11 &\rightarrow&\quad y = 7 \\ x = 3: &\quad 2(3) + y = 11 &\rightarrow&\quad y = 5 \end{aligned}$$
$$\vdots$$

As before, the solutions are pairs: $(1, 9), (2, 7), (3, 5), \ldots$. Notice that one of these pairs, $(3, 5)$, is also a solution to the first equation. Since this pair of values solves both equations at the same time, it is a solution to the system.

Solving Systems of Equations Using Substitution

In Example 3, we found the solution by trial and error. We will now consider two algebraic methods for solving systems of equations. The first method is known as *substitution*. In this approach we solve for one of the variables and then substitute the solution into the other equation.

Example 4   Solve the system

$$\begin{cases} 3x + y = 14 & \text{(equation 1)} \\ 2x + y = 11. & \text{(equation 2)} \end{cases}$$

Solution   Solving equation 1 for y gives

$$y = 14 - 3x \quad \text{(equation 3)}$$

Since a solution to the system satisfies both equations, we substitute this expression for y into equation 2 and solve for x:

$$\begin{aligned} 2x + y &= 11 \\ 2x + \underbrace{14 - 3x}_{y} &= 11 \quad \text{using equation 3 to substitute for } y \text{ in equation 2} \\ 2x - 3x &= 11 - 14 \\ -x &= -3 \\ x &= 3. \end{aligned}$$

To find the corresponding y-value, we can substitute $x = 3$ into either equation. Choosing equation 1, we have $3(3) + y = 14$. Thus $y = 5$, so the (x, y)-pair $(3, 5)$ satisfies both equations, as we now verify:

$$3(3) + 5 = 14 \longrightarrow \quad (3, 5) \text{ solves equation 1}$$
$$2(3) + 5 = 11 \longrightarrow \quad (3, 5) \text{ solves equation 2.}$$

To summarize:

Method of Substitution

- Solve for one of the variables in one of the equations.
- Substitute into the other equation to get an equation in one variable.
- Solve the new equation, then find the value of the other variable by substituting back into one of the original equations.

Deciding Which Substitution Is Simplest

When we use the method of substitution, we can choose which equation to start with and which variable to solve for. For instance, in Example 4, we began by solving equation 1 for y, but this is not the only possible approach. We might instead solve equation 1 for x, obtaining

$$x = \frac{14 - y}{3}. \quad \text{(equation 4)}$$

We can use this equation to substitute for x in equation 2, obtaining

$$2 \underbrace{\left(\frac{14 - y}{3} \right)}_{x} + y = 11 \quad \text{using equation 4 to substitute for } x \text{ in equation 2}$$

$$\frac{2}{3}(14 - y) + y = 11$$
$$2(14 - y) + 3y = 33 \quad \text{multiply both sides by 3}$$
$$28 - 2y + 3y = 33$$
$$y = 33 - 28 = 5.$$

From equation 4, we have $x = (14 - 5)/3 = 3$, so we get the solution $(x, y) = (3, 5)$. This is same pair of values as before, only this time the fractions make the algebra messier. The moral is that we should make our substitution as simple as possible. One way to do this is to look for terms having a coefficient of 1. This is why we chose to solve for y in Example 4.

Solving Systems of Equations Using Elimination

Another technique for solving systems of equations is known as *elimination*. As the name suggests, the goal of elimination is to eliminate one of the variables from the system.

Example 5   We solve the system from Example 4 using elimination instead of substitution:

$$\begin{cases} 3x + y = 14 & \text{(equation 1)} \\ 2x + y = 11. & \text{(equation 2)} \end{cases}$$

Solution

Since equation 2 tells us that $2x + y = 11$, we subtract $2x + y$ from both sides of equation 1:

$$3x + y - (2x + y) = 14 - (2x + y)$$
$$3x + y - (2x + y) = 14 - 11 \quad \text{(equation 5)} \quad \text{because } 2x + y \text{ equals } 11$$
$$3x + y - 2x - y = 3$$
$$x = 3.$$

Notice that we have eliminated the y variable from the system, allowing us to solve for x, obtaining $x = 3$. As before, we substitute $x = 3$ into either equation to find the corresponding value of y. Our solution, $(x, y) = (3, 5)$, is the same answer that we got in Example 4.

Adding and Subtracting Equations

In the last example, we subtracted $2x + y$ from the left-hand side of equation 1 and subtracted 11 from the right-hand side. We can do this because, according to equation 2, the expression $2x + y$ equals 11. Notice that we have, in effect, subtracted equation 2 from equation 1. We write

$$\text{Equation 5} = \text{Equation 1} - \text{Equation 2}.$$

Similarly, we say that we add two equations when we add their left- and right-hand sides, respectively. We also say that we multiply an equation by a constant when we multiply both sides by a constant.

In the next example, we multiply an equation by -2 and then add the result to a second equation.

Example 6

Solve the following system using elimination.

$$\begin{cases} 2x + 7y = -3 & \text{(equation 6)} \\ 4x - 2y = 10. & \text{(equation 7)} \end{cases}$$

Solution

In preparation for eliminating x from the two equations, we multiply equation 6 by -2 in order to make the coefficient of x into -4, the negative of the coefficient in equation 7. This gives equation 8:

$$\underbrace{-4x - 14y}_{-2(2x+7y)} = \underbrace{6}_{-2(-3)} \quad \text{(equation 8} = -2 \times \text{equation 6)}$$

Notice that equation 7 has a $4x$ term and equation 8 has a $-4x$ term. We can eliminate the variable x by adding these two equations:

$$\begin{array}{rrcll} 4x & -2y & = & 10 & \text{(equation 7)} \\ + \quad -4x & -14y & = & 6 & \text{(equation 8)} \\ \hline & -16y & = & 16 & \text{(equation 9} = \text{equation 7} + \text{equation 8)} \end{array}$$

Solving equation 9 we obtain $y = -1$. We substitute $y = -1$ into equation 6 to solve for x:

$$2x + 7(-1) = -3 \quad \text{solving for } x \text{ using equation 6}$$
$$2x = 4$$
$$x = 2.$$

Thus, $(2, -1)$ is the solution. We can verify this solution using equation 7:

$$4 \cdot 2 - 2(-1) = 8 + 2 = 10.$$

In summary:

Method of Elimination

- Multiply one or both equations by constants so that the coefficients of one of the variables are either equal or are negatives of each other.
- Add or subtract the equations to eliminate that variable and solve the resulting equation for the other variable.
- Substitute back into one of the original equations to find the value of the first variable.

Solving Systems of Equations Using Graphs

The number of solutions to a system of linear equations is related to whether their graphs are parallel lines, overlapping lines, or lines that intersect at a single point.

One Solution: Intersecting Lines

In Example 5 we solved the system of equations:

$$\begin{cases} 3x + y = 14 \\ 2x + y = 11. \end{cases}$$

We can also solve this system graphically. In order to graph the lines, we rewrite both equations in slope-intercept form:

$$\begin{cases} y = 14 - 3x \\ y = 11 - 2x. \end{cases}$$

The graphs are shown in Figure 2.28. Each line shows all solutions to one of the equations. Their point of intersection is the common solution of both equations. It appears that these two lines intersect at the point $(x, y) = (3, 5)$. We check that these values satisfy both equations, so the solution to the system of equations is $(x, y) = (3, 5)$.

If it exists, the solution to a system of two linear equations in two variables is the point at which their graphs intersect.

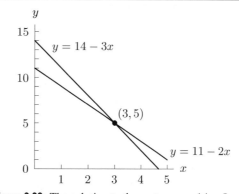

Figure 2.28: The solution to the system $y = 14 - 3x$ and $y = 11 - 2x$ is the point of intersection of these two lines

No Solutions: Parallel Lines

The system in the next example has no solutions.

Example 7   Solve

$$\begin{cases} 6x - 2y = 8 & \text{(equation 10)} \\ 9x - 3y = 6. & \text{(equation 11)} \end{cases}$$

Solution   We can use the process of elimination to eliminate the x terms.

$$18x - 6y = 24 \quad \text{(equation 12 = 3 × equation 10)}$$
$$18x - 6y = 12. \quad \text{(equation 13 = 2 × equation 11)}$$

There are no values of x and y that can make $18x - 6y$ equal to both 12 and 24, so there is *no* solution to this system of equations. To see this graphically, we rewrite each equation in slope-intercept form:

$$\begin{cases} y = -4 + 3x \\ y = -2 + 3x. \end{cases}$$

Since their slopes are both equal $m = 3$, the lines are parallel and they do not intersect. Therefore, the system of equations has no solution. See Figure 2.29.

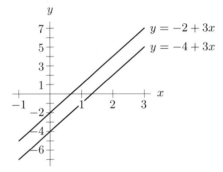

Figure 2.29: The system $y = -4 + 3x$ and $y = -2 + 3x$ has no solution, corresponding to the fact that the graphs of these equations are parallel (non-intersecting) lines

Many Solutions: Overlapping Lines

The system in the next example has many solutions.

Example 8   Solve:

$$\begin{cases} x - 2y = 4 & \text{(equation 14)} \\ 2x - 4y = 8. & \text{(equation 15)} \end{cases}$$

Solution   Notice that equation 15 is really equation 14 in disguise: Multiplying equation 14 by 2 gives equation 15. Therefore, our system of equations is really just the same equation written in two different ways. The set of solutions to the system of equations is the same as the set of solutions to either one of the equations separately. To see this graphically, we rewrite each equation in slope-intercept form:

$$\begin{cases} y = -2 + 0.5x \\ y = -2 + 0.5x. \end{cases}$$

We get the same equation twice, so the two equations represent the same line. The graph of this system is the single line $y = -2 + 0.5x$, and any (x, y) point on the line is a solution to the system. See Figure 2.30. The system of equations has infinitely many solutions.

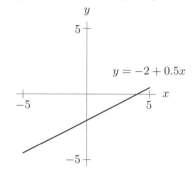

Figure 2.30: The system $x - 2y = 4$ and $2x - 4y = 8$ has infinitely many solutions, since both equations give the same line and every point on the line is a solution

Applications of Systems of Linear Equations

Example 9   At a fabric store, silk costs three times as much as cotton. A customer buys 4 yards of cotton and 1.5 yards of silk, for a total cost of $55. What is the cost per yard of cotton and what is the cost per yard of silk?

Solution   Let x be the cost of cotton and y the cost of silk, in dollars per yard. Then

$$\text{Total cost} = 4 \times \text{Cost per yard of cotton} + 1.5 \times \text{Cost per yard of silk}$$
$$55 = 4x + 1.5y.$$

Also, since the price of silk is three times the price of cotton, we have $y = 3x$. Thus we have the system of equations:

$$\begin{cases} y = 3x \\ 55 = 4x + 1.5y. \end{cases}$$

We use the method of substitution, since the first equation is already in the right form to substitute into the second:

$$55 = 4x + 1.5(3x)$$
$$55 = 4x + 4.5x$$
$$55 = 8.5x$$
$$x = \frac{55}{8.5} = 6.47.$$

Thus cotton costs $6.47 per yard and silk costs $3(6.47) = \$19.41$ per yard.

Example 10   A farmer raises chickens and pigs. His animals together have a total of 95 heads and a total of 310 legs. How many chickens and how many pigs does the farmer have?

Solution   We let x represent the number of chickens and y represent the number of pigs. Each animal has one head, so we know $x + y = 95$. Since chickens have 2 legs and pigs have 4 legs, we know $2x + 4y = 310$. We solve the system of equations:

$$\begin{cases} x + y = 95 \\ 2x + 4y = 310. \end{cases}$$

We can solve this system using the method of substitution or the method of elimination. Using the method of substitution, we solve the first equation for y:

$$y = 95 - x.$$

We substitute this for y in the second equation and solve for x:

$$2x + 4y = 310$$
$$2x + 4(95 - x) = 310$$
$$2x + 380 - 4x = 310$$
$$-2x = 310 - 380$$
$$-2x = -70$$
$$x = 35.$$

Since $x = 35$ and $x + y = 95$, we have $y = 60$. The farmer has 35 chickens and 60 pigs.

Exercises and Problems for Section 2.6

Exercises

Solve the systems of equations in Exercises 1–19.

1. $\begin{cases} x + y = 5 \\ x - y = 7 \end{cases}$

2. $\begin{cases} 3x + y = 10 \\ x + 2y = 15 \end{cases}$

3. $\begin{cases} 3x - 4y = 7 \\ y = 4x - 5 \end{cases}$

4. $\begin{cases} x = y - 9 \\ 4x - y = 0 \end{cases}$

5. $\begin{cases} 2a + 3b = 4 \\ a - 3b = 11 \end{cases}$

6. $\begin{cases} 3w - z = 4 \\ w + 2z = 6 \end{cases}$

7. $\begin{cases} 2p + 3r = 10 \\ -5p + 2r = 13 \end{cases}$

8. $\begin{cases} 5d + 4e = 2 \\ 4d + 5e = 7 \end{cases}$

9. $\begin{cases} 8x - 3y = 7 \\ 4x + y = 11 \end{cases}$

10. $\begin{cases} 4w + 5z = 11 \\ z - 2w = 5 \end{cases}$

11. $\begin{cases} 20n + 50m = 15 \\ 70m + 30n = 22 \end{cases}$

12. $\begin{cases} r + s = -3 \\ s - 2r = 6 \end{cases}$

13. $\begin{cases} y = 20 - 4x \\ y = 30 - 5x \end{cases}$

14. $\begin{cases} 2p + 5q = 14 \\ 5p - 3q = 4 \end{cases}$

15. $\begin{cases} 9x + 10y = 21 \\ 7x + 11y = 26 \end{cases}$

16. $\begin{cases} 7x + 5y = -1 \\ 11x + 8y = -1 \end{cases}$

17. $\begin{cases} 5x + 2y = 1 \\ 2x - 3y = 27 \end{cases}$

18. $\begin{cases} 11v + 7w = 2 \\ 13v + 8w = 1 \end{cases}$

19. $\begin{cases} 3e + 2f = 4 \\ 4e + 5f = -11 \end{cases}$

Determine the points of intersection for Exercises 20–23.

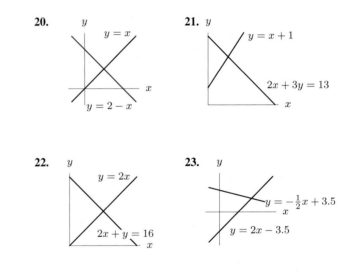

20. $y = x$, $y = 2 - x$

21. $y = x + 1$, $2x + 3y = 13$

22. $y = 2x$, $2x + y = 16$

23. $y = -\frac{1}{2}x + 3.5$, $y = 2x - 3.5$

In Exercises 24–27, solve the system of equations graphically.

24. $\begin{cases} y = 6x - 7 \\ y = 3x + 2 \end{cases}$

25. $\begin{cases} y = -2x + 7 \\ y = 4x + 1 \end{cases}$

26. $\begin{cases} 2x + 5y = 7 \\ -3x + 2y = 1 \end{cases}$

27. $\begin{cases} y = 22 + 4(x - 8) \\ y = 11 - 2(x + 6) \end{cases}$

Problems

Solve the systems of equations in Problems 28–35.

28. $\begin{cases} 2x + 5y = 1 \\ 2y - 3x = 8 \end{cases}$ **29.** $\begin{cases} 7x - 3y = 24 \\ 4y + 5x = 11 \end{cases}$

30. $\begin{cases} 5x - 7y = 31 \\ 2x + 3y = -5 \end{cases}$ **31.** $\begin{cases} 11\alpha - 7\beta = 31 \\ 4\beta - 3\alpha = 2 \end{cases}$

32. $\begin{cases} 3\alpha + \beta = 32 \\ 2\beta - 3\alpha = 1 \end{cases}$ **33.** $\begin{cases} 3x - 2y = 4 \\ 3y - 5x = -5 \end{cases}$

34. $\begin{cases} 3(e + f) = 5e + f + 2 \\ 4(f - e) = e + 2f - 4 \end{cases}$

35. $\begin{cases} 7\kappa - 9\psi = 23 \\ 2\kappa + 3\psi = 1 \end{cases}$

Solve the systems of equations in Problems 36–39 for x and y.

36. $\begin{cases} 2x + 4y = 44 \\ y = \frac{3}{4}x + 6 \end{cases}$ **37.** $\begin{cases} 3x - y = 20 \\ -2x - 3y = 5 \end{cases}$

38. $\begin{cases} 2(x + y) = 5 \\ x = y + 3(x - 3) \end{cases}$

39. $\begin{cases} bx + y = 2b \\ x + by = 1 + b^2, \quad \text{if } b \neq \pm 1 \end{cases}$

40. You want to build a patio. Builder A charges $3 a square foot plus a $500 flat fee, and builder B charges $2.50 a square foot plus a $750 flat fee. For each builder, write an expression relating the cost C to the area s square feet of the patio. Which builder is cheaper for a 200 square foot patio? Which is cheaper for a 1000 square foot patio? For what size patio will both builders charge the same?

41. Two companies sell and deliver sand used as a base for building patios. Company A charges $10 a cubic yard and a flat $40 delivery fee for any amount of sand up to 12 cubic yards. Company B charges $8 a cubic yard and a flat $50 delivery fee for any amount of sand up to 12 cubic yards. You wish to have x cubic yards of sand delivered, where $0 < x \leq 12$. Which company should you buy from?

42. Consider two numbers x and y satisfying the equations $x + y = 4$ and $x - y = 2$.

(a) Describe in words the conditions that each equation places on the two numbers.
(b) Find two numbers x and y satisfying both equations.

43. Find two numbers with sum 17 and difference 12.

44. A motel plans to build small rooms of size 250 ft^2 and large rooms of size 500 ft^2, for a total area of 16,000 ft^2. Also, local fire codes limit the legal occupancy of the small rooms to 2 people and of the large rooms to 5 people, and the total occupancy of the entire motel is limited to 150 people.

(a) Use linear equations to express the constraints imposed by the size of the motel and by the fire code.
(b) Solve the resulting system of equations. What does your solution tell you about the motel?

45. A fast-food fish restaurant serves meals consisting of fish, chips, and hush-puppies.

- One fish, one order of chips, and one pair of hush-puppies costs $2.27.
- Two fish, one order of chips, and one pair of hush-puppies costs $3.26.
- One fish, one order of chips, and two pairs of hush-puppies costs $2.76.

(a) How much should a meal of two fish, two orders of chips, and one pair of hush-puppies cost?
(b) Show that

$$2x + 2y + z = 3(x + y + z) - (x + y + 2z)$$

is an identity in x, y, and z.
(c) Use the identity in part (b) to solve part (a).
(d) Someone says that you do not need the information that "Two fish, one order of chips, and one pair of hush-puppies costs $3.26" to solve part (a). Is this true?

46. Solve the system of equations

$$\begin{cases} 3x + 2y + 5z = 11 \\ 2x - 3y + z = 7 \\ z = 2x \end{cases}$$

Hint: Use the third equation to substitute for z in the other two.

47. For the system

$$\begin{cases} 2x + 3y = 5 \\ 4x + 6y = n, \end{cases}$$

what must be true about n in order for there to be many solutions?

Measures in US recipes are not the same size as in UK recipes. Suppose that a UK cup is u US cups, a UK tablespoon is t US cups, and a UK dessertspoon is d US cups. Problems 48–49 use information from the same web page.[16]

48. According to the web, 3/4 of a US cup is 1/2 a UK cup plus 2 UK tablespoons, and 2/3 of a US cup is 1/2 a UK cup plus 1 UK tablespoon, so

$$\frac{3}{4} = \frac{1}{2}u + 2t$$
$$\frac{2}{3} = \frac{1}{2}u + t.$$

Solve this system and interpret your answer in terms of

the relation between UK and US cups, and between the UK tablespoon and the US cup.

49. According to the web, 1 US cup is 3/4 of a UK cup plus 2 UK dessertspoons, and 1/4 of a US cup is 1/4 of a UK cup minus 1 UK dessertspoon, so

$$1 = \frac{3}{4}u + 2d$$
$$\frac{1}{4} = \frac{1}{4}u - d.$$

Solve this system and interpret your answer in terms of the relation between UK and US cups, and the UK dessertspoon and the US cup.

REVIEW EXERCISES AND PROBLEMS FOR CHAPTER TWO

Exercises

In Exercises 1–4 give the initial value and the slope and explain their meaning in practical terms.

1. The rental charge, C, at a video store is given by $C = 4.29 + 3.99n$, where n is the number of days greater than 2 for which the video is kept.

2. While driving back to college after spring break, a student realizes that his distance from home, D, in miles, is given by $D = 55t + 30$, where t is the time in hours since the student made the realization.

3. The cost, C, in dollars, of making n donuts is given by $C = 250 + n/36$.

4. The number of people, P, remaining in a lecture hall m minutes after the start of a very boring lecture is given by $P = 300 - 19m/3$.

5. A child returns from trick-or-treating with 220 pieces of candy. She eats 5 pieces every day. Write an expression for the number of pieces remaining after t days.

6. A candle of length $L = 12$ inches burns at a rate of 2 inches every 3 hours. How long is the candle after t hours?

Identify the constants b and m in the expression $b + mt$ for the linear functions in Exercises 7–12.

7. $f(t) = 200 + 14t$

8. $g(t) = 77t - 46$

9. $h(t) = t/3$

10. $p(t) = 0.003$

11. $q(t) = \dfrac{2t + 7}{3}$

12. $r(t) = \sqrt{7} - 0.3t\sqrt{8}$

Evaluate the expressions in Exercises 13–16 given that $w(x)$ is a linear function defined by

$$w(x) = 9 + 4x.$$

Simplify your answers.

13. $w(-4)$

14. $w(x - 4)$

15. $w(x + h) - w(x)$

16. $w(9 + 4x)$

Without solving them, say whether the equations in Problems 17–28 have a positive solution, a negative solution, a zero solution, or no solution.

17. $7x = 5$

18. $3x + 5 = 7$

19. $5x + 3 = 7$

20. $5 - 3x = 7$

21. $3 - 5x = 7$

22. $9x = 4x + 6$

23. $9x = 6 - 4x$

24. $9 - 6x = 4x - 9$

25. $8x + 11 = 2x + 3$

26. $11 - 2x = 8 - 4x$

27. $8x + 3 = 8x + 11$

28. $8x + 3x = 2x + 11x$

In Exercises 29–40, say whether the given expression is linear in the indicated variable. Assume all constants are non-zero.

29. $5x - 7 + 2x,\ x$

30. $3x - 2x^2,\ x$

31. $mx + b + c^3,\ x$

32. $mx + b + c^3 x^2,\ x$

33. $P(P - b)(c - P),\ P$

34. $P(P - b)(c - P),\ c$

35. $P(2 + P) - P^2,\ P$

36. $P(2 + P) - 2P^2,\ P$

37. $xy + ax + by + ab,\ x$

38. $xy + ax + by + ab,\ y$

39. $xy + ax + by + ab,\ a$

40. $xy + ax + by + ab,\ b$

[16]http://allrecipes.com/advice/ref/conv/conversions_brit.asp, accessed on May 14, 2003.

Solve the equations in Exercises 41–47.

41. $7 - 3y = -17$ **42.** $13t + 2 = 49$

43. $3t + \dfrac{2(t-1)}{3} = 4$

44. $2(r+5) - 3 = 3(r-8) + 20$

45. $2x + x = 27$

46. $4t + 2(t+1) - 5t = 13$

47. $\dfrac{9}{x-3} - \dfrac{5}{1-x} = 0.$

48. Using simple interest, an investment is worth $I = 1000 + 50t$ dollars after t years. What do 1000 and 50 mean in practical terms? What is the interest rate?

49. The value of a car after t years is $V = 20{,}000 - 1000t$ dollars. What do the 20,000 and 1000 mean in practical terms?

In Exercises 50–53, solve for the indicated variable. Assume all constants are non-zero.

50. $I = Prt$, for r.   **51.** $F = \dfrac{9}{5}C + 32$, for C.

52. $\dfrac{a - cy}{b + dy} + a = 0$, for y, if $c \neq ad$.

53. $\dfrac{Ax - B}{C - B(1 - 2x)} = 3$, for x, if $A \neq 6B$.

In Exercises 54–55, which line has the greater

(a) Slope?   **(b)** y-intercept?

54. $y = 5 - 2x, \quad y = 8 - 4x$

55. $y = 7 + 3x, \quad y = 8 - 10x$

In Exercises 56–57, graph the equation.

56. $y = \dfrac{2}{3}x - 4$ **57.** $4x + 5y = 2$

Write the equations in Exercises 58–63 in the form $y = b + mx$, and identify the values of b and m.

58. $y = \frac{x}{8} - 14$ **59.** $y - 8 = 3(x - 5)$

60. $\dfrac{y - 4}{5} = \dfrac{2x - 3}{3}$ **61.** $4x - 7y = 12$

62. $y = 90$ **63.** $y = \sqrt{8}x$

For Exercises 64–65, put the equation in standard form.

64. $y = 3x - 2$ **65.** $3x = 2y - 1$

In Exercises 66–68, describe in words how the temperature changes with time.

66.

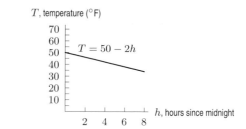

67.

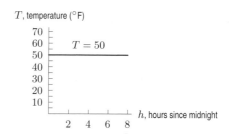

68.

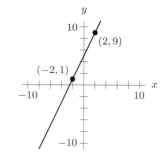

In Exercises 69–71, write an equation for each of the lines whose graph is shown.

69.

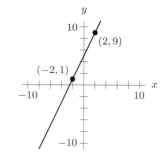

70.

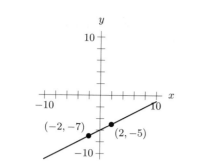

71.

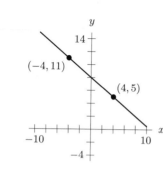

In Exercises 72–83, write an equation in slope-intercept form for the line that

72. Passes through $(4, 6)$ with slope $m = 2$

73. Passes through $(-6, 2)$ with slope $m = -3$

74. Passes through $(-4, -8)$ with slope $m = 1/2$

75. Passes through $(9, 7)$ with slope $m = -2/3$

76. Contains $(6, 8)$ and $(8, 12)$

77. Contains $(-3, 5)$ and $(-6, 4)$

78. Contains $(4, -2)$ and $(-8, 1)$

79. Contains $(5, 6)$ and $(10, 3)$

80. Contains $(4, 7)$ and is parallel to $y = (1/2)x - 3$

81. Contains $(-10, -5)$ and is parallel to $y = -(4/5)x$

82. Passes through $(0, 5)$ and is parallel to $3x + 5y = 6$

83. Passes through $(-10, -30)$ and is perpendicular to $12y - 4x = 8$.

Find formulas for the linear functions described in Exercises 84–91.

84. The graph intercepts the x-axis at $x = 30$ and the y-axis at $y = -80$.

85. $P = s(t)$ describes a population that begins with 8200 members in year $t = 0$ and reaches 12,700 members in year $t = 30$.

86. The total cost C of an international call lasting n minutes if there is a connection fee of \$2.95 plus an additional charge of \$0.35/minute.

87. $w(x)$, where $w(4) = 20, w(12) = -4$

88. The graph of $y = p(x)$ contains the points $(-30, 20)$ and $(70, 140)$.

89. $g(x)$, where $g(5) = 50, g(30) = 25$

90. This function's graph is parallel to the line $y = 20 - 4x$ and contains the point $(3, 12)$.

91. The graph of $f(x)$ passes through $(-1, 4)$ and $(2, -11)$.

Exercises 92–97 give data from a linear function. Find a formula for the function.

92.

Temperature, $y = f(x)$ (°C)	0	5	20
Temperature, x (°F)	32	41	68

93.

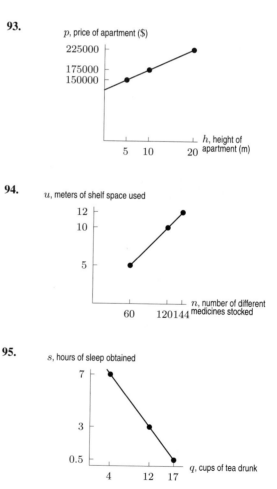

94.

95.

96.

Temperature, $y = f(x)$, (°R)	459.7	469.7	489.7
Temperature, x (°F)	0	10	30

97.

Price per bottle, p ($)		0.50	0.75	1.00
Number of bottles sold, $q = f(p)$		1500	1000	500

In the tables in Exercises 98–103 could the second row be a linear function of the first?

98.

t	1	2	3	4	5
v	5	4	3	2	1

99.

p	−4	−2	0	2	4
q	15	15.5	16	16.5	17

100.

x	0	1	2	3	4
y	0	1	4	9	16

101.

n	3	6	9	12	15
C	5	4	3	4	5

102.

x	1	5	11	19
y	10	8	5	1

103.

x	1	3	9	27
y	−5	−10	−15	−20

Problems

104. Water is added to a barrel at a constant rate for 25 minutes, after which it is full. The quantity of water in the barrel after t minutes is $100 + 4t$ gallons. What do the 100 and 4 mean in practical terms? How much water can the barrel hold?

105. A party facility charges $500 for a banquet room and $20 per person. In addition there is a 20% surcharge on the entire fee. Why would you expect the total cost of the party to be given by a linear function of the number of people, P? Give the function.

106. A design shop offers ceramics painting classes. The fee for the class is $30 and each item that is painted costs $12. There is also a firing fee of $3 for each item and a 7% tax on the cost of the item. Why would you expect the total cost of the class to be given by a linear function of the number of items, i? Give the function.

107. The total cost of ownership for an Epson Stylus Photo R320 printer is given by

$$\text{Total cost} = 200 + 1.00n,$$

where n is the number of 8×10 photos printed. Interpret the constants 200 and 1.00 in practical terms.

108. A ski shop rents out skis for an initial payment of I dollars plus r dollars per day. A skier spends $I + 9r$ dollars on skis for a vacation. What is the meaning of the 9 in this formula?

109. The total number of FDIC-insured banks in the U.S. between 1993 and 2000 can be approximated by $N = 13 - 0.48t$ thousands of banks, where t is the number of years since 1993.[17] What do the numbers 13 and 0.48 represent? If this trend continues, when will there be no FDIC-insured banks in the U.S.?

110. The total area of farmland in the U.S. between 1993 and 2000 can be approximated by $A = 970 - 3.6t$ millions of acres, where t is in years 1993.[18] What do the numbers 970 and 3.6 represent? If this trend continues, when will there be no farmland in the U.S.?

Without solving, decide which of the following statements applies to the solution to each of the equations in Problems 111–130:

(i) $x \leq -1$ (ii) $-1 < x < 0$ (iii) $x = 0$
(iv) $0 < x < 1$ (v) $x \geq 1$ (vi) No solution

Give a reason for your answer.

111. $7x = 4$

112. $4x + 2 = 7$

113. $2x + 8 = 7$

114. $11x + 7 = 2$

115. $11x - 7 = 5$

116. $8 - 2x = 7$

117. $5x + 3 = 7x + 5$

118. $5x + 7 = 2x + 5$

[17]*The World Almanac 2002,* p. 108 (New York: World Almanac Books 2002).
[18]National Agricultural Statistics Service, U.S. Department of Agriculture, 2002.

119. $3 - 4x = -4x - 3$ **120.** $5x + 3 = 8x + 3$

121. $\dfrac{x}{2} = \dfrac{1}{5}$ **122.** $\dfrac{2}{x} = \dfrac{3}{5}$

123. $\dfrac{x + 8}{2} = 4$ **124.** $\dfrac{x + 8}{3} = 2$

125. $\dfrac{10}{5 + x} = 1$ **126.** $\dfrac{10}{5 + x} = 2$

127. $\dfrac{2x + 5}{2x + 6} = 1$ **128.** $\dfrac{2x + 2}{3x + 5} = 1$

129. $\dfrac{1}{x} - 2 = 2$ **130.** $3 - \dfrac{1}{x} = 3$

In Problems 131–136, find possible formulas for the linear functions described if $f(t)$ is a linear function with y-intercept r and slope s. Your formulas will involve the positive constants r and s.

131. This function has the same slope as f, but its y-intercept is 3 units higher.

132. This function has the same y-intercept as f, but its graph is three times as steep.

133. The axis intercepts of this function's graph are twice as far from the origin as the intercepts of the graph of f.

134. This function's y-intercept is 1 unit lower than the y-intercept of f, but its graph is 1 unit steeper.

135. This function's graph is half as steep as the graph of f, and it passes through the origin.

136. This function has the same y-intercept as f, but its graph never crosses the t-axis.

Find formulas for the linear functions described in Problems 137–138.

137. The graph of $y = v(x)$ intersects the graph of $u(x) = 1 + x^3$ at $x = -2$ and $x = 3$.

138. The graph of f intersects the graph of $y = 0.5x^3 - 4$ at $x = -2$ and $x = 4$.

139. Your favorite Chinese restaurant charges $3 per dish. You leave a 15% tip. Find an expression for your cost when you order d dishes and say whether it is linear in d.

140. You buy a $15 meal and leave a $t\%$ tip. Find an expression for your total cost and say whether it is linear in t.

141. Your personal income taxes are 25% of your adjusted income. To calculate your adjusted income you subtract $1000 for each child you have from your income, I. You have five children. Find an expression for the tax in terms of I and say whether it is linear in I.

142. Two parents take their N children to the movies. Adult tickets are $9 each and child tickets are $7. They buy a $3 box of popcorn for each child and pay $5 parking. Find an expression for the cost of the excursion and say whether it is linear in N.

143. A person departs Tucson driving east on I-10 at 65 mph. One hour later a second person departs in the same direction at 75 mph. Both vehicles stop when the second person catches up.

(a) Write an expression for the distance, D, in miles, between the two vehicles as a function of the time t in hours since the first one left Tucson.

(b) For what values of t does the expression make sense in practical terms?

144. A pond with vertical sides has a depth of 2 ft and a surface area of 10 ft^2. If the pond is full of water and evaporation causes the water level to drop at the rate of 0.3 inches/day, write an expression that represent the volume of water in the pond after d days. For what values of d is your expression valid?

145. A 200 gallon container contains 100 gallons of water. Water pours in at 1 gallon per minute, but drains out at the rate of 5 gallon per minute. How much water is in the container after t minutes? Is the container emptying or filling? When is the container empty or full?

146. While on a European vacation, a student wants to be able to convert from degrees Celsius (°C) to degrees (°F). One day the temperature is 20°C, which she is told is equivalent to 68°F. Another day, the temperature is 25°C, which she is told is equivalent to 77°F.

(a) Using the information she is given, write a linear equation that converts temperatures from Celsius to Fahrenheit.

(b) Using the formula from part (a), convert the temperature of

(i) 10°C to Fahrenheit.

(ii) 86°F to Celsius.

147. The road to the summit of Mt. Haleakala, located in Haleakala National Park in Hawaii, holds the world's record for climbing to the highest elevation in the shortest distance.[19] You can drive from sea level to the 10,023 ft summit over a distance of 35 miles, passing through five distinct climate zones. The temperature drops about 3°F per 1000 foot rise. The average temperature at Park Headquarters (7000 foot elevation) is 53°F.[20]

(a) Write a linear equation to represent the average temperature, T in °F, in Haleakala National Park as a function of the elevation, E in ft.

---

[19] www.hawaiiweb.com/maui/html/sites/haleakala_national_park.html.

[20] www.mauidownhill.com/haleakala/facts/haleakalaweather.html.

(b) Using the formula from part (a), what would you expect the average temperature to be at

   (i) Sun Visitor Center, at 9745 feet?

   (ii) Sea level?

148. When the carnival comes to town, a group of students wants to attend. The cost of admission and going on 3 rides is $12.50, while the cost of admission and going on 6 rides is $17.

   (a) Write a linear equation to represent the cost, C, of admission and going on n rides at the carnival.

   (b) Write your equation from part (a) in slope-intercept form.

   (c) What is the meaning of the slope and the intercept of the equation in part (b)?

149. A small band would like to sell CDs of its music. The business manager says that it costs $320 to produce 100 CDs and $400 to produce 500 CDs.

   (a) Write a linear equation to represent the cost, $\$C$, of producing n CDs.

   (b) Find the slope and the intercept of the equation, and interpret their meaning.

   (c) How much does it cost to produce 750 CDs?

   (d) If the band can afford to spend $500, how many CDs can it produce?

150. The cost of owning a timeshare consists of two parts: the initial cost of buying the timeshare, and the annual cost of maintaining it (the maintenance fee), which is paid each year in advance. Figure 2.31 shows the cumulative cost of owning a particular California timeshare over a 10 year period from the time it was bought. The first maintenance fee was paid at the time of purchase. What was the maintenance fee? How much did the timeshare cost to buy?

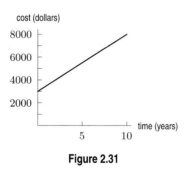

Figure 2.31

151. The unit of currency in the Republic of South Africa is the rand. Its value in dollars is shown in the Figure 2.32

(for the years 1990, 1991, 1993, 1994, and 2001), together with a line that approximates the points well.

   (a) Estimate the value of the rand in 1990 and in 2001, and write an equation for the line shown in Figure 2.32.

   (b) Where does this line cross the horizontal axis? What is the practical meaning of this? Is the answer realistic?

   (c) Use the equation in part (a) to predict the value of the rand in 1988. Its actual value was 44 cents. How well does your prediction agree with this?

   (d) Use the equation in part (a) to predict when the rand was worth $1.00. The actual year was 1981. How well does your prediction agree with this?

   (e) In general, is the equation in part (a) a good predictor of either the past or the future value of the rand?

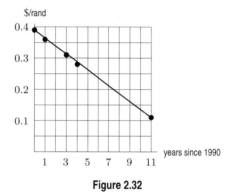

Figure 2.32

152. A company charges 19 cents to send a Short Message Service (SMS) message over a cell phone. It is considering introducing an SMS happy-hour in the evening, during which SMSs cost 9 cents each. The company's SMS profits under the current scheme are $-25{,}000 + 0.19x$ dollars, where x is the number of SMSs sent during the happy-hour time. To launch the happy-hour costs $10,000, so the profits under the happy-hour scheme are $-35{,}000 + 0.09x$ dollars. Under the current scheme, 70,000 SMSs are sent during the happy-hour time.

   (a) What are the company's current profits on SMSs during happy-hour time?

   (b) Use graphs to determine how many SMSs must be sent for the company to make more profits under the happy-hour scheme.

   (c) Use your graph from part (b) to determine how many SMSs must be sent for the company to make a positive profit on them in the happy-hour scheme.

Problems 153–155 involve a simple mathematical model of the behavior of songbirds. This model relates d, the amount of time in hours a songbird spends each day defending its territory, to s, the amount of time it spends singing (looking for mates). The more time a bird devotes to one of these behaviors, the less time it can devote to the other.[21]

153. Suppose that a particular species must consume 5 calories for each hour spent singing and 10 calories for each hour spent defending its territory. If this species consumes 60 calories per day on these activities, find a linear equation relating s and d and sketch its graph, placing d on the vertical axis. Say what the s- and d-intercepts of the graph tell you about the bird.

154. A second species must consume 4 calories for each hour spent singing and 12 calories for each hour spent defending its territory. If this species also consumes 60 calories per day on these activities, find a linear equation relating s and d and sketch its graph, and sketch its graph, placing d on the vertical axis. Say what the intercepts tell you about the bird.

155. What does the point of intersection of the graphs you drew for Questions 153 and 154 tell you about the two bird species?

For Problems 156–158, use the equation

$$F = \frac{9}{5}C + 32,$$

which relates temperatures in degrees Celsius (°C) to degrees Fahrenheit (°F).

156. (a) By solving the equation for C, find a formula which converts temperatures in degrees Fahrenheit (°F) to degrees Celsius (°C).
(b) Typical oven cooking temperatures are between 350°F and 450°F. What are these temperatures in °C?

157. A rule of thumb used by some cooks[22] is: For the range of temperatures used for cooking, the number of degrees Fahrenheit is twice the number of degrees Celsius.

(a) Write an equation, in terms of F and C, that represents this rule of thumb.
(b) For what values of F and C does this rule of thumb give the correct conversion temperature?
(c) Typical oven cooking temperatures are between 350°F (176.7°C) and 450°F (232.2°C). Use graphs to show that the maximum error in using this rule of thumb occurs at the upper end of the temperature range.

158. A convenient way to get a feel for temperatures measured in Celsius is to know[23] that room temperature is about 21°C.

(a) Show that, according to this, room temperature is about 70°F.
(b) Someone says: "So if 21°C is about 70°F, then 42°C is about 140°F." Is this statement accurate?

Problems 159–161 concern laptop computer batteries. As the computer runs, the total charge on the battery, measured in milliampere-hours or mAh, goes down. A typical new laptop battery has a capacity of 4500 mAh when fully charged, and under normal use will discharge at a constant rate of about 0.25 mAh/s.

159. Find an expression for $f(t)$, the amount of charge (in mAh) remaining on a new battery after t seconds of normal use, assuming that the battery is charged to capacity at time $t = 0$.

160. If fully charged at time $t = 0$, the charge remaining on an older laptop battery after t seconds of use is given by $g(t) = 3200 - 0.4t$. What do the parameters of this function tell you about the old laptop battery as compared to a new battery?

161. Given that $f(t_1) = 0$ and $g(t_2) = 0$, evaluate the expression t_1/t_2. What does the expression tells you about the two batteries?

---

[21]This model neglects other behaviors such as foraging, nest building, caring for young, and sleeping.
[22]Graham and Rosemary Haley, *Haley's Hints* Rev. Ed. (Canada: 3H Productions, 1999), p.355.
[23]Graham and Rosemary Haley, *Haley's Hints* Rev. Ed. (Canada: 3H Productions, 1999), p.351.

Chapter Three

Powers and Power Functions

Contents

3.1   POWER FUNCTIONS

After linear functions, we consider functions where the independent variable is raised to a power. For example, the area of a circle, A, is a function of its radius, r, given by

$$A = \pi r^2.$$

Here A is proportional to r^2, with constant of proportionality π. Since r is raised to the power 2, this is not a linear function.

A **power function** of x is a function that can be given by

$$y = f(x) = kx^p, \qquad \text{for constants } k \text{ and } p.$$

We call k the coefficient and p the exponent.

Example 1   The stopping distance in feet of an Alfa Romeo is proportional to the square of its speed, v mph, at the time the brakes are applied, and is given by the power function

$$f(v) = 0.036v^2.$$

(a) Identify the constants k and p.
(b) Find the stopping distances of an Alfa Romeo going at 35 mph and at 140 mph (its top speed). Which stopping distance is larger? Explain your answer in algebraic and practical terms.

Solution   (a) This is a power function $f(v) = kv^p$ with $k = 0.036$ and $p = 2$.
(b) At 35 mph the stopping distance is

$$f(35) = 0.036(35^2) = 44.1 \text{ ft.}$$

At 140 mph it is

$$f(140) = 0.036(140^2) = 705.6 \text{ ft.}$$

This is almost as long as two football fields (including the end zones). The stopping distance for 140 mph is larger. This makes sense algebraically, because $140 > 35$, so $140^2 > 35^2$. It also makes sense in practical terms, since a faster car takes longer to stop. See Figure 3.1.

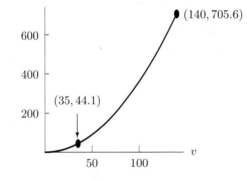

Figure 3.1: Stopping distance of an Alfa Romeo

Graphs of Positive Integer Powers: $y = x^3, x^5, x^7 \ldots,$ and $y = x^2, x^4, x^6 \ldots$

In Figure 3.1 we graphed the function $f(v) = 0.036v^2$ for positive values of v. Figure 3.2 shows the graph of $y = x^2$ for positive and negative values of x. Since the square of a negative number is positive, the graph is always above the x-axis (except at $x = 0$). The graphs of all power functions with p a positive even integer have the same characteristic $\bigcup$-shape and are symmetric about the y-axis. For instance, the graphs of $y = x^2$ and $y = x^4$ in Figure 3.2 are similar in shape, although the graph of $y = x^4$ is flatter near the origin and steeper away from the origin than the graph of $y = x^2$.

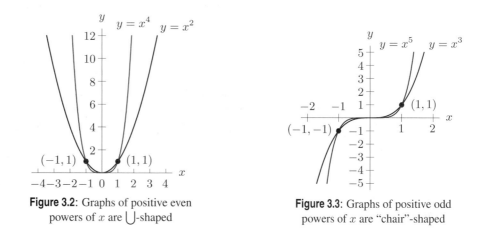

Figure 3.2: Graphs of positive even powers of x are $\bigcup$-shaped

Figure 3.3: Graphs of positive odd powers of x are "chair"-shaped

Odd powers have a different shape because the cube (or fifth power, etc.) of a negative number is negative, so the graph must be below the x-axis for negative x. The graphs of power functions with p a positive odd integer resemble the side view of a chair and are symmetric about the origin. Figure 3.3 shows the graphs of $y = x^3$ and $y = x^5$. The graph of $y = x^5$ is flatter near the origin and steeper far from the origin than the graph of $y = x^3$.

Negative Powers

In Example 1 we saw that the output value for an input value of 140 mph was larger than the output value for an input value of 35, because squaring a larger number gives a larger value. In the next example we consider what happens when the exponent p is negative.

Example 2   The weight, w, in pounds, of an astronaut r thousand miles from the center of the earth is given by

$$w = 2880r^{-2}.$$

(a) Is this a power function? Identify the constants k and p.
(b) How much does the astronaut weigh at the earth's surface, 4000 miles from the center? How much does the astronaut weigh 1000 miles above the earth's surface? Which weight is smaller?

Solution   (a) This is a power function $w = kr^p$ with $k = 2880$ and $p = -2$.
(b) At the earth's surface we have $r = 4$, since r is measured in thousands of miles, so

$$w = 2880(4^{-2}) = \frac{2880}{4^2} = 180 \text{ lb.}$$

So the astronaut weighs 180 lb at the earth's surface. At 1000 miles above the earth's surface we have $r = 5$, so

$$w = 2880(5^{-2}) = \frac{2880}{5^2} = 115.2 \text{ lb.}$$

So the astronaut weighs about 115 lb when she is 1000 miles above the earth's surface. Notice that this is smaller than her weight at the surface. For this function, larger input values give smaller output values.

Example 3   Rewrite the expression for the function in Example 2 in a form that explains why larger inputs give smaller outputs.

Solution   If we write

$$w = 2880r^{-2} = \frac{2880}{r^2},$$

we see that the r^2 is in the denominator, so inputting a larger value of r means dividing by a larger number, which results in a smaller output. This can be seen from the graph in Figure 3.4, which shows the function values decreasing as you move from left to right.

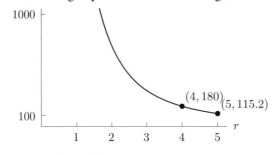

Figure 3.4: Weight of an astronaut

Graphs of Negative Integer Powers: $y = x^{-1}, x^{-3}, x^{-5}, \ldots$ and $y = x^{-2}, x^{-4}, x^{-6}, \ldots$

For negative powers, if we rewrite

$$y = x^{-1} = \frac{1}{x}$$

and

$$y = x^{-2} = \frac{1}{x^2},$$

then we can see from the form of the expression that as $x > 0$ increases, the denominator increases and the value of the function decreases. We can also see this numerically in Table 3.1.

Table 3.1   *Negative powers of x get small as x gets large*

x	0	10	20	30	40	50
$x^{-1} = 1/x$	Undefined	0.1	0.05	0.033	0.025	0.02
$x^{-2} = 1/x^2$	Undefined	0.01	0.0025	0.0011	0.0006	0.0004

The graphs of power functions with odd negative powers, $y = x^{-3}, x^{-5}, \ldots$ resemble the graph of $y = x^{-1} = 1/x$, and are symmetric about the origin. The graphs of power functions with even negative powers, $y = x^{-4}, x^{-6}, \ldots$ are similar in shape to the graph of $y = x^{-2} = 1/x^2$, and are symmetric about the y-axis. See Figures 3.5 and 3.6. Notice that the graphical behavior matches the numerical behavior we see in Table 3.1. If the exponent p is negative, then as x increases in the first quadrant, $y = x^p$ decreases.

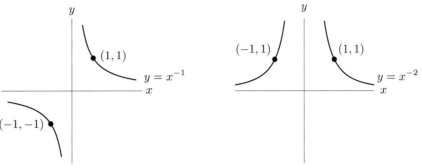

Figure 3.5: Graph of $y = x^{-1} = 1/x$ **Figure 3.6**: Graph of $y = x^{-2} = 1/x^2$

Inverse Proportionality

We often describe a power function $y = kx^p$ by saying that y is proportional to x^p. When the exponent is negative, there is another way of describing the function that comes from expressing it as a fraction. For example, if

$$y = 2x^{-3} = \frac{2}{x^3},$$

we say that y is *inversely proportional* to x^3. In general, if

$$y = kx^{-n} = \frac{k}{x^n}, \quad n \text{ positive,}$$

we say y is inversely proportional to x^n.

Fractional Powers

So far we have looked only at integer values of p, but it is also possible to have fractional exponents.

Example 4   For a certain species of animal, the bone length, L, in cm, is given by

$$L = 7\sqrt{A},$$

where A is the cross-sectional area of the bone in cm^2.

(a) Is L a power function of A? If so, identify the values of k and p.
(b) Find the bone lengths for $A = 36$ cm^2 and $A = 100$ cm^2. Which bone length is larger? Explain your answer in algebraic terms.

Solution   (a) Since $\sqrt{A} = A^{1/2}$, we have

$$L = 7A^{1/2}.$$

So this is a power function of A with $k = 7$ and $p = 1/2$. See Figure 3.7.
(b) For $A = 36$ cm^2, the bone length is

$$L = 7\sqrt{36} = 42 \text{ cm.}$$

For $A = 100$ cm^2, the bone length is

$$L = 7\sqrt{100} = 70 \text{ cm.}$$

The bone length for $A = 100$ cm^2 is larger. This makes sense algebraically, because $100 > 36$, so $\sqrt{100} > \sqrt{36}$.

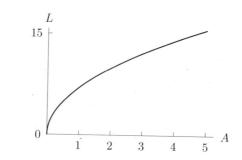

Figure 3.7: Bone length as a function of cross-sectional area

Graphs of Positive Fractional Powers: $y = x^{1/2}, x^{1/3}, x^{1/4}, \ldots$

Figure 3.8 shows the graphs of $y = x^{1/2}$ and $y = x^{1/4}$. These graphs have the same shape, although $y = x^{1/4}$ is steeper near the origin and flatter away from the origin than $y = x^{1/2}$. The same can be said about the graphs of $y = x^{1/3}$ and $y = x^{1/5}$ in Figure 3.9. In general, if p is a positive integer, then the graph of $y = x^{1/p}$ resembles the graph of $y = x^{1/2}$ if p is even; if p is odd, the graph resembles the graph of $y = x^{1/3}$. Notice the difference between the way the graphs of $y = x^{1/2}$ and $y = x^{1/3}$ bend and the way the graphs of $y = x^2$ and x^3 bend. For example, the graph of $y = x^2$ bends up, but the graph of $y = x^{1/2}$ bends to the right.

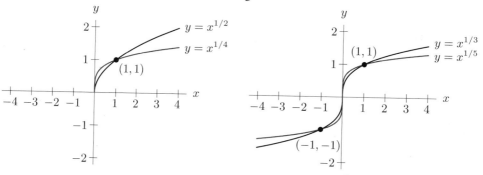

Figure 3.8: The graphs of $y = x^{1/2}$ and $y = x^{1/4}$ **Figure 3.9**: The graphs of $y = x^{1/3}$ and $y = x^{1/5}$

The Special Cases $p = 0$ and $p = 1$

We have seen that in general a power function is not linear. But if $p = 0$ or $p = 1$, then the power functions are linear. For example, the graph of $y = x^0 = 1$ is a horizontal line through the point $(1, 1)$. The graph of $y = x^1 = x$ is a line through the origin with slope 1. See Figures 3.10 and fig3gpk11.

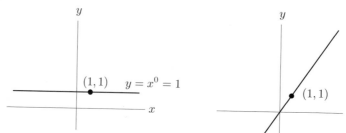

Figure 3.10: Graph of $y = x^0 = 1$ **Figure 3.11**: Graph of $y = x^1 = x$

Exercises and Problems for Section 3.1

Exercises

Are the functions in Exercises 1–6 power functions?

1. $y = 14x^{12}$

2. $y = 12 \cdot 14^x$

3. $y = 3x^3 + 2x^2$

4. $y = 2/(x^3)$

5. $y = x^3/2$

6. $y = \sqrt{4x^4}$

In Exercises 7–10, find

(a) $f(0)$

(b) $f(2)$

(c) $f(a)$

(d) $f(x+1)$

7. $f(x) = 3x^2$

8. $f(x) = 4x^{-2}$

9. $f(x) = 5x^3$

10. $f(x) = 2x^3$

In Exercises 11–15, identify the exponent and the constant of proportionality for each power function.

11. The area of a square of side x is $A = x^2$.

12. The perimeter of a square of side x is $P = 4x$.

13. The side of a cube of volume V is $x = \sqrt[3]{V}$.

14. The circumference of a circle of radius r is $C = 2\pi r$.

15. The surface area of a sphere of radius r is $S = 4\pi r^2$.

16. The area, A, of a rectangle whose length is 3 times its width is given by $A = 3w^2$, where w is its width.

   (a) Identify the coefficient and exponent of this power function.

   (b) If the width is 5 cm, what is the area of the rectangle?

17. The volume, V, of a cylinder whose radius is 5 times its height is given by $V = \frac{1}{5}\pi r^3$, where r is the radius.

   (a) Identify the coefficient and exponent of this power function.

   (b) If the radius is 2 cm, what is the volume?

   (c) If the height is 0.8 cm, what is the volume?

18. A ball dropped into a hole reaches a depth $d = 4.9t^2$ meters, where t is the time in seconds since it was dropped.

   (a) Identify the coefficient and exponent of this power function.

   (b) How deep is the ball after 2 seconds?

   (c) If the ball hits the bottom of the hole after 4 seconds, how deep is the hole?

In Exercises 19–22,

   (a) Is y proportional, or is it inversely proportional, to a positive power of x?

   (b) Make a table of values showing corresponding values for y when x is 1, 10, 100, and 1000.

   (c) Use your table to determine whether y increases or decreases as x gets larger.

19. $y = 2x^2$

20. $y = 3\sqrt{x}$

21. $y = \dfrac{1}{x}$

22. $y = \dfrac{5}{x^2}$

For the graphs of power functions $f(x) = kx^p$ in Exercises 23–27, is

(a) $p > 1$ **(b)** $p = 1$ **(c)** $0 < p < 1$

(d) $p = 0$ **(e)** $p < 0$

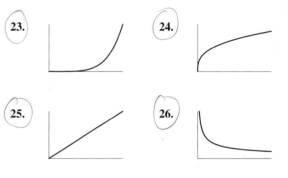

23.   **24.**

25.   **26.**

27.

Problems

Problems 28–30 describe power functions of the form $y = kx^p$, for p an integer. Is p

(a) Even or odd? **(b)** Positive or negative?

28.

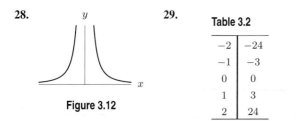

Figure 3.12

29.

Table 3.2

−2	−24
−1	−3
0	0
1	3
2	24

30. The graph of $y = f(x)$ gets closer to the x-axis as x gets large. For $x < 0$, $y < 0$, and for $x > 0$, $y > 0$.

In Problems 31–34, a and b are positive constants. If $a > b$ then which is larger?

31. a^4, b^4

32. $a^{1/4}, b^{1/4}$

33. a^{-4}, b^{-4}

34. $a^{-1/4}, b^{-1/4}$

35. A student takes a part-time job to earn \$2400 for summer travel. The number of hours, h, the student has to work is inversely proportional to the wage, w, in dollars per hour, and is given by

$$h = \frac{2400}{w}.$$

(a) How many hours does the student have to work if the job pays \$4 an hour? What if it pays \$10 an hour?

(b) How do the number of hours change as the wage goes up from \$4 an hour to \$10 an hour? Explain your answer in algebraic and practical terms.

(c) Is the wage, w, inversely proportional to the number hours, h? Express w as a function of h.

36. The weight, in pounds, of an astronaut at a distance r, in thousands of miles from the Earth's center, is given by Weight $= 2880/r^2$. Use this expression to support the statement "An astronaut who is 250,000 miles from the Earth's center is almost weightless."

37. If a ball is dropped from a high window, the distance, D, in feet, it falls is proportional to the square of the time, t, in seconds, since it was dropped and is given by

$$D = 16t^2.$$

How far has the ball fallen after three seconds and after five seconds? Which distance is larger? Explain your answer in algebraic terms.

38. A city's electricity consumption, E, in gigawatt-hours per year, is given by

$$E = \frac{0.15}{p^{3/2}},$$

where p is the price in dollars per kilowatt-hour charged.

(a) Is E a power function of p? If so, identify the exponent and the constant of proportionality.

(b) What is the electricity consumption at a price of \$0.16 per kilowatt-hour? At a price of \$0.25 per kilowatt hour? Explain the change in electricity consumption in algebraic terms.

39. The surface area of a mammal is given by $f(M) = kM^{2/3}$, where M is the body mass, and the constant of proportionality k is a positive number that depends on the body shape of the mammal. Is the surface area larger for a mammal of body mass 60 kilograms or for a mammal of body mass 70 kilograms? Explain your answer in algebraic terms.

40. The radius, r, in cm, of a sphere of volume V cm^3 is approximately $r = 0.620 \sqrt[3]{V}$.

(a) Graph the radius function, r, for volumes from 0 to 40 cm^3.

(b) Use your graph to estimate the volume of a sphere of radius 2 cm.

41. Plot the expressions $x^2 \cdot x^3$, x^5, and x^6, on three separate graphs in the window $-1 < x < 1$, $-1 < y < 1$. Does it appear from the graphs that $x^2 \cdot x^3 = x^5 = x^{2+3}$ or $x^2 \cdot x^3 = x^6 = x^{2 \cdot 3}$?

42. Plot the expressions $-x^4$ and $(-x)^4$ on the same graph in the window $-1 < x < 1$, $-1 < y < 1$. Is $-x^4 = (-x)^4$?

3.2   EXPRESSIONS FOR POWER FUNCTIONS

In the last section we saw that the value of the exponent p in a power function $f(x) = kx^p$ affects the behavior of the function. To recognize the value of k and p, some algebraic simplification is sometimes necessary.[1]

Example 1   Is the given function a power function? If so, identify the coefficient k and the exponent p.

(a) $f(x) = \dfrac{2}{x^3}$ (b) $g(x) = 4x + 2$ (c) $h(x) = \dfrac{5x}{2}$ (d) $j(x) = \dfrac{3x^5}{12x^6}$

Solution   (a) We have
$$f(x) = \frac{2}{x^3} = 2x^{-3},$$
so this is a power function with $k = 2$ and $p = -3$.

(b) We cannot rewrite $4x + 2$ in the form kx^p, since there are two terms. This is not a power function.

(c) We have
$$h(x) = \frac{5x}{2} = \frac{5}{2}x^1,$$
so $k = 5/2$ and $p = 1$.

(d) We have
$$j(x) = \frac{3x^5}{12x^6} = \frac{1}{4x} = \frac{1}{4}x^{-1}.$$
We see that $k = 1/4$ and $p = -1$.

Recognizing the value of the exponent p helps us see the behavior of a function.

Example 2   For each power function, identify the coefficient k, and the exponent, p. (Assume that all constants are positive.) Which graph (I)–(IV) gives the best fit?

(a) Volume of a sphere as a function of its radius r:
$$V = \frac{4}{3}\pi r^3.$$

(b) Gravitational force exerted by Earth on an object as a function of its distance r from the earth's center:
$$F = \frac{GmM}{r^2}, \quad G, m, M \text{ constants.}$$

(c) Period of a pendulum as a function of its length l:
$$P = 2\pi\sqrt{\frac{l}{g}}, \quad g \text{ constant.}$$

(d) Pressure of a quantity of gas as a function of its volume V (at a fixed temperature T):
$$P = \frac{nRT}{V}, \quad n, R, T \text{ constants.}$$

(e) Energy as a function of mass m,
$$E = mc^2, \quad c \text{ a constant (the speed of light).}$$

---

[1] See the Tools sections beginning on page 153 for a review of how to manipulate power expressions.

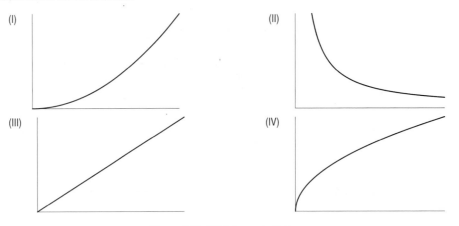

Figure 3.13: Which graph fits?

Solution

(a) The volume V is a power function of r, with $k = \frac{4}{3}\pi$ and exponent $p = 3$. Recall that the graph of a power function with p a positive odd integer resembles the side view of a chair. In the first quadrant, a graph such as (I) satisfies this requirement.

(b) Writing this as

$$F = \frac{GmM}{r^2} = GmMr^{-2}$$

we see that F is a constant times r^{-2}, so it is a power function of r with $k = GmM$ and exponent $p = -2$. Since the exponent is negative, the function decreases, and graph (II) fits best.

(c) Writing this as

$$P = 2\pi\sqrt{\frac{l}{g}} = 2\pi\frac{\sqrt{l}}{\sqrt{g}} = \frac{2\pi}{\sqrt{g}}l^{1/2},$$

we see that P is a power function of l with $k = \frac{2\pi}{\sqrt{g}}$ and exponent $p = 1/2$. Since the exponent is a proper fraction, the graph is increasing and bends downward. Thus, graph (IV) fits best.

(d) Writing this as

$$P = \frac{nRT}{V} = nRTV^{-1},$$

we see that P is a constant times V^{-1}, so it is a power function of V with $k = nRT$ and exponent $p = -1$. Since the exponent is negative, as in part (b), graph (II) again fits best.

(e) Writing this as

$$E = c^2m^1,$$

we see that E is a power function of m with coefficient $k = c^2$ and exponent $p = 1$. We recognize this as an equation of a line. Thus graph (III) fits best.

Example 3   Without evaluating, say which is larger, $f(10)$ or $f(5)$.

(a) $f(A) = 7A^3$ (b) $f(x) = \sqrt{\dfrac{25}{x^3}}$ (c) $f(t) = \dfrac{-4}{\sqrt[3]{t}}$

(d) $f(u) = \sqrt{3u^3}$ (e) $f(r) = \dfrac{r^2}{5r^{3/2}}$

Solution

(a) Since $10 > 5$, we have $10^3 > 5^3$. Multiplying both sides of this inequality by the positive number 7 does not change the direction of the inequality, so $f(10) > f(5)$.

(b) We have
$$f(x) = \sqrt{\frac{25}{x^3}} = \frac{\sqrt{25}}{\sqrt{x^3}} = \frac{5}{x^{3/2}} = 5x^{-3/2}.$$
Since the exponent is negative, we have $10^{-3/2} < 5^{-3/2}$, so $f(10) < f(5)$.

(c) We have
$$f(t) = -4t^{-1/3},$$
with coefficient -4 and exponent $-1/3$. Since the exponent is negative, we have $10^{-1/3} < 5^{-1/3}$. Multiplying both sides by the negative number -4 reverses the direction of this inequality, so $f(10) > f(5)$ (both numbers are negative, and $f(10)$ is closer to zero).

(d) We have
$$f(u) = \sqrt{3u^3} = \sqrt{3}\sqrt{u^3} = 3^{1/2}u^{3/2}.$$
So the coefficient is $3^{1/2}$ and the exponent is $3/2$. Since both the coefficient and the exponent are positive, $f(10) > f(5)$.

(e) We have
$$f(r) = \frac{r^2}{5r^{3/2}} = \frac{1}{5}r^{2-3/2} = \frac{1}{5}r^{1/2}.$$
So the coefficient is $\frac{1}{5}$ and the exponent is $1/2$. Both these numbers are positive, so $f(10) > f(5)$.

Example 4   The surface area of a closed cylinder of radius r and height h is
$$S = 2\pi r^2 + 2\pi rh.$$

(a) Write the expression as a power function of r or h, if possible.
(b) If the height is twice the radius, write S as a power function of r.

Solution   (a) We cannot write this expression as a power function of h or r, because there are two terms which cannot be combined into either the form $S = kr^p$ or the form $S = kh^p$.
(b) Since $h = 2r$, we have
$$S = 2\pi r^2 + 2\pi r \cdot 2r$$
$$= 2\pi r^2 + 4\pi r^2$$
$$= 6\pi r^2.$$
Thus, S is a power function of r, with $k = 6\pi$, and $p = 2$.

Exercises and Problems for Section 3.2

Exercises

In Exercises 1–10 write the expression as a constant times a power of a variable. Identify the coefficient and the exponent.

1. $3\sqrt{p}$

2. $\sqrt{2b}$

3. $\frac{4}{\sqrt[4]{z}}$

4. $\frac{\sqrt{x}}{\sqrt[3]{x}}$

5. $\frac{x^4}{4\sqrt{x^2}}$

6. $\left(\frac{1}{5\sqrt{x}}\right)^3$

7. $\sqrt[3]{\frac{8}{x^6}}$

8. $(2\sqrt{x})x^2$

9. $\sqrt{\frac{\sqrt{3}}{4}s}$

10. $\sqrt{\sqrt{4t^3}}$

In Exercises 11–24, can the expression be written in the form kx^p? If so, give the values of k and p.

11. $\dfrac{2}{3\sqrt{x}}$

12. $\dfrac{1}{3x^2}$

13. $(7x^3)^2$

14. $\sqrt{\dfrac{49}{x^5}}$

15. $(3x^2)^3$

16. $\left(\dfrac{2}{\sqrt{x}}\right)^3$

17. $y = \dfrac{1}{5x}$

18. $(x^2 + 3x^2)^2$

19. $3 + x^2$

20. $\sqrt{9x^5}$

21. $\left(\dfrac{1}{2\sqrt{x}}\right)^3$

22. $(x^2 + 4x)^2$

23. $\left((-2x)^2\right)^3$

24. $\sqrt[3]{x/8}$

In Exercises 25–30, a and b are positive numbers and $a > b$. Which is larger, $f(a)$ or $f(b)$?

25. $f(x) = \dfrac{x^3}{3}$

26. $f(t) = \dfrac{\sqrt{5}}{t^7}$

27. $f(r) = \sqrt{\dfrac{12}{r}}$

28. $f(t) = -3t^2\left(-2t^3\right)^5$

29. $f(v) = -5\left((-2v)^2\right)^3$

30. $f(x) = 2\sqrt[3]{x} + 3\sqrt[9]{x^3}$

Problems

In Problems 31–42, what is the exponent of the power function? Which of (I)–(IV) in Figure 3.14 best fits its graph? Assume all constants are positive.

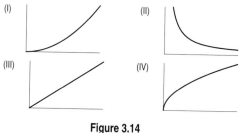

(I)   (II)   (III)   (IV)

Figure 3.14

31. The heart mass, H, of a mammal as a function of its body mass, B:
$$H = kB.$$

32. The number of species, N, on an island as a function of the area, A, of the island:
$$N = k\sqrt[3]{A}.$$

33. The energy, E, of a swimming dolphin as a function of the speed, v of the dolphin:
$$E = av^3.$$

34. The strength, S, of a beam as a function of its thickness, h:
$$S = bh^2.$$

35. The average velocity, v, on a trip over a fixed distance d as a function of the time of travel, t:
$$v = \dfrac{d}{t}.$$

36. The surface area, S, of a mammal as a function of the body mass, B:
$$S = kB^{2/3}.$$

37. The number of animal species, N, of a certain body length as a function of the body length, L:
$$N = \dfrac{A}{L^2}.$$

38. The circulation time, T, of a mammal as a function of its body mass, B:
$$T = M\sqrt[4]{B}.$$

39. The weight, W, of plaice (a type of fish) as a function of the length, L, of the fish:
$$W = \dfrac{a}{b} \cdot L^3.$$

40. The surface area, s, of a person with weight w and height h as a function of w, if height h is fixed:
$$s = 0.01w^{0.25}h^{0.75}.$$

41. The judged loudness, J, of a sound as a function of the actual loudness L:
$$J = aL^{0.3}.$$

42. The blood mass, M of a mammal as a function of its body mass, B:
$$M = kB.$$

43. The perimeter of a rectangle of length l and width w is

$$P = 2l + 2w.$$

(a) Write the expression as a power function of l or w, if possible.

(b) If the length is three times the width, write P as a power function of w and give the values of the coefficient k and the exponent p.

44. A window is in the shape of a square with a semicircle on top. If the side of the square is l ft. then the area of the glass sheet in the window is

$$A = l^2 + \frac{\pi}{2}\left(\frac{l}{2}\right)^2 \text{ ft}^2.$$

(a) Is A a power function of l? If so, identify the coefficient k and the exponent p.

(b) Without computing the area, say which area is larger, when $l = 1.5$ ft. or $l = 2.5$ ft. Explain your answer in algebraic terms.

45. The side length of an equilateral triangle of area A is

$$s = \sqrt{\frac{4}{\sqrt{3}}A}.$$

Write s as a power function of A and identify the coefficient and the exponent.

46. A certificate of deposit is worth $P(1 + r)^t$ dollars after t years, where r is the annual interest rate expressed as a decimal, and P is the amount initially deposited. In each part below, state which investment will be worth more.

(a) Investment A, in which $P = \$7000$, $r = 4\%$, and $t = 5$ years or investment B, in which $P = \$6500$, $r = 6\%$ and $t = 7$ years.

(b) Investment A, in which $P = \$10,000$, $r = 2\%$, and $t = 10$ years or investment B, in which $P = \$5000$, $r = 4\%$, and $t = 10$ years.

(c) Investment A, in which $P = \$10,000$, $r = 3\%$, and $t = 30$ years or investment B, in which $P = \$15,000$, $r = 4\%$, and $t = 15$ years.

47. A town's population in thousands in 20 years is given by $15(1 + x)^{20}$, where x is the growth rate per year. What is the population in 20 years if the growth rate is

(a) 2%?   (b) 7%?   (c) −5%?

48. An astronaut r thousand miles from the center of the earth weighs $2880/r^2$ lbs, and the surface of the earth is 4000 miles from the center.

(a) If the astronaut is h miles above the surface of the earth, express r as a function of h.

(b) Express her weight w in pounds as a function of h.

The variables in Problems 49–58 are all positive. What is the effect of increasing a on the value of the expression? Does the value increase, decrease, or remain unchanged?

49. $\dfrac{(ax)^2}{a^2}$

50. $(ax)^2 + a^2$

51. $\dfrac{(ax)^2}{a^3}$

52. a^x

53. $\dfrac{1}{a^{-x}}$

54. $x + a^{-1}$

55. $a^0 x$

56. $x + \dfrac{1}{a^0}$

57. $\dfrac{a^4}{(ax)^3}$

58. $\dfrac{(ax)^{1/3}}{\sqrt{a}}$

3.3 EQUATIONS INVOLVING POWERS

In Example 1 on page 126 we considered the stopping distance of an Alfa Romeo traveling at v mph, which is given, in feet, by

$$\text{Stopping distance} = 0.036v^2.$$

Suppose we know the stopping distance and want to find v. Then we must solve an equation.

Example 1   At the scene of an accident involving an Alfa Romeo, skid marks show that the stopping distance is 270 feet. How fast was the car going when the brakes were applied?

Solution   Substituting 270 for the stopping distance, we have the equation

$$270 = 0.036v^2.$$

To find the speed of the car, we solve for v. First, we isolate v^2 by dividing both sides by 0.036:

$$v^2 = \frac{270}{0.036} = 7500.$$

Taking the square root of both sides, we get

$$v = \pm\sqrt{7500}.$$

Since the speed of the car is a positive number, we choose the positive square root as the solution:

$$v = \sqrt{7500} = 86.603 \text{ mph.}$$

We see that the car was going nearly 87 mph at the time the brakes were applied.

Example 2   A cube-shaped box has volume 10 ft^3. What is the length of its sides?

Solution   If s is the side length in feet then the volume is s^3, so we must solve the equation

$$s^3 = 10.$$

Taking the cube root of both sides, we get

$$s = 10^{1/3} = 2.15 \text{ ft.}$$

Example 3   An astronaut's weight in pounds is inversely proportional to the square of her distance, r, in thousands of miles, from the earth's center and is given by

$$\text{Weight} = f(r) = \frac{2880}{r^2}.$$

Find the distance from the earth's center at the point when the astronaut's weight is 100 lbs.

Solution   Since the weight is 100 lbs, the distance from the earth's center is the value of r such that $100 = f(r)$, so

$$100 = \frac{2880}{r^2}$$
$$100r^2 = 2880$$
$$r^2 = 28.8$$
$$r = \pm\sqrt{28.8} = \pm 5.367 \text{ thousand miles.}$$

We choose the positive square root since r is a distance. When the astronaut weighs 100 lbs, she is approximately 5367 miles from the earth's center (which is about 1367 miles from the earth's surface).

How Many Solutions are There?

Notice that in Examples 1 and 3 the equation had two solutions, and in Example 2 the equation had only one solution. In general, how many solutions can we expect for the equation $x^p = a$? In what follows we assume the exponent p can be either a positive or negative integer, but not zero.

Two Solutions

When we have an equation like $x^2 = 5$ or $x^4 = 16$, in which an even power of x is set equal to a positive number, we have two solutions. For instance, $x^4 = 16$ has two solutions, $x = 2$ and $x = -2$, because $(-2)^4 = 2^4 = 16$. See Figure 3.15.

In general, if p is an even integer (not zero) and a is positive, the equation

$$x^p = a$$

has two solutions, $x = a^{1/p}$ and $x = -a^{1/p}$.

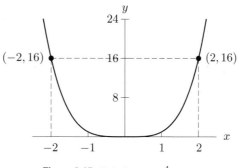

Figure 3.15: Solutions to $x^4 = 16$

Example 4   Find all solutions for each equation.

(a) $x^2 = 64$ (b) $x^4 = 64$ (c) $x^{-2} = 64$ (d) $x^{-4} = 64$

Solution   (a) We take square roots of both sides. Because 64 has two square roots, there are two solutions, 8 and -8.

(b) We take the fourth root of both sides and get

$$x = \pm(64)^{1/4} = \pm(2^6)^{1/4} = \pm 2^{6/4} = \pm 2^{4/4} \cdot 2^{2/4} = \pm 2\sqrt{2}.$$

(c) Multiplying both sides by x^2 we get

$$1 = 64x^2$$
$$\frac{1}{64} = x^2$$
$$x = \pm\frac{1}{8}.$$

(d) We can use the same method as in part (c), or we can simply raise both sides to the $-1/4$ power:

$$x = \pm 64^{-1/4} = \pm\frac{1}{64^{1/4}} = \pm\frac{1}{2\sqrt{2}}.$$

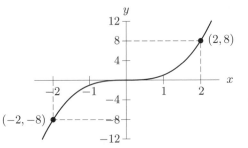

Figure 3.16: Solutions to $x^3 = 8$ and $x^3 = -8$

One Solution

When we have an equation like $x^3 = 8$ or $x^5 = -2$, in which an odd power of x is set equal to a number (positive or negative), we have one solution. For instance, $x^3 = 8$ has the solution $x = 2$, because $2^3 = 8$, and $x^3 = -8$ has the solution $x = -2$, because $(-2)^3 = -8$. See Figure 3.16.

In general, if p is odd,

$$x^p = a$$

has one solution, $x = a^{1/p}$. There is another situation when the equation $x^p = a$ has one solution, and that is if $a = 0$. In that case there is just the one solution $x = 0$.

Example 5   Find all solutions for each equation.

    (a) $x^5 = -773$ (b) $x^4 = 773$ (c) $x^8 = 0$

Solution   (a) Since $p = 5$ is odd, there is one solution: $x = \sqrt[5]{-773} = -3.781$.
    (b) Since $p = 4$ is even and $a = 773$ is positive, there are two solutions, $x = \sqrt[4]{773} = 5.273$ and
         $x = -\sqrt[4]{773} = -5.273$.
    (c) Since $a = 0$, there is one solution, $x = 0$.

No Solutions

Since we cannot take the square root of a negative number, the equation $x^2 = -64$ has no solutions. In general, if p is even and a is negative, then the equation

$$x^p = a$$

has no solutions. There is another situation when the equation $x^p = a$ has no solution, and that is if p is negative and $a = 0$.

Example 6   How many solutions are there to each of the following equations?

    (a) $-773 = 22x^3$ (b) $-773 = 22x^2$
    (c) $773 = 22x^2$ (d) $0 = 22x^2$

Solution   (a) Since the exponent is odd, there is one solution.
    (b) Since the exponent is even and $-773/22$ is a negative number, there are no solutions.
    (c) Since the exponent is even and $773/22$ is a positive number, there are two solutions.
    (d) Since $0/22$ is 0, there is one solution.

Example 7

A $5000 certificate of deposit at an annual interest rate of r yields $5000(1 + r)^{10}$ after 10 years. What does the equation $5000(1 + r)^{10} = 10{,}000$ represent? How many solutions are there? Which solutions make sense in practical terms?

Solution

We know that $5000(1 + r)^{10}$ represents the value of the certificate after 10 years. So the solution to the equation

$$5000(1 + r)^{10} = 10{,}000$$

represents the interest rate at which the certificate will grow to $10,000 in 10 years. To solve the equation, we divide both sides by 5000 to get

$$(1 + r)^{10} = \frac{10{,}000}{5000} = 2,$$

then take 10th roots to get

$$1 + r = \pm \sqrt[10]{2}$$
$$r = -1 \pm \sqrt[10]{2}.$$

Since $\sqrt[10]{2} = 1.0718$, there are two possible values for r,

$$r = -1 + 1.0718 = 0.0718 \quad \text{and} \quad r = -1 - 1.0718 = -2.0718.$$

Since r represents an interest rate, it must be positive, so only the first solution makes sense. The interest rate needed to double the value in 10 years is $r = 0.0718$, or 7.18%.

Equations Involving Fractional Powers

For the equation $x^{n/m} = a$, we can raise both sides to the mth power to get an equation with integer powers, $x^n = a^m$. However, raising both sides of an equation to an even power can sometimes produce an equation with more solutions than the original one, so it is important to check your solutions at the end.

Example 8

Solve each of the following equations:
(a) $\sqrt{t} + 9 = 21$
(b) $\sqrt{t} + 21 = 9$
(c) $2A^{1/5} = 10$

Solution

(a) We solve for $\sqrt{t}$ and then square both sides:

$$\sqrt{t} + 9 = 21$$
$$\sqrt{t} = 12$$
$$(\sqrt{t})^2 = (12)^2$$
$$t = 144.$$

We check that 144 is a solution: $\sqrt{144} + 9 = 12 + 9 = 21$.

(b) Proceeding as before, we solve for $\sqrt{t}$ and then square both sides:

$$\sqrt{t} + 21 = 9$$
$$\sqrt{t} = -12$$
$$(\sqrt{t})^2 = (-12)^2$$
$$t = 144.$$

However, in this case 144 is not a solution: $\sqrt{144} + 21 = 12 + 21 = 33 \neq 9$. We could have noticed this without solving the equation, since $\sqrt{t}$ is always positive, so $\sqrt{t} + 21$ cannot be 9.

(c) The first step is to isolate the $A^{1/5}$. We then raise both sides of the equation to the fifth power:

$$2A^{1/5} = 10$$
$$A^{1/5} = 5$$
$$\left(A^{1/5}\right)^5 = 5^5$$
$$A = 3125.$$

Checking $A = 3125$ in the original equation we get

$$2(3125)^{1/5} = 2(5^5)^{1/5} = 2 \cdot 5 = 10.$$

Exercises and Problems for Section 3.3

Exercises

In Exercises 1–21, solve the equation for the variable.

1. $x^3 = 50$

2. $2x^2 = 8.6$

3. $4 = x^{-1/2}$

4. $4w^3 + 7 = 0$

5. $z^2 + 5 = 0$

6. $2b^4 - 11 = 81$

7. $\sqrt{a} - 2 = 7$

8. $3\sqrt[3]{x} + 1 = 16$

9. $\sqrt{y - 2} = 11$

10. $\sqrt{2y - 1} = 9$

11. $\sqrt{3x - 2} + 1 = 10$

12. $(x + 1)^2 + 4 = 29$

13. $(3c - 2)^3 - 50 = 100$

14. $4x^4 = 18x^2$

15. $2p^3 = 7p$

16. $\frac{1}{4}t^3 = t$

17. $16 - \frac{1}{L^2} = 0$

18. $\sqrt{r^2 + 144} = 13$

19. $\frac{1}{\sqrt[3]{x}} = -3$

20. $2\sqrt{x} = \frac{1}{3}x$

21. $12 = \sqrt{\dfrac{z}{5\pi}}$

In Exercises 22–26, solve the equation for the indicated variable. Assume all other letters represent non-zero constants.

22. $y = kx^2$, for x

23. $A = \frac{1}{2}\pi r^2$, for r

24. $L = kB^2D^3$, for D

25. $yx^2 = 3(xy)^3$, for x

26. $w = 4\pi\sqrt{\dfrac{x}{t}}$, for x.

Without solving them, say whether the equations in Exercises 27–42 have

(i)   One positive solution   (ii)   One negative solution
(iii) One solution at $x = 0$ (iv) Two solutions
(v)   Three solutions   (vi)   No solution

Give a reason for your answer.

27. $x^3 = 5$

28. $x^5 = 3$

29. $x^2 = 5$

30. $x^2 = 0$

31. $x^2 = -9$

32. $x^3 = -9$

33. $x^4 = -16$

34. $x^7 = -9$

35. $x^{1/3} = 2$

36. $x^{1/3} = -2$

37. $x^{1/2} = 12$

38. $x^{1/2} = -12$

39. $x^{-1} = 4$

40. $x^{-2} = 4$

41. $x^{-3} = -8$

42. $x^{-6} = -\frac{1}{64}$

Problems

43. A city's electricity consumption, E, in gigawatt-hours per year, is given by $E = 0.15p^{-3/2}$, where p is the price in dollars per kilowatt-hour charged. What does the solution to the equation $0.15p^{-3/2} = 2$ represent? Find the solution.

44. The surface area, S, in cm^2, of a mammal of mass M kg, is given by $S = kM^{2/3}$, where k depends on the body shape of the mammal. For people, assume that $k = 1095$.

    (a) Find the body mass of a person whose surface area is 21,000 cm^2.
    (b) What does the solution to the equation $1095M^{2/3} = 30{,}000$ represent?
    (c) Express M in terms of S.

45. Solve each of the following geometric formulas for the radius r.

    (a) The circumference of a circle of radius r: $C = 2\pi r$.
    (b) The area of a circle of radius r: $A = \pi r^2$.
    (c) The volume of a sphere of radius r: $V = (4/3)\pi r^3$.
    (d) The volume of a cylinder of radius r and height h: $V = \pi r^2 h$.
    (e) The volume of a cone of radius at the open end r and height h: $V = (1/3)\pi r^2 h$.

46. The volume of a cone of base radius r and height h is $(1/3)\pi r^2 h$, and the volume of a sphere of radius r is $(4/3)\pi r^3$. Suppose a particular sphere of radius r has the same volume as a particular cone of base radius r.

    (a) Write an equation expressing this situation.
    (b) What is the height of the cone in terms of r?

47. How can you tell immediately that the equation $x + 5\sqrt{x} = -4$ has no solutions?

In Problems 48–59, decide for what values of the constant A the equation has

(a) The solution $t = 0$ **(b)** A positive solution
(c) A negative solution

48. $t^3 = A$ **49.** $t^4 = A$ **50.** $(-t)^3 = A$

51. $(-t)^4 = A$ **52.** $t^3 = A^2$ **53.** $t^4 = A^2$

54. $t^4 = -A^2$ **55.** $t^3 = -A^2$ **56.** $t^3 + 1 = A$

57. $At^2 + 1 = 0$ **58.** $At^2 = 0$ **59.** $A^2 t^2 + 1 = 0$

In Problems 60–63, decide for what value(s) of the constant A (if any) the equation has

(a) The solution $x = 1$ **(b)** A solution $x > 1$
(c) No solution

60. $4x^2 = A$ **61.** $4x^3 = A$

62. $-4x^2 = A$ **63.** $4x^{-2} = A$

64. Figure 3.17 shows two points of intersection of the graphs of $y = 1.3x^3$ and $y = 120x$.

    (a) Use a graphing calculator to find the x-coordinate of the non-zero point of intersection accurate to one decimal place.
    (b) Use algebra to find the non-zero point of intersection accurate to three decimal places.

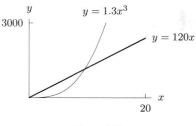

Figure 3.17

65. Figure 3.18 shows two points of intersection of the graphs of $y = 0.2x^5$ and $y = 1000x^2$.

    (a) Use a graphing calculator to find the x-coordinate of the non-zero point of intersection accurate to one decimal place.
    (b) Use algebra to find the non-zero point of intersection accurate to three decimal places.

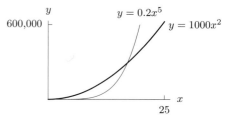

Figure 3.18

3.4 MODELING WITH POWER FUNCTIONS

In this section we see how to find values of p and k that make the power function $y = kx^p$ fit a given situation.

Example 1

The period of a pendulum is the amount of time for the pendulum to make one full swing and return to its starting point. The period, P, of a pendulum is proportional to the square root of its length, L.

(a) The pendulum in a grandfather clock is 3 feet long and has a period of 1.924 seconds. Find the constant of proportionality and write P in terms of L.

(b) The length of Foucault's pendulum, built in 1851 in the Pantheon in Paris, was 197 feet. Find the period of Foucault's pendulum.

Solution

(a) Since P is proportional to the square root of L, we have

$$P = k\sqrt{L} \quad \text{or} \quad P = kL^{1/2}.$$

To find the constant of proportionality, we substitute $L = 3$ and $P = 1.924$:

$$P = kL^{1/2}$$
$$1.924 = k(3^{1/2})$$
$$k = \frac{1.924}{3^{1/2}} = 1.111.$$

The power function for the period of a pendulum in terms of its length is

$$P = 1.111L^{1/2}.$$

(b) We substitute $L = 197$ and see that $P = 1.111(197^{1/2}) = 15.59$ seconds. The period of Foucault's pendulum was about 15.59 seconds.

Example 2

Find a formula for y in terms of x.

(a) The quantity y is proportional to the square of x, with constant of proportionality -0.68.

(b) The quantity y is proportional to the 4th power of x, and $y = 150$ when $x = 2$.

(c) The quantity y is inversely proportional to the cube of x, and $y = 5$ when $x = 3$.

Solution

(a) We have $y = kx^2$, with $k = -0.68$. The formula is

$$y = -0.68x^2.$$

(b) We have $y = kx^4$. We substitute $y = 150$ and $x = 2$ and solve for k:

$$y = kx^4$$
$$150 = k \cdot (2^4)$$
$$k = 150/(2^4) = 9.375.$$

The formula is

$$y = 9.375x^4.$$

(c) We have $y = k/x^3$. We substitute $y = 5$ and $x = 3$ and solve for k:

$$y = \frac{k}{x^3}$$
$$5 = \frac{k}{3^3}$$
$$k = 5 \cdot 3^3 = 135.$$

The formula is

$$y = \frac{135}{x^3}.$$

Example 3   The Dubois formula relates a person's surface area, S, in m^2, to weight, w, in kg, and height, h, in cm, by
$$S = 0.01w^{0.25}h^{0.75}.$$

(a) What is the surface area of a person who weighs 65 kg and is 160 cm tall?
(b) If weight is proportional to height with $w = 0.4h$, write S in terms of h and simplify your answer. Is surface area proportional to height?

Solution   (a) Substituting $w = 65$ and $h = 160$, we have
$$S = 0.01(65^{0.25})(160^{0.75}) = 1.277 \text{ m}^2.$$

(b) We substitute $w = 0.4h$ and simplify using the properties of exponents:
$$S = 0.01(0.4h)^{0.25}(h^{0.75})$$
$$= 0.01(0.4^{0.25})(h^{0.25})(h^{0.75})$$
$$= 0.01(0.4^{0.25})(h^{0.25+0.75})$$
$$= 0.01(0.4^{0.25})(h^1)$$
$$= 0.00795h.$$

Since $S = 0.00795h = kh$ for $k = 0.00795$, we see that, if weight is proportional to height, then surface area is also proportional to height.

What if We Do Not Know the Power?

If one quantity is proportional to a second quantity, we have
$$\text{First quantity} = k \times \text{Second quantity}.$$

We can solve for the constant of proportionality k to obtain
$$k = \frac{\text{First quantity}}{\text{Second quantity}}.$$

This tells us that the *ratio* of the first quantity to the second quantity is the constant value, k. Similarly, if one quantity y is proportional to a power of another quantity, x, then we have
$$y = kx^p, \qquad \text{for some } p.$$

We can solve for k to obtain
$$k = \frac{y}{x^p}.$$

In this case, the ratio of y to the pth power of x is a constant.

If y is proportional to x^p, then the ratio y/x^p is constant.

Example 4   Using Table 3.3, show that y could be inversely proportional to x^2.

Table 3.3

x	1	2	3	4	5	6
y	21,600	5400	2400	1350	864	600

Solution   We need to show that $y = kx^{-2}$, so we need to calculate y/x^{-2} to see if this ratio is constant. We have:

$$\frac{21{,}600}{1^{-2}} = 21{,}600 \qquad \frac{5400}{2^{-2}} = 21{,}600 \qquad \frac{2400}{3^{-2}} = 21{,}600$$

$$\frac{1350}{4^{-2}} = 21{,}600 \qquad \frac{864}{5^{-2}} = 21{,}600 \qquad \frac{600}{6^{-2}} = 21{,}600.$$

We see that the ratio is constant: 21,600. Therefore $y = 21{,}600x^{-2}$.

Example 5   Table 3.4 relates the weight, y, of plaice (a type of fish) to its length, x.[2] Determine whether this data supports the hypothesis that, for this type of fish,

(a) Weight is approximately proportional to length.
(b) Weight is approximately proportional to the cube of length.

Table 3.4   *Are length and weight of fish proportional?*

Length (cm), x	33.5	34.5	35.5	36.5	37.5	38.5	39.5	40.5	41.5
Weight (gm), y	332	363	391	419	455	500	538	574	623

Solution   (a) To see if weight y is proportional to length x, we see if the ratios y/x are approximately constant:
$$\frac{332}{33.5} = 9.910, \qquad \frac{363}{34.5} = 10.522, \qquad \frac{391}{35.5} = 11.014.$$
Since the ratios do not appear to be approximately constant, we conclude that y is not proportional to x.

(b) To see if weight y is proportional to the cube of length x, we see if the ratios y/x^3 are approximately constant. To three decimal places, we have

$$\frac{332}{(33.5)^3} = 0.009, \qquad \frac{363}{(34.5)^3} = 0.009, \qquad \frac{391}{(35.5)^3} = 0.009, \qquad \frac{419}{(36.5)^3} = 0.009,$$

and so on for the other ratios. To three decimal places, all the ratios y/x^3 are the same, so weight is approximately proportional to the cube of length, with constant of proportionality 0.009.

---

[2] Adapted from "On the Dynamics of Exploited Fish Populations", by R. J. H. Beverton and S. J. Holt, *Fishery Investigations*, Series II, 19, 1957.

The Behavior of Power Functions

Understanding power functions can help us predict what happens to the value of the output variable when we change the input. If y is directly proportional to x, then doubling the value of x doubles the value of y, tripling the value of x triples the value of y, and so on. What happens if y is proportional to a power of x?

Example 6

What happens to the stopping distance for the Alfa Romeo in Example 1 on page 126:

(a) If its speed is doubled from $v = 20$ mph to 40 mph? From 30 mph to 60 mph?

(b) If the speed is tripled from 10 mph to 30 mph? From 25 mph to 75 mph?

$f(v) = 0.036 v^2$

Solution

(a) If the speed doubles from $v = 20$ mph to 40 mph, the stopping distance increases from $0.036(20)^2 = 14.4$ ft to $0.036(40)^2 = 57.6$ ft. Taking ratios, we see that the stopping distance increases by a factor of 4:

$$\frac{57.6 \text{ ft}}{14.4 \text{ ft}} = 4.$$

Likewise, if the speed doubles from $v = 30$ mph to 60 mph, the stopping distance increases from 32.4 ft to 129.6 ft. Once again, this is a fourfold increase:

$$\frac{129.6 \text{ ft}}{32.4 \text{ ft}} = 4.$$

(b) If the speed triples from $v = 10$ mph to 30 mph, the stopping distance increases from 3.6 ft to 32.4 ft. If the speed triples from 25 mph to 75 mph, the stopping distance increases from 22.5 ft to 202.5 ft. In both cases the stopping distance increases by a factor of 9:

$$\frac{32.4 \text{ ft}}{3.6 \text{ ft}} = \frac{202.5 \text{ ft}}{22.5 \text{ ft}} = 9.$$

From the last example, we see that, in one case, doubling the speed quadruples the stopping distance, while tripling the speed multiplies the stopping distance by 9. This result is true for all speeds of the Alpha Romeo, as we show algebraically in Examples 7 and 8.

Example 7

For the Alfa Romeo in Example 1 on page 126, what does the expression $0.036(2v)^2$ represent in terms of stopping distance? Write this expression as a constant times $0.036v^2$. What does your answer tell you about how stopping distance changes if you double the car's speed?

Solution

The expression $0.036v^2$ represents the stopping distance of an Alfa Romeo traveling at v mph, so the expression $0.036(2v)^2$ represents the stopping distance if the car's speed is doubled from v to $2v$. We have

$$\text{New stopping distance} = 0.036(2v)^2 = 0.036 \cdot 2^2 v^2 = 4(0.036v^2)$$
$$= 4 \cdot \text{ Old stopping distance.}$$

So if the car's speed is doubled, its stopping distance is multiplied by 4.

Example 8

What happens to the stopping distance if the speed of the Alfa Romeo in Example 1 on page 126 is tripled?

Solution

If the original speed is v, then after being tripled the speed is $3v$, so

$$\text{New stopping distance} = 0.036(3v)^2 = 0.036 \cdot 3^2 v^2 = 9(0.036v^2)$$
$$= 9 \cdot \text{ Old stopping distance.}$$

So if the car's speed is tripled, its stopping distance is multiplied by 9.

Example 9   An astronaut's weight, w, in pounds, is inversely proportional to the square of her distance, r, in thousands of miles, from the earth's center. The astronaut weighs 180 lbs on the earth's surface and the radius of the earth is approximately 4000 miles.

(a) Find the constant of proportionality and write w in terms of r.

(b) As r doubles, what happens to w?

Solution   (a) We have $w = k/r^2$, with $k > 0$. We substitute the values $w = 180$ and $r = 4$ and solve for k:

$$180 = \frac{k}{4^2}$$
$$k = 180(4^2) = 2880.$$

The formula is

$$w = \frac{2880}{r^2}.$$

(b) If r is doubled,

$$\text{New weight} = \frac{2880}{(2r)^2} = \frac{2880}{4r^2}$$
$$= \frac{1}{4} \cdot \frac{2880}{r^2} = \frac{1}{4} \cdot \text{Old weight.}$$

This makes sense: as the distance from earth goes up, the gravitational pull of the earth decreases, so the weight of the object decreases.

Exercises and Problems for Section 3.4

Exercises

In Exercises 1–4, write a formula for y in terms of x if y satisfies the given conditions.

1. Proportional to the 5th power of x, and $y = 744$ when $x = 2$.

2. Proportional to the cube of x, with constant of proportionality -0.35.

3. Proportional to the square of x, and $y = 1000$ when $x = 5$.

4. Proportional to the 4th power of x, and $y = 10.125$ when $x = 3$.

5. Find a formula for s in terms of t if s is proportional to the square root of t, and $s = 100$ when $t = 50$.

6. If A is inversely proportional to the cube of B, and $A = 20.5$ when $B = -4$, write A as a power function of B.

7. Suppose c is directly proportional to the square of d. If $c = 50$ when $d = 5$, find the constant of proportionality and write the formula for c in terms of d. Use your formula to find c when $d = 7$.

8. Suppose c is inversely proportional to the square of d. If $c = 50$ when $d = 5$, find the constant of proportional-

ity and write the formula for c in terms of d. Use your formula to find c when $d = 7$.

In Exercises 9–12, write a formula representing the function.

9. The strength, S, of a beam is proportional to the square of its thickness, h.

10. The energy, E, expended by a swimming dolphin is proportional to the cube of the speed, v, of the dolphin.

11. The radius, r, of a circle is proportional to the square root of the area, A.

12. Kinetic energy, K, is proportional to the square of velocity, v.

In Exercises 13–16, what happens to y when x is doubled? Here k is a positive constant.

13. $y = kx^3$

14. $y = \dfrac{k}{x^3}$

15. $xy = k$

16. $\dfrac{y}{x^4} = k$

Problems

17. The volume V of a sphere is a function of its radius r given by

$$V = f(r) = \frac{4}{3}\pi r^3.$$

  (a) Find $\dfrac{f(2r)}{f(r)}$.   **(b)** Find $\dfrac{f(r)}{f(\frac{1}{2}r)}$.

  (c) What do you notice about your answers to (a) and (b)? Explain this result in terms of sphere volumes.

18. The gravitational force F exerted on an object of mass m at a distance r from the Earth's center is given by

$$F = g(r) = kmr^{-2}, \qquad k, m \text{ constant.}$$

  (a) Find $\dfrac{g\left(\frac{r}{10}\right)}{g(r)}$.   **(b)** Find $\dfrac{g(r)}{g(10r)}$.

  (c) What do you notice about your answers to (a) and (b)? Explain this result in terms of gravitational force.

In Problems 19–22, find possible formulas for the power functions.

19.

x	0	1	2	3
y	0	2	8	18

20.

x	1	2	3	4
y	4	16	36	64

21.

x	−2	−1	1	2
y	−16	−1	−1	−16

22.

x	−2	−1	1	2
y	8/5	1/5	−1/5	−8/5

23. A square of side x has area x^2. By what factor does the area change if the length is

  (a) Doubled?   **(b)** Tripled?

  (c) Halved?   **(d)** Multiplied by 0.1?

24. A cube of side x has volume x^3. By what factor does the volume change if the length is

  (a) Doubled?   **(b)** Tripled?

  (c) Halved?   **(d)** Multiplied by 0.1?

25. If the radius of a circle is halved, what happens to its area?

26. If the side length of a cube is increased by 10%, what happens to its surface area?

27. If the side length of a cube is increased by 10%, what happens to its volume?

28. The thrust, T, in pounds, of a ship's propeller is proportional to the square of the propeller speed, R, in rotations per minute, times the fourth power of the propeller diameter, D, in feet.[3]

  (a) Write a formula for T in terms of R and D.

  (b) If $R = 300D$ for a certain propeller, is T a power function of D?

  (c) If $D = 0.25\sqrt{R}$ for a different propeller, is T a power function of R?

29. Poiseuille's law gives the rate of flow, R, of a gas through a cylindrical pipe in terms of the radius of the pipe, r, for a fixed drop in pressure between the two ends of the pipe.

  (a) Find a formula for Poiseuille's Law, given that the rate of flow is proportional to the fourth power of the radius.

  (b) If $R = 400 \text{ cm}^3/\text{sec}$ in a pipe of radius 3 cm for a certain gas, find a formula for the rate of flow of that gas through a pipe of radius r cm.

  (c) What is the rate of flow of the gas in part (b) through a pipe with a 5 cm radius?

30. The circulation time of a mammal (that is, the average time it takes for all the blood in the body to circulate once and return to the heart) is proportional to the fourth root of the body mass of the mammal.

  (a) Write a formula for the circulation time, T, in terms of the body mass, B.

  (b) If an elephant of body mass 5230 kilograms has a circulation time of 148 seconds, find the constant of proportionality.

  (c) What is the circulation time of a human with body mass 70 kilograms?

31. When an aircraft takes off, it accelerates until it reaches its takeoff speed V. In doing so it uses up a distance R of the runway, where R is proportional to the square of the takeoff speed. If V is measured in mph, and R is measured in feet then the constant of proportionality is 0.1639.[4]

  (a) A Boeing 747-400 aircraft has a takeoff speed of about 210 miles per hour. How much runway does it need?

---

[3] Gillner, Thomas C., *Modern Ship Design*, (US Naval Institute Press, 1972).
[4] Adapted from H. Tennekes, *The Simple Science of Flight,* (Cambridge: MIT Press, 1996).

(b) What would the constant of proportionality be if R was measured in meters, and V was measured in meters per second?

32. Biologists estimate that the number of animal species of a certain body length is inversely proportional to the square of the body length.[5] Write a formula for the number of animal species, N, of a certain body length in terms of the length, L. Are there more species at large lengths or at small lengths? Explain.

33. Figure 3.19 shows the life span of birds and mammals in captivity as a function of their body size.[6]

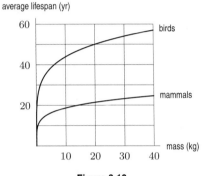

Figure 3.19

(a) Which mammals live longer in captivity, large ones or small ones? If a bird and a mammal have the same size, which has the greater life span in captivity?

(b) What would you expect the life span to be for a body size of 0 kg? For each of the body sizes 10, 20, 30, 40 kg, estimate the life span for birds and mammals, and then use these estimates to plot bird life span as a function of mammal life span. Use this graph to support the statement that in captivity "birds tend to live more than twice as long as mammals of the same size".

(c) The graphs come from the experimentally derived estimates $L_M = 11.8W^{0.20}$ and $L_B = 28.3W^{0.19}$, where L_M and L_B are the life spans of the mammals and birds, and W is body size.

  (i) Solve each formula for W and show that $L_B = 28.3\left(\dfrac{L_M}{11.8}\right)^{0.95}$. What has this to do with the graph you produced in part (b)?

  (ii) Clearly $L_M = L_B$ when $W = 0$. Is there another value of W that makes $L_M = L_B$? What does this mean in terms of the life spans of birds and mammals? Is this realistic?

34. If z is proportional to a power of y and y is proportional to a power of x, is z proportional to a power of x?

35. If z is proportional to a power of x and y is proportional to the same power of x, is $z + y$ proportional to a power of x?

36. If z is proportional to a power of x and y is proportional to a power of x, is zy proportional to a power of x?

37. If z is proportional to a power of x and y is proportional to a different power of x, is $z + y$ proportional to a power of x?

REVIEW EXERCISES AND PROBLEMS FOR CHAPTER THREE

Exercises

In Exercises 1–12, is the function a power function? If it is a power function, write it in the form $y = kx^p$ and give the values of k and p.

In Exercises 13–24, write each expression as a constant times a power of the variable, and state the base, exponent and coefficient.

1. $y = \dfrac{3}{x^2}$

2. $y = 5\sqrt{x}$

3. $y = \dfrac{3}{8x}$

4. $y = 2^x$

5. $y = \dfrac{5}{2\sqrt{x}}$

6. $y = (3x^5)^2$

7. $y = \dfrac{2x^2}{10}$

8. $y = 3 \cdot 5^x$

9. $y = (5x)^3$

10. $y = \dfrac{8}{x}$

11. $y = \dfrac{x}{5}$

12. $y = 3x^2 + 4$

13. $\dfrac{-1}{7w}$

14. $(-2t)^3$

15. $\dfrac{125v^5}{25v^3}$

16. $\dfrac{4}{3}\pi r^3$

17. $(4x^3)(3x^{-2})$

18. $-z^4$

19. $\dfrac{3}{x^2}$

20. $\dfrac{8}{-2/x^6}$

21. $3(-4r)^2$

22. $\dfrac{\frac{5}{2}}{10t^5}$

23. $(\pi a)(\pi a)$

24. $\pi a + \pi a$

---

[5] *US News & World Report*, August 18, 1997, p. 79.

[6] Adapted from K. Schmidt-Nielsen, *Scaling, Why is Animal Size so Important?* p. 147 (Cambridge: CUP, 1984).

In Exercises 25–30, solve for x, assuming all constants are non-zero.

25. $\dfrac{50}{x^3} = 2.8$ **26.** $\dfrac{5}{x^2} = \dfrac{8}{x^3}$

27. $\dfrac{12}{\sqrt{x}} = 3$ **28.** $\dfrac{1}{\sqrt{x-3}} = \dfrac{5}{4}$

29. $\dfrac{100}{(x-2)^2} = 4$ **30.** $\dfrac{A}{Bx^n} = C$

31. Solve the equation $\dfrac{2\pi\sqrt{L}}{C^2} = R$

(a) For L **(b)** For C

In Exercises 32–37 say without solving how many solutions the equation has. Assume that a is a positive constant.

32. $x^3 = a$ **33.** $2t^2 = a$

34. $x^{1/2} = a$ **35.** $s^6 = -a$

36. $3t^{1/5} = a$ **37.** $x^5 = -a$

Problems

38. Heap's Law says that the number of different vocabulary words V in a typical English text of length n words is approximately

$$V = Kn^{\beta}, \quad \text{for constants } K \text{ and } \beta.$$

Figure 3.20 shows a typical graph of this function for specific values of K and β.

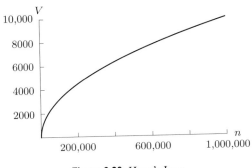
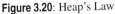

Figure 3.20: Heap's Law

(a) For the function graphed, which of the following is true: $\beta < 0$ or $0 < \beta < 1$ or $\beta > 1$?
(b) Assuming $\beta = 0.5$ in Figure 3.20, is K closer to 1, 10, or 100?

Without solving them, say whether the equations in Problems 39–52 have a positive solution $x = a$ such that

(i) $a > 1$ (ii) $a = 1$
(iii) $0 < a < 1$ (iv)   No positive solution

Give a reason for your answer.

39. $x^3 = 2$ **40.** $x^2 = 5$

41. $x^7 = \frac{1}{8}$ **42.** $x^3 = 0.6$

43. $2x^3 = 5$ **44.** $5x^2 = 3$

45. $2x^5 = 3$ **46.** $3x^5 = 3$

47. $x^{-1} = 9$ **48.** $x^{-3} = \frac{1}{4}$

49. $8x^{-5} = 3$ **50.** $5x^{-8} = -3$

51. $3x^{-5} = 8$ **52.** $-8x^{-5} = -8$

In Problems 53–55, demonstrate a sequence of operations which could be used to solve $4x^2 = 16$. Begin with the step given.

53. Take the square root of both sides.

54. Divide both sides of the equation by 4.

55. Divide both sides of the equation by 16.

56. Which of the following steps is the appropriate next step to solve the equation $x^3 + 8 = 64$?

(a) Take the cube root of both sides of the equation
(b) Subtract 8 from both sides of the equation.

57. Which of the following equations have the same solutions as the equation $9x^2 = 81$?

(a) $3x = 9$ **(b)** $9x = \pm 9$

(c) $3x = \pm 9$ **(d)** $x^2 = 9$

58. The energy, E, in foot-pounds, delivered by an ocean wave is proportional[7] to the length, L, of the wave times the square of its height, h.

   (a) Write a formula for E in terms of L and h.
   (b) A 30-foot high wave of length 600 feet delivers 4 million foot-pounds of energy. Find the constant of proportionality and give its units.
   (c) If the height of a wave is one-fourth the length, find the energy E in terms of the length L.
   (d) If the length is 5 times the height, find the energy E in terms of the height h.

59. Poiseuille's law tells us that the rate of flow, R, of a gas through a cylindrical pipe is proportional to the fourth power of the radius, r, of the pipe, given a fixed drop in pressure between the two ends of the pipe. For a certain gas, if the rate of flow is measured in cm^3/sec and the radius is measured in cm, the constant of proportionality is 4.94.

   (a) If the rate of flow of this gas through a pipe is 500 cm^3/sec, what is the radius of the pipe?
   (b) Solve for the radius r in terms of the rate of flow R.
   (c) Is r proportional to a power of R? If so, what power?

60. The energy, E, in foot-pounds, delivered by an ocean wave is proportional[8] to the length, L, in feet, of the wave times the square of its height, h, in feet, with constant of proportionality 7.4.

   (a) If a wave is 50 ft long and delivers 40,000 ft-lbs of energy, what is its height?
   (b) For waves that are 20 ft long, solve for the height of the wave in terms of the energy. Put the answer in the form $h = kE^p$ and give the values of the coefficient k and the exponent p.

61. A quantity P is inversely proportional to the cube of a quantity R. Solve for R in terms of P. Is R inversely proportional or proportional to a positive power of P? What power?

62. The thrust, T, in pounds, of a ship's propeller is proportional to the square of the propeller speed, R, in rotations per minute, times the fourth power of the propeller diameter, D, in feet.[9]

   (a) Write a formula for T in terms of R and D.
   (b) Solve for the propeller speed R in terms of the thrust T and the diameter D. Write your answer in the form $R = CT^n D^m$ for some constants C, n, and m. What are the values of n and m?
   (c) Solve for the propeller diameter D in terms of the thrust T and the speed R. Write your answer in the

form $D = CT^n R^m$ for some constants C, n, and m. What are the values of n and m?

63. Table 3.5 shows the weight and diameter of various different Sassafras trees.[10] The researchers who collected the data theorize that weight, w, is related to diameter, d, by a power model

$$w = kd^s$$

for some constants k and s.

Table 3.5 *Weight and diameter of Sassafras trees*

Diameter, d (cm)	5.6	6.5	11.8	16.7	23.4
Weight, w (kg)	5.636	7.364	30.696	76.730	169.290

   (a) Plot the data with w on the vertical axis and d on the horizontal axis.
   (b) Does your plot support the hypothesis that $s = 1$? Why or why not?
   (c) What would a graph of the relationship look like if $s = 2$? What would it look like if $s = 3$? Does it look possible that $s = 2$ or 3 from your data plot?
   (d) Instead of looking at the plot to decide whether $s = 2$ is a possibility, look at the data. Add a row to Table 3.5 that shows the ratios w/d^2. What is the overall trend for the numbers in this table: increasing, decreasing, or roughly the same? What should the trend be if $s = 2$ is the correct value for the hypothesis?
   (e) To decide if $s = 3$ is a possibility, add a row to Table 3.5 that shows the ratio w/d^3. How does this table behave differently from the one in part (d)? What do you conclude about the correct value of s for the hypothesis?
   (f) Make your best estimate of the exponent s and the constant of proportionality k.

64. One of Kepler's three laws of planetary motion states that the square of the period, P, of a body orbiting the sun is proportional to the cube of its average distance, d, from the sun. The earth has a period of 365 days and its distance from the sun is approximately 93,000,000 miles.

   (a) Find a formula that gives P as a function of d.
   (b) The planet Mars has an average distance from the sun of 142,000,000 miles. What is the period in earth days for Mars?

---

[7]Gillner, Thomas C., *Modern Ship Design*, (US Naval Institute Press, 1972).
[8]Gillner, Thomas C., *Modern Ship Design*, (US Naval Institute Press, 1972).
[9]Gillner, Thomas C., *Modern Ship Design*, (US Naval Institute Press, 1972).
[10]http://www.yale.edu/fes519b/totoket/allom/allom.htm

TOOLS FOR CHAPTER 3

POSITIVE INTEGER POWERS AND THE EXPONENT RULES

Repeated addition leads to multiplication. For example,

$$\underbrace{2 + 2 + 2 + 2 + 2}_{\text{5 terms in sum}} = 5 \times 2.$$

Similarly, repeated multiplication leads to *exponentiation*. For example,

$$\underbrace{2 \times 2 \times 2 \times 2 \times 2}_{\text{5 factors in product}} = 2^5.$$

Here, 2 is called the *base*, and 5 is called the *exponent* or *power* of 2. Notice that 2^5 is not the same as 5^2, because $2^5 = 32$ and $5^2 = 25$.

In general, if a is any number and n is a positive integer, then we define exponentiation as an abbreviation for repeated multiplication:

$$\underbrace{a \cdot a \cdot a \cdots a}_{n \text{ factors}} = a^n.$$

Notice that $a^1 = a$, because here we have only 1 factor of a. For example, $5^1 = 5$. We call a^2 the *square* of a and we call a^3 the *cube* of a.

Multiplication with a Common Base

When we multiply, we can add the exponents to get a more compact form. For example, $5^2 \cdot 5^3 = (5 \cdot 5) \cdot (5 \cdot 5 \cdot 5) = 5^{2+3} = 5^5$. In general, when we multiply with a common base,

$$a^n \cdot a^m = \underbrace{a \cdot a \cdot a \cdots a}_{n \text{ factors}} \underbrace{a \cdot a \cdot a \cdots a}_{m \text{ factors}} = \underbrace{a \cdot a \cdot a \cdots a}_{n + m \text{ factors}} = a^{n+m}.$$

Thus,

$$a^n \cdot a^m = a^{n+m}.$$

Example 1 Write with a single exponent.

(a) $q^5 \cdot q^7$ (b) $6^2 \cdot 6^3$ (c) $2^n \cdot 2^m$
(d) $3^n \cdot 3^4$ (e) $(x+y)^2(x+y)^3$

Solution

(a) Following the rule $a^n \cdot a^m = a^{n+m}$, we know that $q^5 \cdot q^7 = q^{5+7} = q^{12}$.
(b) $6^2 \cdot 6^3 = 6^{2+3} = 6^5$.
(c) $2^n \cdot 2^m = 2^{n+m}$.
(d) $3^n \cdot 3^4 = 3^{n+4}$.
(e) $(x+y)^2(x+y)^3 = (x+y)^{2+3} = (x+y)^5$.

Division with a Common Base

When we multiply exponentials with a common base we add the exponents. When we divide exponentials with a common base, we subtract the exponents. For example, when we divide 5^6 by 5^2, we get

$$\frac{5^6}{5^2} = \frac{\overbrace{5 \cdot 5 \cdot 5 \cdot 5 \cdot 5 \cdot 5}^{6 \text{ factors of } 5}}{\underbrace{5 \cdot 5}_{2 \text{ factors of } 5}} = \frac{\overbrace{\cancel{5} \cdot \cancel{5}}^{2 \text{ factors of } 5 \text{ cancel}} \cdot \overbrace{5 \cdot 5 \cdot 5 \cdot 5}^{6 - 2 = 4 \text{ factors of } 5 \text{ are left after canceling}}}{\underbrace{\cancel{5} \cdot \cancel{5}}_{2 \text{ factors of } 5 \text{ cancel}}} = \underbrace{5 \cdot 5 \cdot 5 \cdot 5}_{6 - 2 = 4 \text{ factors}} = 5^{6-2} = 5^4.$$

When we divide one product of fives by another, we can cancel them (divide 5 by 5 to get 1), leaving the number of fives in the numerator minus the number of fives in the denominator. More generally, if $n > m$,

$$\frac{a^n}{a^m} = \frac{\overbrace{a \cdot a \cdot a \cdot a \cdots a}^{n \text{ factors of } a}}{\underbrace{a \cdot a \cdots a}_{m \text{ factors of } a}} = \frac{\overbrace{\cancel{a} \cdot \cancel{a} \cdots \cancel{a}}^{m \text{ factors of } a \text{ cancel}} \cdot \overbrace{a \cdots a}^{n - m \text{ factors of } a \text{ are left after canceling}}}{\underbrace{\cancel{a} \cdot \cancel{a} \cdots \cancel{a}}_{m \text{ factors of } a \text{ cancel}}} = \underbrace{a \cdot a \cdot a \cdots a}_{n - m \text{ factors}} = a^{n-m}.$$

Thus,

$$\frac{a^n}{a^m} = a^{n-m}, \text{ if } n > m.$$

Example 2   Write with a single exponent.

(a) $\dfrac{q^7}{q^5}$ (b) $\dfrac{6^7}{6^3}$ (c) $\dfrac{3^n}{3^4}$, where $n > 4$

(d) $\dfrac{\pi^5}{\pi^3}$ (e) $\dfrac{(c+d)^8}{(c+d)^2}$

Solution

(a) Since $\dfrac{a^n}{a^m} = a^{n-m}$, we know that $\dfrac{q^7}{q^5} = q^{7-5} = q^2$.

(b) $\dfrac{6^7}{6^3} = 6^{7-3} = 6^4$.

(c) $\dfrac{3^n}{3^4} = 3^{n-4}$.

(d) $\dfrac{\pi^5}{\pi^3} = \pi^{5-3} = \pi^2$.

(e) $\dfrac{(c+d)^8}{(c+d)^2} = (c+d)^{8-2} = (c+d)^6$.

Raising a Power to a Power

When we take a number written in exponential form and raise it to a power, we multiply the exponents. For example

$$(5^2)^3 = 5^2 \cdot 5^2 \cdot 5^2 = 5^{2+2+2} = 5^6.$$

More generally,

The m factors of a are multiplied n times, giving a total of $m \cdot n$ factors of a

$$(a^m)^n = \underbrace{(a \cdot a \cdot a \cdots a)}_{m \text{ factors of } a}{}^n = \overbrace{\underbrace{(a \cdot a \cdot a \cdots a)}_{m \text{ factors of } a}\underbrace{(a \cdot a \cdot a \cdots a)}_{m \text{ factors of } a} \cdots \underbrace{(a \cdot a \cdot a \cdots a)}_{m \text{ factors of } a}} = a^{m \cdot n}.$$

Thus,

$$(a^m)^n = a^{m \cdot n}.$$

Example 3   Write with a single exponent:

(a) $(7^n)^m$ (b) $(q^7)^5$ (c) $\left((x+y)^2\right)^3$

Solution   (a) Since $(a^m)^n = a^{m \cdot n}$, we know that $(7^n)^m = 7^{nm}$.
(b) $(q^7)^5 = q^{7 \cdot 5} = q^{35}$.
(c) $\left((x+y)^2\right)^3 = (x+y)^{2 \cdot 3} = (x+y)^6$.

Multiplication with a Common Exponent

Sometimes, we want to multiply two exponentials with different bases which have the same exponent. For example, when we multiply $5^2 \cdot 4^2$ we can change the order of the factors and rewrite it as $5^2 \cdot 4^2 = (5 \cdot 5) \cdot (4 \cdot 4) = 5 \cdot 5 \cdot 4 \cdot 4 = (5 \cdot 4) \cdot (5 \cdot 4) = (5 \cdot 4)^2 = 20^2$. Sometimes, we want to use this process in reverse: $10^2 = (2 \cdot 5)^2 = 2^2 \cdot 5^2$.

In general,

n factors of a n factors of b

$$(a \cdot b)^n = \underbrace{(a \cdot b)(a \cdot b)(a \cdot b) \cdots (a \cdot b)}_{n \text{ factors of } (a \cdot b)} = \overbrace{(a \cdot a \cdot a \cdots a)}^{n \text{ factors of } a} \cdot \overbrace{(b \cdot b \cdot b \cdots b)}^{n \text{ factors of } b} = a^n \cdot b^n.$$

Since we can rearrange the order using the commutative property of multiplication

Thus,

$$(ab)^n = a^n b^n.$$

Example 4   Write with a single exponent:

(a) $c^4 d^4$ (b) $2^n \cdot 3^n$ (c) $(x^2 + y^2)^5 (c-d)^5$

Solution   (a) Since $a^n b^n = (ab)^n$, we have $c^4 d^4 = (cd)^4$.
(b) $2^n \cdot 3^n = (2 \cdot 3)^n = 6^n$.
(c) $(x^2 + y^2)^5 (c-d)^5 = \left((x^2 + y^2)(c-d)\right)^5$.

Example 5   Write without parentheses:

(a) $(qp)^7$ (b) $(3x)^n$

Solution   (a) Since $(ab)^n = a^n b^n$, we have $(qp)^7 = q^7 p^7$.
(b) $(3x)^n = 3^n x^n$.

Division with a Common Exponent

Division of two powers with the same exponent works the same way as multiplication. For example,

$$\frac{6^4}{3^4} = \frac{6 \cdot 6 \cdot 6 \cdot 6}{3 \cdot 3 \cdot 3 \cdot 3} = \frac{6}{3} \cdot \frac{6}{3} \cdot \frac{6}{3} \cdot \frac{6}{3} = \left(\frac{6}{3}\right)^4 = 2^4 = 16.$$

Or, from the other side,

$$\left(\frac{4}{5}\right)^3 = \frac{4}{5} \cdot \frac{4}{5} \cdot \frac{4}{5} = \frac{4 \cdot 4 \cdot 4}{5 \cdot 5 \cdot 5} = \frac{4^3}{5^3}.$$

More generally,

$$\left(\frac{a}{b}\right)^n = \underbrace{\left(\frac{a}{b}\right) \cdot \left(\frac{a}{b}\right) \cdot \left(\frac{a}{b}\right) \cdots \left(\frac{a}{b}\right)}_{n \text{ factors of } a/b} = \frac{\overbrace{a \cdot a \cdot a \cdots a}^{n \text{ factors of } a}}{\underbrace{b \cdot b \cdot b \cdots b}_{n \text{ factors of } b}} = \frac{a^n}{b^n}.$$

Thus,

$$\left(\frac{a}{b}\right)^n = \frac{a^n}{b^n}.$$

Example 6   Write with a single exponent:

(a) $\dfrac{q^7}{p^7}$ (b) $\dfrac{9x^2}{y^2}$ (c) $\dfrac{(x^2 + y^2)^5}{(a + b)^5}$

Solution   (a) Since $\dfrac{a^n}{b^n} = \left(\dfrac{a}{b}\right)^n$, we have $\dfrac{q^7}{p^7} = \left(\dfrac{q}{p}\right)^7$.

(b) $\dfrac{9x^2}{y^2} = \left(\dfrac{3x}{y}\right)^2$.

(c) $\dfrac{(x^2 + y^2)^5}{(a + b)^5} = \left(\dfrac{x^2 + y^2}{a + b}\right)^5$.

Example 7   Write without parentheses.

(a) $\left(\dfrac{c}{d}\right)^{12}$ (b) $\left(\dfrac{2u}{3v}\right)^3$

Solution

(a) Since $\left(\dfrac{a}{b}\right)^n = \dfrac{a^n}{b^n}$, we have $\left(\dfrac{c}{d}\right)^{12} = \dfrac{c^{12}}{d^{12}}$.

(b) $\left(\dfrac{2u}{3v}\right)^3 = \dfrac{(2u)^3}{(3v)^3} = \dfrac{2^3 u^3}{3^3 v^3} = \dfrac{8u^3}{27v^3}$.

Exponent Rules

We summarize the results of this section as follows.

Expressions with a Common Base

If m and n are positive integers, with $n > m$ in the second equation,

1. $a^n \cdot a^m = a^{n+m}$ 2. $\dfrac{a^n}{a^m} = a^{n-m}$ 3. $(a^m)^n = a^{m \cdot n}$

Expressions with a Common Exponent

If n is a positive integer,

1. $(ab)^n = a^n b^n$ 2. $\left(\dfrac{a}{b}\right)^n = \dfrac{a^n}{b^n}$

Common Mistakes

Be aware of the following notational conventions:

$$ab^n = a(b^n), \qquad \text{but, in general, } ab^n \neq (ab)^n,$$
$$-b^n = -(b^n), \qquad \text{but, in general, } -b^n \neq (-b)^n,$$
$$-ab^n = (-a)(b^n).$$

For example, $-2^4 = -(2^4) = -16$, but $(-2)^4 = (-2)(-2)(-2)(-2) = +16$.

Example 8   Evaluate the following expressions for $x = -2$ and $y = 3$:

(a) $(xy)^4$ (b) $-xy^2$ (c) $(x+y)^2$
(d) x^y (e) $-4x^3$ (f) $-y^2$

Solution

(a) $(-2 \cdot 3)^4 = (-6)^4 = (-6)(-6)(-6)(-6) = 1296$.
(b) $-(-2) \cdot (3)^2 = 2 \cdot 9 = 18$.
(c) $(-2+3)^2 = (1)^2 = 1$.
(d) $(-2)^3 = (-2)(-2)(-2) = -8$.
(e) $-4(-2)^3 = -4(-2)(-2)(-2) = 32$.
(f) $-(3)^2 = -9$.

Exercises on Positive Integer Powers and the Exponent Rules

Evaluate the expressions in Exercises 1–4 without using a calculator.

1. $3 \cdot 2^3$

2. -3^2

3. $(-2)^3$

4. $5^1 \cdot 1^4 \cdot 3^2$

In Exercises 5–8, write the expression in the form x^n, assuming $x \neq 0$.

5. $x^3 \cdot x^5$

6. $(x^4 \cdot x)^2$

7. $\dfrac{x^3 \cdot x^4}{x^2}$

8. $\left(\dfrac{x^5}{x^2}\right)^4$

In Exercises 9–19, write with a single exponent.

9. $\left((a+b)^2\right)^5$

10. $(x+y+z)^{21}(u+v+w)^{21}$

11. $(a+b)^2(a+b)^5$

12. $\dfrac{(a+b)^5}{(a+b)^2}$

13. $\dfrac{a^b}{3^b}$

14. $\dfrac{9^2}{q^4}$

15. $2^n 2^2$

16. $A^{n+3} B^n B^3$

17. $B^a B^{a+1}$

18. $\dfrac{2^a 3^a}{6^b}$

19. $(x^2+y)^3(x+y^2)^3$

Without a calculator, decide whether the quantities in Exercises 20–27 are positive or negative.

20. $(-4)^3$

21. -4^3

22. $(-3)^4$

23. -3^4

24. -16^{33}

25. $(-12)^{15}$

26. $(-23)^{42}$

27. -31^{66}

ZERO, NEGATIVE, AND FRACTIONAL POWERS

The natural definition for exponentiation as an abbreviation for multiplication holds for positive integers only. For example, 4^5 means 4 multiplied by itself 5 times, but we cannot use the same definition for 4^0 or 4^{-1} or 4^{-2}. It makes sense to choose definitions for exponents like $0, -1, -2$ which are consistent with the exponent rules.

The exponent rule for division, $\dfrac{a^n}{a^m} = a^{n-m}$, is true for positive integers n and m, with $n > m$. We can also make it true for $n = m$ and $n < m$ if we define a^0 and a^{-n} in the right way.

Zero Powers

To define a^0, we assume the exponent rule for division, $\dfrac{a^n}{a^m} = a^{n-m}$, applies for $n = m$. For example, if $a \neq 0$,

$$\frac{a^2}{a^2} = a^{2-2} = a^0.$$

But $\dfrac{a^2}{a^2} = 1$, so we define

$$a^0 = 1,$$

if $a \neq 0$.

Negative Powers

To define a^{-1}, we assume the exponent rule for division, $\dfrac{a^n}{a^m} = a^{n-m}$, applies for $n < m$. For example, if $a \neq 0$,

$$\frac{a^0}{a^1} = a^{0-1} = a^{-1}.$$

But $\dfrac{a^0}{a^1} = \dfrac{1}{a}$, so we define

$$a^{-1} = \frac{1}{a},$$

if $a \neq 0$. Similarly, we can show that it makes sense to define

$$a^{-2} = \frac{1}{a^2}.$$

More generally, we define

$$a^{-n} = \frac{1}{a^n},$$

if $a \neq 0$.

A negative exponent tells us to take the reciprocal, *not* to make the number negative.

Example 9   Evaluate:

(a) 5^0 (b) 3^{-2} (c) 2^{-1} (d) $(-2)^{-3}$ (e) $\left(\dfrac{2}{3}\right)^{-1}$

Solution   (a) Any nonzero number to the zero power is one, so $5^0 = 1$.

(b) Using

$$a^{-n} = \frac{1}{a^n},$$

with $n = 2$ gives

$$3^{-2} = \frac{1}{3^2} = \frac{1}{9}.$$

(c) We have

$$2^{-1} = \frac{1}{2^1} = \frac{1}{2}.$$

(d) We have

$$(-2)^{-3} = \frac{1}{(-2)^3} = \frac{1}{(-2)\cdot(-2)\cdot(-2)} = \frac{1}{-8} = -\frac{1}{8}.$$

(e) We have

$$\left(\frac{2}{3}\right)^{-1} = \frac{1}{\left(\frac{2}{3}\right)} = \frac{3}{2}.$$

Example 10   Evaluate for $r = -1$ and $s = 7$:

(a) $(5r)^{-3}$ (b) $-\dfrac{rs^{-2}}{5^0}$ (c) -5^r

(d) $2s^r$ (e) $(174s^4 r^{12})^0$

Solution   (a) Substituting -1 for r gives

$$(5 \cdot (-1))^{-3} = (-5)^{-3} = \frac{1}{(-5)^3} = \frac{1}{-125} = -\frac{1}{125}.$$

(b) Substituting -1 for r and 7 for s gives

$$-\frac{(-1) \cdot 7^{-2}}{5^0} = -\frac{-1}{7^2 \cdot 5^0} = -\frac{-1}{49 \cdot 1} = \frac{1}{49}.$$

(c) Substituting -1 for r gives

$$-5^{-1} = -\frac{1}{5^1} = -\frac{1}{5}.$$

(d) Substituting -1 for r and 7 for s gives

$$2(7)^{-1} = 2 \cdot \frac{1}{7} = \frac{2}{7}.$$

(e) Any nonzero number raised to the zero power is 1.   So

$$(174s^4 r^{12})^0 = 1.$$

The definitions are summarized in the following box.

Definitions For Exponentiation

If a is any number and n is a positive integer:
- $a^0 = 1$
- $a^{-n} = \dfrac{1}{a^n}$

With these definitions, the exponent rule for division, $\dfrac{a^n}{a^m} = a^{n-m}$ for $n > m$, now applies to $n = m$ and to $n < m$, so to all integers n and m.

Example 11   Are the expressions 2^{-n}, $\dfrac{1}{2^n}$ and $\left(\dfrac{1}{2}\right)^n$ equivalent?

Solution   First we make a table of the expressions for various different values of n.

Table 3.6   *Values of 2^{-n}, $1/2^n$, and $(1/2)^n$*

n	-1	0	1	2
2^{-n}	$2^{-(-1)} = 2$	$2^0 = 1$	$2^{-1} = 1/2$	$2^{-2} = 1/4$
$\dfrac{1}{2^n}$	$\dfrac{1}{2^{-1}} = 2$	$\dfrac{1}{2^0} = 1$	$\dfrac{1}{2^1} = 1/2$	$\dfrac{1}{2^2} = 1/4$
$\left(\dfrac{1}{2}\right)^n$	$\left(\dfrac{1}{2}\right)^{-1} = 2$	$\left(\dfrac{1}{2}\right)^0 = 1$	$\left(\dfrac{1}{2}\right)^1 = 1/2$	$\left(\dfrac{1}{2}\right)^2 = 1/4$

The expressions have the same value for each value of n chosen, so it seems that they could be equivalent. We can show that they are equivalent using the exponent rules:

$$2^{-n} = \frac{1}{2^n} = \frac{1^n}{2^n} = \left(\frac{1}{2}\right)^n.$$

Example 12   Rewrite with only positive exponents. Assume all variables are positive.

(a) $\dfrac{1}{3x^{-2}}$ (b) $\left(\dfrac{x}{y}\right)^{-3}$ (c) $\dfrac{3r^{-2}}{(2r)^{-4}}$ (d) $\dfrac{(a+b)^{-2}}{(a+b)^{-5}}$

Solution   (a) We have

$$\frac{1}{3x^{-2}} = \frac{1}{3\cdot\frac{1}{x^2}} = \frac{1}{\frac{3}{x^2}} = \frac{x^2}{3}.$$

(b) We have

$$\left(\frac{x}{y}\right)^{-3} = \frac{1}{\left(\frac{x}{y}\right)^3} = \frac{1}{\frac{x^3}{y^3}} = \frac{y^3}{x^3}.$$

(c) We have

$$\frac{3r^{-2}}{(2r)^{-4}} = \frac{3\cdot\frac{1}{r^2}}{\frac{1}{(2r)^4}} = \frac{\frac{3}{r^2}}{\frac{1}{16r^4}} = \frac{3}{r^2}\cdot\frac{16r^4}{1} = \frac{48r^4}{r^2} = 48r^2.$$

(d) We have

$$\frac{(a+b)^{-2}}{(a+b)^{-5}} = (a+b)^{-2+5} = (a+b)^3.$$

Fractional Powers and Roots

We chose definitions for zero and negative exponents to be consistent with the exponent rules. To define $a^{1/n}$, we follow the same procedure. The exponent rule, $(a^m)^n = a^{m\cdot n}$, is true for positive integers n and m. We can also make it true for fractions $1/n$ if we define $a^{1/n}$ in the right way.

To define $a^{1/n}$, we assume the exponent rule $(a^m)^n = a^{m\cdot n}$ applies to fractions. For $a > 0$ we have

$$(a^{1/2})^2 = a^{(1/2)\cdot 2} = a^1 = a.$$

Thus, $a^{1/2}$ is a number which when squared gives a, so we define

$$a^{1/2} = \sqrt{a} = \text{the \textbf{square root} of } a.$$

$$a^{1/3} = \sqrt[3]{a} = \text{the \textbf{cube root} of } a.$$

Similarly, we define

$$a^{1/n} = \sqrt[n]{a} = \text{the } \textbf{n}^{\textbf{th}} \textbf{ root} \text{ of } a.$$

We review the definition of roots:

Let n be an integer greater than 1:
- For $a \geq 0$,

    $\sqrt{a}$ is the positive number whose square is a.

    $\sqrt[n]{a}$ is the positive number whose n^{th} power is a.

- For $a < 0$,

    If n is even, $\sqrt[n]{a}$ is not a real number.

    If n is odd, $\sqrt[n]{a}$ is the negative number whose n^{th} power is a.

For example, $\sqrt{49} = 7$ because $7^2 = 49$. Likewise, $\sqrt[3]{125} = 5$ because $5^3 = 125$, and $\sqrt[5]{32} = 2$ because $2^5 = 32$. Similarly, $\sqrt[3]{-27} = -3$ because $(-3)^3 = -27$, and $\sqrt{-9}$ is not a real number, because the square of no real number is negative. We summarize the definitions:

Definitions For Exponentiation

If a is any number and m and n are positive integers:[11]
- $a^{1/n} = \sqrt[n]{a}$ (if n is even, we assume $a \geq 0$.)
- $a^{m/n} = \sqrt[n]{a^m} = (\sqrt[n]{a})^m$ (if n is even, we assume $a \geq 0$.)

With these definitions, the rules for exponents given for positive integer exponents apply also when the exponent is fractional.

Example 13   Evaluate the following:

(a) $25^{1/2}$ (b) $9^{-1/2}$ (c) $8^{1/3}$ (d) $27^{-1/3}$

Solution

(a) We have
$$25^{1/2} = \sqrt{25} = 5.$$

(b) We have
$$9^{-1/2} = \frac{1}{9^{1/2}} = \frac{1}{\sqrt{9}} = \frac{1}{3}.$$

(c) Since $2^3 = 8$, we have
$$8^{1/3} = \sqrt[3]{8} = 2.$$

(d) Since $3^3 = 27$, we have
$$27^{-1/3} = \frac{1}{27^{1/3}} = \frac{1}{\sqrt[3]{27}} = \frac{1}{3}.$$

The definition of rational exponents allows us to evaluate simple fractional power expressions without a calculator.

---

[11] When we write a fractional power, we assume that the base is restricted to the values for which the power is defined.

Example 14   Find

(a) $64^{2/3}$

(b) $9^{-3/2}$

Solution   (a) The exponent $2/3$ tells us to square and to take the cube root. We can do these two operations in any order. We have

$$64^{2/3} = \sqrt[3]{64^2} = \sqrt[3]{4096} = 16.$$

It is easier to take the cube root first, and then square the result as follows:

$$64^{2/3} = \left(64^{1/3}\right)^2 = \left(\sqrt[3]{64}\right)^2 = 4^2 = 16.$$

(b) The exponent $-3/2$ tells us to take the reciprocal, to cube, and to take the square root. We can do these three operations in any order. We choose to do this the easiest way as follows:

$$9^{-3/2} = \frac{1}{9^{3/2}} = \frac{1}{(\sqrt{9})^3} = \frac{1}{3^3} = \frac{1}{27}.$$

Calculator Note: Some calculators will not compute $a^{m/n}$ for $m \neq 1$ when a is negative, even if n is odd. For example, although $(-1)^{2/3}$ is defined, a calculator may display "error."

Example 15   Evaluate, if possible:

(a) $(-216)^{2/3}$

(b) $(-625)^{3/4}$

Solution   (a) To find $(-216)^{2/3}$, we can first evaluate $(-216)^{1/3} = -6$, and then square the result. This gives $(-6)^2 = 36$.

(b) We conclude that $(-625)^{3/4}$ is not a real number since $(-625)^{1/4}$ is an even root of a negative number.

Example 16   Write each of the following as an equivalent expression in the form x^n and give the value for n.

(a) $\dfrac{1}{x^3}$

(b) $\sqrt[5]{x}$

(c) $(\sqrt[3]{x})^2$

(d) $\sqrt{x^5}$

(e) $\dfrac{1}{\sqrt[4]{x}}$

(f) $\left(\dfrac{1}{\sqrt{x}}\right)^3$

Solution   (a) We have

$$\frac{1}{x^3} = x^{-3} \quad \text{so} \quad n = -3.$$

(b) We have

$$\sqrt[5]{x} = x^{1/5} \quad \text{so} \quad n = 1/5.$$

(c) We have

$$(\sqrt[3]{x})^2 = (x^{1/3})^2 = x^{2/3} \quad \text{so} \quad n = 2/3.$$

(d) We have

$$\sqrt{x^5} = (x^5)^{1/2} = x^{5/2} \quad \text{so} \quad n = 5/2.$$

(e) We have

$$\frac{1}{\sqrt[4]{x}} = \frac{1}{x^{1/4}} = x^{-1/4} \quad \text{so} \quad n = -1/4.$$

(f) We have

$$\left(\frac{1}{\sqrt{x}}\right)^3 = \left(\frac{1}{x^{1/2}}\right)^3 = (x^{-1/2})^3 = x^{-3/2} \quad \text{so} \quad n = -3/2.$$

Exercises on Zero, Negative, and Fractional Powers

Evaluate the expressions in Exercises 1–8 without using a calculator.

1. 3^0

2. 0^3

3. $4^{1/2}$

4. 5^{-2}

5. 9^{-1}

6. $25^{-1/2}$

7. $\left(\dfrac{4}{9}\right)^{-1/2}$

8. $\left(\dfrac{64}{27}\right)^{-1/3}$

In Exercises 9–13, write the expression as an equivalent expression in the form x^n and give the value for n.

9. $\dfrac{1}{\sqrt{x}}$

10. $\dfrac{1}{x^5}$

11. $\sqrt{x^3}$

12. $\left(\sqrt[3]{x}\right)^5$

13. $1/(1/x^{-2})$

Without a calculator, decide whether the quantities in Exercises 14–21 are positive or negative.

14. 17^{-1}

15. $(-5)^{-2}$

16. -5^{-2}

17. $(-4)^{-3}$

18. $(-73)^0$

19. -48^0

20. $(-47)^{-15}$

21. $(-61)^{-42}$

Chapter Four

More on Functions

Contents

4.1 DOMAIN AND RANGE

Some functions have restrictions on the possible inputs they can take.

Example 1   In Example 1 on page 144, we saw that the period, P, in seconds, of a pendulum of length L, in feet, is given by the function
$$P = f(L) = 1.111L^{1/2}.$$
(a) For what values of L is the expression for $f(L)$ defined?
(b) What values of L make sense for pendulums?

Solution   (a) Since $L^{1/2} = \sqrt{L}$, the expression for f is a constant times the square root of L. Since we can only take square roots of numbers that are positive or zero, $f(L)$ is defined only for $L \geq 0$.
(b) A real pendulum would have to have positive length, so we should exclude $L = 0$, leaving $L > 0$ (in practice, there are probably further restrictions on the length).

In general, we make the following definition:

> If $Q = f(t)$, then the **domain** of f is the set of allowable input values, t.

How do we Know Which Inputs are Allowable?

In Example 1 we saw two different ways of deciding which inputs are allowable: looking at the algebraic expression defining a function or considering the situation that the function is modeling. In general, if f is defined by an algebraic expression and no other information is given, we allow all inputs x for which $f(x)$ is defined, and only those inputs. For example, we exclude any input that leads to dividing by zero or taking the square root of a negative number.

Example 2   For each of the following functions, find the domain.
(a) $f(x) = \dfrac{1}{x-2}$ (b) $g(x) = \dfrac{1}{x} - 2$ (c) $h(x) = \sqrt{1-x}$ (d) $k(x) = \sqrt{x-1}.$

Solution   (a) Since the expression for $f(x)$ is an algebraic fraction, we watch out for inputs that lead to dividing by zero. The denominator is zero when $x = 2$,
$$f(2) = \frac{1}{2-2} = \frac{1}{0} = \text{Undefined}.$$
Any other value of x is allowable, so the domain is all real numbers except $x = 2$. We sometimes write this as
$$\text{Domain: all real } x \neq 2.$$
(b) Here we have an x in the denominator, so $x = 0$ is a problem:
$$g(0) = \frac{1}{0} - 2 = \text{Undefined}.$$
Any other input is allowable, so
$$\text{Domain: all real } x \neq 0.$$

(c) Here we need to make sure that we are taking the square root of a number which is either zero or positive. So we must have $1 - x \geq 0$, which means $x \leq 1$. So

$$\text{Domain: } x \leq 1$$

(d) This time the order of subtraction under the square root is reversed, so we want $x - 1 \geq 0$, which means $x \geq 1$. So

$$\text{Domain: } x \geq 1.$$

We can work out more complicated examples by following through the steps in the calculation of the value of the function, and checking at each stage whether the step gives an undefined value.

Example 3   Find the domain of

$$f(x) = \frac{5}{2 - \sqrt{x - 2}}.$$

Solution   We need to know for what values of x the expression for f is defined, so we consider one by one the operations used in forming the expression.

(a) Subtract 2 from x. This does not restrict the domain, since we can subtract 2 from any number.
(b) Take the square root of $x - 2$. We cannot take the square root of a negative number, so the result of step (a) cannot be negative. Thus, $x \geq 2$.
(c) Subtract $\sqrt{x - 2}$ from 2. This does not restrict the domain, since we can subtract any number from 2.
(d) Divide 5 by $2 - \sqrt{x - 2}$. We cannot divide by 0, so $\sqrt{x - 2}$ cannot equal 2. Therefore, $x - 2$ cannot equal 4, and x cannot equal 6.
(e) Putting steps (c) and (d) together, we see that the domain is $x \geq 2$, $x \neq 6$—that is, the domain is all numbers greater than or equal to 2 except for 6.

Determining the Domain from the Context

If a function is being used to model a particular situation, then we only allow inputs that make sense in the situation.

Example 4   Find the domain of

(a) the function $f(A) = A/350$ used to calculate the number of gallons of paint needed to cover an area A, as in Example 9 on page 43
(b) the function f in Example 3 on page 41 giving the cost of purchasing n stamps,

$$f(n) = 0.39n$$

Solution   (a) Although the algebraic expression for f has all real numbers A as its domain, in this case we say f has domain $A \geq 0$ because the area painted must not be negative.
(b) We cannot purchase a negative number of stamps, nor can we purchase a fractional number of stamps. Therefore, the domain is all integers greater than or equal to 0.

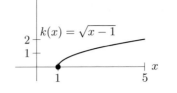

Figure 4.1: The domain of $k(x) = \sqrt{x-1}$ is $x \geq 1$

Visualizing the Domain on the Graph

An input value a for a function f corresponds to a point $(a, f(a))$ on the graph, so the domain of f is all values on the input axis for which there is a corresponding point on the graph. For example, Figure 4.1 shows the graph of of $k(x) = \sqrt{x-1}$, which we saw in Example 2(d) has domain $x \geq 1$. The graph has no portion to the left of $x = 1$, starts at $(1, 0)$, and has points corresponding to every number to the right $x = 1$.

Example 5   Find the domain of the function
$$G = 0.75 - 0.3h$$
giving the amount of gasoline, G gallons, in a portable electric generator h hours after it starts.

Solution   Considered algebraically, the expression for G makes sense for all values of h. However, in the context, the only values of h that make sense are those from when the generator starts, at $h = 0$, until it runs out of gas, where $G = 0$. Figure 4.2 shows that the allowable values of h are $0 \leq h \leq 2.5$.

We can also calculate algebraically the value of h that makes $G = 0$ by solving the equation $0.3h = 0.75$, so when $h = 0.75/0.3 = 2.5$.

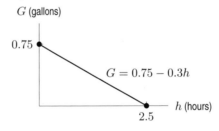

Figure 4.2: Gasoline runs out when $h = 2.5$

Example 6   Estimate the domain of the function in Figure 4.3, assuming that the entire graph is shown.

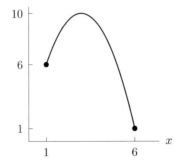

Figure 4.3: Find the domain and range of this function

Solution   Since the graph only has points with x-coordinates between 1 and 6, these are the only x-values for which $f(x)$ is defined, so the domain is $1 \leq x \leq 6$.

The Range of a Function

If the equation $f(t) = -3$ has a solution, then the number -3 is an output value of the function f. In general,

---

If $Q = f(t)$, then the **range** of f is the set of output values, Q. Thus, it is the set of Q-values for which the equation

$$Q = f(t)$$

has a solution in t.

Example 7   Find the range of the function $f(x) = 5 - x$

Solution   A number y is in the range if the equation $y = f(x)$ can be solved for x. In this case the equation can be solved no matter what y is:

$$y = 5 - x$$
$$y + x = 5$$
$$x = 5 - y.$$

Since y can be any number, we have

Range: all real numbers.

Example 8   Find the range of (a) $f(x) = 3x + 5$ (b) $g(x) = 5$

Solution   (a) The equation

$$3x + 5 = k$$

always has a solution, no matter what the value of k, because we can always subtract 5 and divide through by 3. So the range of $f(x) = 3x + 5$ is all real numbers.
(b) This function has output 5 no matter what the input, so its range consists of the single number 5.

In general, if m is not zero, the range of a linear function $f(x) = b + mx$ is all real numbers.

Visualizing the Range on a Graph

If $y = f(x)$, then there is a point on the graph of f with coordinates (x, y), so the range is the set of all values on the vertical axis with a corresponding point on the graph.

Example 9   For the function

$$g(x) = \frac{1}{x-1},$$

(a) Find the domain and range from the graph
(b) Verify your answer algebraically.

Solution   (a) In Figure 4.4, we see that there is no point (x, y) on the graph with $x = 1$ or with $y = 0$. Thus, the domain is all real numbers $x \neq 1$, and the range is all real numbers $y \neq 0$.

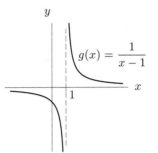

Figure 4.4: Domain and range of $g(x) = 1/(x-1)$

(b) The expression for $g(x)$ is undefined when $x = 1$, because $1/((1) - 1) = 1/0$ and division by 0 is undefined. It is defined for all other values of x, so

Domain: all real x, $x \neq 1$.

To find the range, we check to see if the equation $y = 1/(x-1)$ can be solved for x. Provided $y \neq 0$, the equation can be solved as follows:

$$y = \frac{1}{x-1}$$
$$\frac{1}{y} = x - 1$$
$$x = 1 + \frac{1}{y}.$$

If $y = 0$, then the equation has no solution, since 1 divided by a real number is never zero. Thus we have

Range: all real y, $y \neq 0$.

Example 10   Estimate the range of the function in Figure 4.3, assuming that the entire graph is shown.

Solution   The range is the set of outputs of the function, which are the numbers that occur as y-coordinates of points on the graph. The range of this function is $1 \leq y \leq 10$.

Domain and Range of a Power Function

The domain of a power function $f(x) = kx^p$, with $k \neq 0$, is the values of x for which x^p is defined.

Example 11   Graph the following functions and give their domain.

(a) $g(x) = x^3$ (b) $h(x) = x^{-2}$ (c) $f(x) = x^{1/2}$ (d) $k(x) = x^{-1/3}$

Solution   See Figure 4.5.

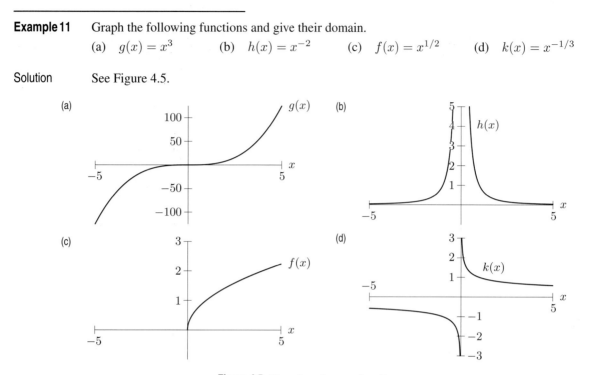

Figure 4.5: Domains of power functions

(a) Since we can cube any number, the domain of g is all real numbers.
(b) We have

$$h(x) = x^{-2} = \frac{1}{x^2}.$$

Since we cannot divide by 0, we cannot input 0 into h. Any other number produces an output, so the domain of h is all numbers except 0. Notice that on the graph of h, there is no point with $x = 0$.
(c) The domain of f is all non-negative numbers. Notice that the graph of f has no points with negative x-coordinates.
(d) We can take the cube root of any real number, but we cannot divide by 0, so the domain of k is all real numbers except 0.

In general, we have the following rules for determining the domain of a power function:

- x^p is defined if x and p are positive
- Negative powers of x are not defined at $x = 0$
- Fractional powers $x^{n/m}$, where n/m is a fraction in lowest terms and m is even, are not defined if $x < 0$.

We can also use these rules to decide whether a number a is in the range of a power function $f(x) = kx^p$. We need to know if there is a number x such that $f(x) = a$, so we want to solve

$$kx^p = a.$$

A solution to this equation, if it exists, is

$$x = \left(\frac{a}{k}\right)^{1/p}.$$

So to decide if x exists we need to decide if the right-hand side, $(a/k)^{1/p}$, is defined.

Example 12   Find the range, and explain your answer in terms of equations.

(a) $f(x) = 2x^2$ (b) $g(x) = 5x^3$ (c) $h(x) = \dfrac{1}{x}$ (d) $k(x) = \dfrac{-3}{x^2}$.

Solution

(a) The equation $2x^2 = a$ is equivalent to $x^2 = a/2$, which has no solutions if a is negative. So negative numbers are not in the range of f. On the other hand, $x^2 = a/2$ does have solutions if a is positive or zero. So

$$\text{Range} = \text{all } a \text{ such that } a \geq 0.$$

We can see this from the graph of $y = 2x^2$ in Figure 4.6. There is a point on the graph corresponding to every non-negative y-value.

(b) The equation $5x^3 = a$ is equivalent to $x^3 = a/5$, which has a solution for all numbers a. Therefore

$$\text{Range} = \text{all real numbers}.$$

We also see this from the graph of $y = 5x^3$ in Figure 4.7. There is a point on the graph corresponding to every number on the vertical axis.

(c) The equation $x^{-1} = a$ has the solution $x = a^{-1}$ for all a except $a = 0$, so the range is all real numbers except 0. Figure 4.8 shows that there is a point on the graph for every y-value except 0.

(d) The equation $-3x^{-2} = a$ is equivalent to $x^2 = -3/a$. The right hand side must be positive for this to have solutions, so a must be negative. There is no solution if $a = 0$, since a is in the denominator. So the range is all negative numbers. Figure 4.9 shows that there are points on the graph for all negative y-values, and no points if $y \geq 0$.

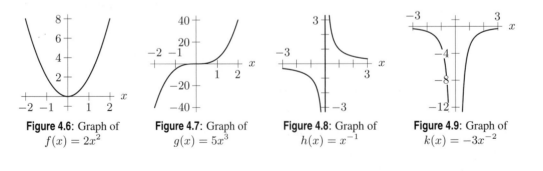

Figure 4.6: Graph of $f(x) = 2x^2$ **Figure 4.7:** Graph of $g(x) = 5x^3$ **Figure 4.8:** Graph of $h(x) = x^{-1}$ **Figure 4.9:** Graph of $k(x) = -3x^{-2}$

Exercises and Problems for Section 4.1

Exercises

For the functions in Exercises 1–6,
  (a) List the algebraic operations in order of evaluation. What restrictions does each operation place on the domain of the function?
  (b) Give the function's domain.

1. $y = \dfrac{2}{x-3}$

2. $y = \sqrt{x-5} + 1$

3. $y = 4 - (x-3)^2$

4. $y = \dfrac{7}{4-(x-3)^2}$

5. $y = 4 - (x-3)^{1/2}$

6. $y = \dfrac{7}{4-(x-3)^{1/2}}$

In Exercises 7–10:
  (a) Find the values of a for which $f(x) = a$ has a solution
  (b) Give the range of f.

7. $f(x) = \dfrac{5x+7}{2}$

8. $f(x) = \dfrac{2}{5x+7}$

9. $f(x) = 2(x+3)^2$

10. $f(x) = 5 + 2(4x+3)^2$

In Exercises 11–22, find
  (a) The domain.   (b) The range.

11. $m(x) = 9 - x$

12. $y = x^2$

13. $y = 7$

14. $y = x^2 - 3$

15. $f(x) = x - 3$

16. $y = 5x - 1$

17. $f(x) = \dfrac{1}{\sqrt{x-4}}$

18. $y = \sqrt{x} + 1$

19. $y = \sqrt{x+1}$

20. $y = \dfrac{1}{x-2}$

21. $f(x) = \dfrac{1}{x+1} + 3$

22. $y = \sqrt{2x-4}$

Problems

23. A restaurant is open from 2 pm to 2 am each day, and a maximum of 200 clients can fit inside. If $f(t)$ is the number of clients in the restaurant t hours after 2 pm each day,
    (a) What is reasonable domain for f?
    (b) What is a reasonable range for f?

24. A car's average gas mileage, G, is a function $f(v)$ of the average speed driven, v. What is a reasonable domain for $f(v)$?

In Exercises 25–26, assume the entire graph is shown. Estimate:
  (a) The domain of the function.
  (b) The range of the function.

25.
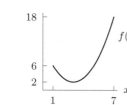

26.
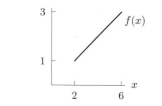

27. The value, V, of a car is a function of its age, a, in years. Find and interpret:
    (a) The domain   (b) The range

28. The cost, $\$C$, of producing x units of a product is given by the function $C = 2000 + 4x$, up to a cost of $\$10{,}000$. Find and interpret:
    (a) The domain.   (b) The range.

29. For what values of k does the equation $5 - 3x = k$ have a solution? What does your answer say about the range of the function $f(x) = 5 - 3x$?

30. For what values of k does the equation $-1 = k$ have solution? What does your answer tell you about the range of the function $f(x) = -1$?

4.2 UNDERSTANDING THE STRUCTURE OF FUNCTIONS

A function can be thought of as an operation that takes an input and produces an output. Sometimes it helps in understanding how a function works if we break it down into a series of simpler operations.

Example 1   Scientists measure temperature in degrees Celsius (°C). The freezing point of water is 0°C, which is the same as 32°F. Also, each degree increase (or decrease) in Celsius corresponds to a 1.8° increase (or decrease) in Fahrenheit. If the temperature is $T°$F then the temperature in degrees Celsius is given by

$$f(T) = \frac{1}{1.8}(T - 32).$$

Describe the steps in evaluating this function and explain the significance of each step in terms of temperature scales.

Solution   Starting with the input T, we compute the output $f(T)$ in the following two steps
- subtract 32, giving $T - 32$
- multiply by $1/1.8$, giving $(1/1.8)(T - 32)$

The first step subtracts the freezing point, and gives the number of degrees Fahrenheit above freezing. The second step converts that into the number of degrees Celsius above freezing. Since freezing point is 0°C, this is the temperature in degrees Celsius.

Example 2   As a spherical balloon inflates, its radius grows at the rate of 2 in/sec. Express the volume of the balloon as a function of the time t in seconds since it started inflating.

Solution   We calculate V from t in two steps. First we calculate the radius r. Since the radius grows at the rate of 2 in/sec, after t seconds it is given by

$$r = 2t.$$

Next we calculate the volume using the formula

$$V = \frac{4}{3}\pi r^3.$$

We put these two steps together by putting $r = 2t$ in the formula for the volume,

$$V = \frac{4}{3}\pi(2t)^3 = \frac{4}{3}\pi(8t^3) = \frac{32}{3}\pi t^3.$$

Composition of Functions

Each of the two steps in Example 1 is itself a function:
- the "subtract 32" function, $u = h(T) = T - 32$
- the "multiply by $1/1.8$" function, $g(u) = \frac{1}{1.8}u$.

To obtain $f(T)$, we take the output of h and make it the input or g. So we put $u = h(T) = T - 32$ in $g(u)$ to get

$$f(T) = g(u) = g(h(T))$$
$$= \frac{1}{1.8}u = \frac{1}{1.8}(T - 32).$$

This process of inputting one function into another to form a third function is called *composing* the functions.

> We call the function $f(t) = g(h(t))$ the **composition** of g with h. The function $g(h(t))$ is defined by using the output of the function h as the input to f.

Example 3   Express the function
$$V = f(t) = \frac{32}{3}\pi t^3$$
in Example 2 as a composition of two functions.

Solution   The output $f(t)$ is calculated in two steps:
- the "multiply by 2" function: $r = h(t) = 2t$
- the volume function, $V = g(t) = (4/3)\pi r^3$.

Thus
$$f(t) = g(h(t)).$$

Substitution

In composing functions, it is often useful to work with an intermediate variable that represents the output of the first function and the input of the second function. We express the composition $g(h(x))$ in two steps,
$$y = g(u) \quad \text{and} \quad u = h(x).$$
We call this *substituting $u = h(x)$ into g*.

Example 4   Find $y = g(h(x))$.
(a) $g(x) = x^5, h(x) = x + 3$ (b) $g(x) = x + 3, h(x) = x^5$
(c) $g(x) = \sqrt{x}, h(x) = 2x + 1$ (d) $g(x) = 2x + 1, h(x) = \sqrt{x}$

Solution   (a) We introduce the variable u to represent the output of h and the input of g. We have $y = u^5$ and $u = x + 3$, so
$$y = u^5 = (x + 3)^5.$$
Therefore $g(h(x)) = (x + 3)^5$.
(b) This time $y = u + 3$ and $u = x^5$, so
$$y = u + 3 = x^5 + 3.$$
Therefore $g(h(x)) = x^5 + 3$. Note that this is different from the answer in part (a), because although we use the same two functions, we compose them in a different order.
(c) We have $y = \sqrt{u}$ and $u = 2x + 1$, so
$$y = \sqrt{u} = \sqrt{2x + 1}.$$
Therefore $g(h(x)) = \sqrt{2x + 1}$.
(d) We have $y = 2u + 1$ and $u = \sqrt{x}$, so
$$y = 2u + 1 = 2\sqrt{x} + 1.$$
Therefore $g(h(x)) = 2\sqrt{x} + 1$. Again, this is different from part (c) because the functions are composed in a different order.

Substitution is a useful method for breaking a function down into smaller steps.

Example 5   Decompose each of the following functions by defining a new variable u as a function of x so that when we write y as a function of u, composing these two functions gives the original function.

(a) $y = (2x+1)^5$ (b) $y = \dfrac{1}{\sqrt{x^2+1}}$ (c) $y = 2(5-x^2)^3 + 1$

Solution   (a) We define $u = 2x + 1$ as the inside function and $y = u^5$ as the outside function. Then, by substitution,
$$y = u^5 = (2x+1)^5.$$

(b) We define $u = x^2 + 1$ as the inside function and $y = 1/\sqrt{u}$ as the outside function. Then, by substitution,
$$y = \frac{1}{\sqrt{u}} = \frac{1}{\sqrt{x^2+1}}.$$

Notice that we instead could have defined $u = \sqrt{x^2+1}$ and $y = 1/u$. There are other possibilities as well. There are often multiple ways to decompose a function.

(c) One possibility is to define $u = 5 - x^2$ as the inside function and $y = 2u^3 + 1$ as the outside function. Then, by substitution,
$$y = 2u^3 + 1 = 2(5-x^2)^3 + 1.$$

Again, there are other possible answers.

Inside and Outside Functions

When we compose $y = u^5$ and $u = x + 3$ to obtain $y = (x+3)^5$, we call $y = u^5$ the *outside function* and $u = x+3$ the *inside function*, as a reminder that we are substituting the inside function into the outside function. We recognize that the functions
$$y = (2x+1)^3, \qquad y = (x^2+1)^3, \qquad y = (5-2x)^3$$

have something in common. In each case, the expression on the right is of the form
$$(\text{expression})^3,$$

so we can write the function as a composition in which the outside function is $y = u^3$. Similarly, the functions
$$y = (2x+1)^3, \qquad y = \sqrt{2x+1}, \qquad y = \frac{1}{(2x+1)^2}$$

can be expressed as a composition in which the inside function is
$$u = 2x + 1.$$

Horizontal Shifts

Example 6   A ball thrown in the air has height $h(t) = 90t - 16t^2$ feet after t seconds (see Figure 4.10). A second ball is thrown in exactly the same way 2 seconds later and has height $g(t)$ after t seconds.

(a) How does the graph of g compare with the graph of h?
(b) Write an expression for g in terms of t.

Solution   (a) Since the second ball follows the same motion as the first ball, only 2 seconds later, the graph of g has the same shape as the graph of h, but is shifted 2 seconds to the right (see Figure 4.10).

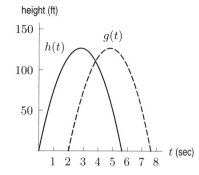

Figure 4.10: Height of two balls thrown at different times

(b) The second ball has the same height at t seconds as the first ball had 2 seconds earlier, at time $t - 2$ seconds. So to calculate $g(t)$, we first subtract 2 from t, giving $t - 2$, then input the result into h:

$$g(t) = h(t - 2) = 90(t - 2) - 16(t - 2)^2.$$

In the previous example, g is the composition of h with the "subtract 2" function. The effect of this composition is to shift the graph 2 units to the right.

Example 7   Sketch the graph of each function and give its domain.

(a) $y = \dfrac{1}{x}$ (b) $y = \dfrac{1}{x - 1}$ (c) $y = \dfrac{1}{x + 3}$

Solution   (a) See Figure 4.11(a). Since $1/x$ is defined for all x except $x = 0$, we have

Domain: all real x, $x \neq 0$.

(b) See Figure 4.11(b). Notice that it is the same shape as the graph of $y = 1/x$, shifted 1 unit to the right. Since the original graph has no y-value at $x = 0$, the new graph has no y value at $x = 1$, which makes sense because $1/(x - 1)$ is defined for all x except $x = 1$, so

Domain: all real x, $x \neq 1$.

(c) See Figure 4.11(c), which has the same shape as the graph of $y = 1/x$, shifted 3 units to the left. This time there is no y-value for $x = -3$, and $1/(x + 3)$ is defined for all x except $x = -3$, so

Domain: all real x, $x \neq -3$.

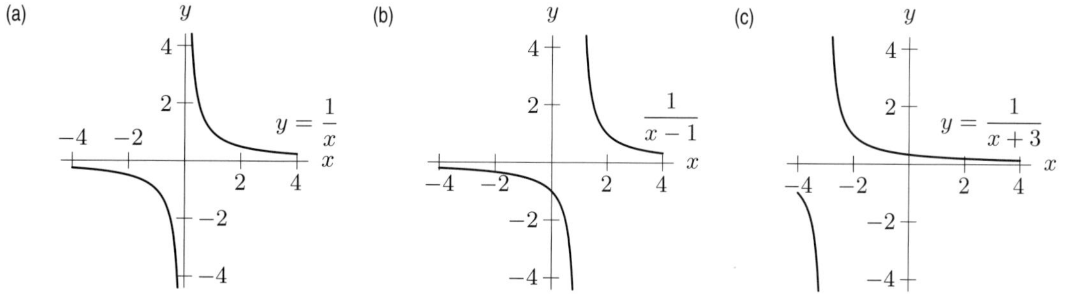

Figure 4.11: Horizontal shifts of $y = 1/x$

In general, we have

For a function $y = f(x)$ and a constant k, the graph of

$$y = f(x - k)$$

is the graph of f shifted k units to the right. (A negative shift to the right is a shift to the left, so if k is negative we get a shift to the left of $|k|$ units.)

Vertical Shifts

Example 8   Repeat Example 6, with the difference that this time the second ball is thrown in the air at exactly the same time as the first ball, but from a position 20 feet higher.

Solution   Since the second ball leaves at the same time but 20 feet higher, the graph showing its height $g(t)$ at time t is the same shape as the graph showing the height $h(t)$ of the first ball, but shifted up by 20. See Figure 4.12. To calculate $g(t)$, we calculate the height of the first ball, $f(t)$, and then add 20:

$$g(t) = h(t) + 2 = 90t - 16t^2 + 20.$$

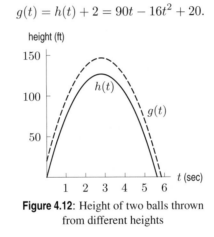

Figure 4.12: Height of two balls thrown from different heights

The difference between this situation and horizontal shifts is that we add (or subtract) a constant to the output, not the input. In Example 8 we compose the "add 20" function with h. Notice that the order of composition is different from the order in Example 6. In that example, h was the outside function, whereas in this example h is the inside function. The effect of adding 20 to the output is to shift the graph up by 20.

Example 9   Sketch the graph and give the range of each function.
   (a) $y = x^2$ (b) $y = x^2 + 1$ (c) $y = x^2 - 3$

Solution   (a) See Figure 4.13(a). Since x^2 is never negative, we have

$$\text{Range: all real } y \geq 0.$$

(b) For every x-value, the y-value for $y = x^2 + 1$ is one unit larger than the y-value for $y = x^2$. Since all the y-coordinates are increased by 1, the graph shifts vertically up by 1 unit. See Figure 4.13(b). Since $x^2 \geq 0$ for all x, we have $x^2 + 1 \geq 1$ for all x, so

$$\text{Range: all real } y \geq 1.$$

(c) The y-coordinates are 3 units smaller than the corresponding y-coordinates of $y = x^2$, so the graph is shifted *down* by 3 units. See Figure 4.13(c). We have $x^2 - 3 \geq -3$ for all x, so

$$\text{Range: all real } y \geq -3.$$

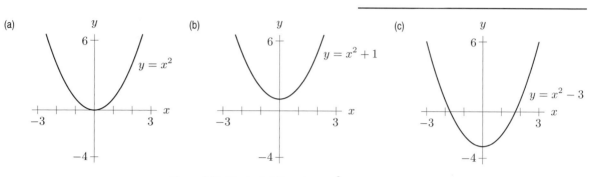

Figure 4.13: Vertical shifts of $y = x^2$

In general, we have

> For a function $y = f(x)$ and a constant k, the graph of
>
> $$y = f(x) + k$$
>
> is the graph of $y = f(x)$ shifted up by k units. (A negative shift up is a shift down, so if k is negative, we get a shift down by $|k|$ units.)

Using Vertical and Horizontal Shifts

Example 10   Figure 4.14 shows the graph of $y = \sqrt{x}$. Use shifts of this function to graph each of the following:

(a) $y = \sqrt{x} + 1$ (b) $y = \sqrt{x - 1}$ (c) $y = \sqrt{x} - 2$ (d) $y = \sqrt{x + 3}$

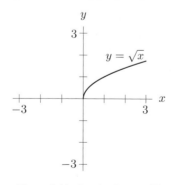

Figure 4.14: Graph of $y = \sqrt{x}$

Solution   (a) Since 1 is added to the expression $\sqrt{x}$, the graph is a vertical shift up 1 unit of the graph of $y = \sqrt{x}$. See Figure 4.15(a).

(b) Since $x - 1$ is substituted for x in the expression $\sqrt{x}$, the graph is a horizontal shift of the graph of $y = \sqrt{x}$. Substituting $x = 1$ gives $\sqrt{1-1} = \sqrt{0}$, so we see that $x = 1$ in the new graph gives the same value as $x = 0$ in the old graph. The graph is shifted to the right 1 unit. See Figure 4.15(b).

(c) Since 2 is subtracted from the expression $\sqrt{x}$, the graph is a vertical shift down 2 units of the graph of $y = \sqrt{x}$. See Figure 4.15(c).

(d) Since $x + 3$ is substituted for x in the expression $\sqrt{x}$, the graph is a horizontal shift of the graph of $y = \sqrt{x}$. Substituting $x = -3$ gives $\sqrt{-3+3} = \sqrt{0}$, so we see that $x = -3$ in the new graph gives the same value as $x = 0$ in the old graph. The graph is shifted to the left 3 units. See Figure 4.15(d).

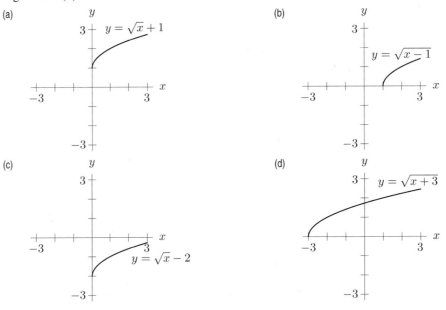

Figure 4.15: Horizontal and vertical shifts of $y = \sqrt{x}$

Example 11   Tickets are being sold online for an event. After the tickets become available, the first three days are a private sale. The public sale is from noon day 3 to noon day 8, and tickets sell at a constant rate during this time. Let $N = f(t)$ represent the total number of tickets sold by day t, and assume 1500 tickets have sold at the start of the public sale (day 3) and 2500 have been sold by the end of the public sale (day 8).

(a) Sketch a graph of $N = f(t)$ for the public sale, and give the domain and range.

(b) If 600 additional tickets sell during the private phase, find a formula, $v(t)$, for ticket sales in terms of $f(t)$. Sketch the graph of $v(t)$ and give the domain and range.

(c) If, instead, there is a delay of two days between the private and public sales, find a formula, $h(t)$, for ticket sales in terms of $f(t)$. Sketch the graph of $h(t)$ and give the domain and range.

Solution   (a) Since tickets sell at a constant rate, the graph is a line segment with one endpoint at $t = 3$, $N = 1500$ and the other endpoint at $t = 8$, $N = 2500$. See Figure 4.16(a). We see that

Domain is $3 \leq t \leq 8$ and   Range is $1500 \leq N \leq 2500$.

(b) Every value of the output variable, number of tickets sold, is increased by 600. Since this variable is on the vertical axis, the graph will shift up by 600 units.

$$\text{At } t = 3, \text{ the number of tickets sold is } f(3) + 600 = 1500 + 600 = 2100.$$
$$\text{At } t = 8, \text{ the number of tickets sold is } f(8) + 600 = 2500 + 600 = 3100.$$

For any t, we have
$$v(t) = f(t) + 600.$$

The graph is shifted vertically up by 600 units. See Figure 4.16(b). We see that

$$\text{Domain is } 3 \leq t \leq 8 \quad \text{and} \quad \text{Range is } 2100 \leq N \leq 3100.$$

(c) The public sale will take place two days later, from day 5 to day 10. The change is in the input variable t and the graph will shift to the right by two days. See Figure 4.16(c). We see that

$$\text{Domain is } 5 \leq t \leq 10 \quad \text{and} \quad \text{Range is } 1500 \leq N \leq 2500.$$

Notice that

day 5 under new schedule   corresponds to   day 3 under old schedule,
day 10 under new schedule   corresponds to   day 8 under old schedule,
day t under new schedule   corresponds to   day $t - 2$ under old schedule.

We see that
$$h(t) = f(t - 2).$$

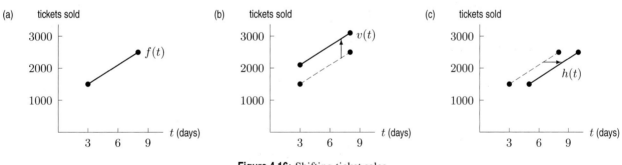

Figure 4.16: Shifting ticket sales

Example 12   (a) Find a formula for the linear function $f(t)$ in Example 11(a).
(b) Use the vertical shift to find a formula for $v(t)$ in Example 11(b).
(c) Use the horizontal shift to find a formula for $h(t)$ in Example 11(c).

Solution   (a) We find the equation of the line between two points. The slope is:

$$\text{Slope} = \frac{\text{Change in } N}{\text{Change in } t} = \frac{2500 - 1500}{8 - 3} = \frac{1000}{5} = 200.$$

Using the point-slope form with the point $t = 3$, $N = 1500$, we have

$$N = 1500 + 200(t - 3)$$
$$= 1500 + 200t - 600$$
$$= 900 + 200t.$$

The formula is $N = f(t) = 900 + 200t$ on domain $3 \le t \le 8$.

(b) Using the vertical shift, we have:

$$v(t) = f(t) + 600$$
$$= (900 + 200t) + 600$$
$$= 1500 + 200t.$$

The vertical shift does not change the domain, so the formula is $v(t) = 1500 + 200t$ on domain $3 \le t \le 8$.

(c) Using the horizontal shift, we have:

$$h(t) = f(t - 2)$$
$$= 900 + 200(t - 2)$$
$$= 900 + 200t - 400$$
$$= 500 + 200t.$$

The horizontal shift shifts the domain, so the formula is $h(t) = 500 + 200t$ on domain $5 \le t \le 10$.

Exercises and Problems for Section 4.2

Exercises

Write the expressions in Exercises 1–8 in the form $y = kh(x)^p$ for some function $h(x)$.

1. $y = \dfrac{(2x + 1)^5}{3}$

2. $y = \dfrac{17}{(1 - x^3)^4}$

3. $y = \sqrt{5 - x^3}$

4. $y = \dfrac{2}{\sqrt{1 + \dfrac{1}{x}}}$

5. $y = 100 \left(1 + \sqrt{x}\right)^4$

6. $y = \left(1 + x + x^2\right)^3$

7. $y = 5\sqrt{12 - \sqrt[3]{x}}$

8. $y = 0.5(x + 3)^2 + 7(3 + x)^2$

In Exercises 9–14, the graph of the function $g(x)$ is a horizontal and/or vertical shift of the graph of $f(x) = x^3$, shown in Figure 4.17. For each of the shifts described, sketch the graph of $g(x)$ and find a formula for $g(x)$.

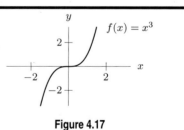

Figure 4.17

9. Shifted vertically up 3 units.

10. Shifted vertically down 2 units.

11. Shifted horizontally to the left 1 unit.

12. Shifted horizontally to the right 2 units.

13. Shifted vertically down 3 units and horizontally to the left 1 unit.

14. Shifted vertically up 2 units and horizontally to the right 4 units.

In Exercises 15–18, the graph of the function $g(x)$ is a horizontal and/or vertical shift of the graph of $f(x) = 5 - x$, shown in Figure 4.18. For each of the shifts described, sketch the graph of $g(x)$ and find a formula for $g(x)$.

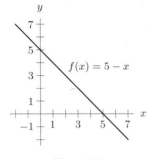

Figure 4.18

15. Shifted vertically down 3 units.

16. Shifted vertically up 1 unit.

17. Shifted horizontally to the right 2 units.

18. Shifted horizontally to the left 4 units.

Each of the graphs in Exercises 19–24 is the graph of $y = x^2$ shifted horizontally and/or vertically. Find a formula for the function.

19.

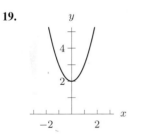

20.

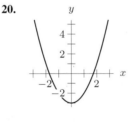

21.

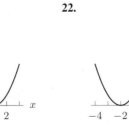

22.

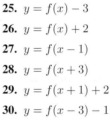

23.

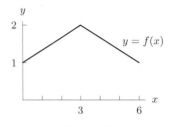

24.

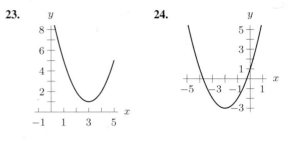

Problems 25–30 refer to the graph of $y = f(x)$ in Figure 4.19. Sketch the graph of each function.

Figure 4.19

25. $y = f(x) - 3$

26. $y = f(x) + 2$

27. $y = f(x - 1)$

28. $y = f(x + 3)$

29. $y = f(x + 1) + 2$

30. $y = f(x - 3) - 1$

In Exercises 31–38, use substitution to compose the two functions.

31. $y = u^4$ and $u = x + 1$

32. $y = 5u^3$ and $u = 3 - 4x$

33. $w = r^2 + 5$ and $r = t^3$

34. $p = 2q^4$ and $D = 5p - 1$

35. $w = 5s^3$ and $q = 3 + 2w$

36. $P = 3q^2 + 1$ and $q = 2r^3$

37. $y = u^2 + u + 1$ and $u = x^2$

38. $y = 2u^2 + 5u + 7$ and $u = 3x^3$

Problems

In Problems 39–40, the two functions share either an inside function or an outside function. Which is it? Describe the common function.

39. $y = (2x + 1)^3$ and $y = \dfrac{1}{\sqrt{2x + 1}}$

40. $y = \sqrt{5x - 2}$ and $y = \sqrt{x^2 + 4}$

In Problems 41–43, decompose the function by defining a new variable u as a function of x so that when we write y as a function of u, composing these two functions gives the original function.

41. $y = \sqrt{x^2 + 1}$

42. $y = 5(x - 2)^3$

43. $y = 3x^3 - 2$

In Problems 44–45, find

(a) $f(g(x))$ **(b)** $g(f(x))$

44. $f(x) = x^3$ and $g(x) = 5 + 2x$

45. $f(x) = x^3 + 1$ and $g(x) = \sqrt{x}$

46. Using $f(t) = 3t^2$ and $g(t) = 2t + 1$, find

  (a) $f(g(t))$ **(b)** $g(f(t))$ **(c)** $f(f(t))$ **(d)** $g(g(t))$

47. If $f(g(x)) = 5(x^2 + 1)^3$ and $g(x) = x^2 + 1$, find $f(x)$.

48. Give three different composite functions with the property that the outside function raises the inside function to the third power.

49. Give a formula for a composite function with the property that the outside function takes the square root and the inside function multiplies by 5 and adds 2.

50. Give a formula for a composite function with the property that the inside function takes the square root and the outside function multiplies by 5 and adds 2.

51. The rate R at which the drug level in the body changes when an intravenous line is used is a function of the amount Q of the drug in the body. For a certain drug, we have $R = 25 - 0.08Q$. The quantity Q of the drug is a function of time t with $Q = \sqrt{t}$ over a fixed time period. Express the rate R as a function of time t.

52. The function $H = f(t)$ gives the temperature, $H°$F of an object t minutes after it is taken out of the refrigerator and left to sit in a room. Give a formula in terms of $f(t)$ for the temperature of the object as a function of time if the situation is modified as described:

  (a) The object is taken out of the refrigerator 5 minutes later. (Give the formula in terms of $f(t)$ and give a reasonable domain for your function.)

  (b) Both the refrigerator and the room are $10°$F colder.

53. A sand dune is eroding over time. The height, h in cm, of the sand dune as a function of year, t, is given by $h = f(t)$. If the formula is modified as shown to create a new model showing the height of the sand dune, describe how the situation has changed.

  (a) $h = f(t + 30)$

  (b) $h = f(t) + 50$

54. A line has equation $y = x$.

  (a) Find the new equation if the line is shifted vertically up by 5 units.

  (b) Find the new equation if the line is shifted horizontally to the left by 5 units.

  (c) Compare your answers to (a) and (b) and explain your result graphically.

55. If we compose the two functions $w = f(s)$ and $q = g(w)$ using substitution, what is the input variable of the resulting function? What is the output variable?

4.3 INVERSE FUNCTIONS

In Section 4.2 we saw how to break a function down into a sequence of simpler operations. In this section we consider how to use this idea to undo the operation of a function.

Inverse Operations

An inverse operation is used to undo an operation. Performing an operation followed by its inverse operation gets us back to where we started. For example, the inverse operation of "add 5" is "subtract 5." If we start with any number x and perform these operations in order, we get back to where we started:

$$ x \quad \rightarrow \quad x + 5 \quad \rightarrow \quad (x + 5) - 5 \quad \rightarrow \quad x $$

$$ \text{(add 5)} \qquad\qquad \text{(subtract 5)} $$

Example 1   State in words the inverse operation.

(a)   Add 12   (b)   Multiply by 7
(c)   Raise to the 5th power   (d)   Take the cube root

Solution

(a) To undo adding 12, we subtract 12. The inverse operation is "Subtract 12".
(b) To undo multiplying by 7, we divide by 7. The inverse operation is "Divide by 7". (We could also say that the inverse operation is to multiply by 1/7, which is equivalent to dividing by 7.)
(c) To undo raising an expression to the 5th power, we raise to the $1/5^{\text{th}}$ power. The inverse operation is to "Raise to the $1/5^{\text{th}}$ power" or, equivalently, "Take the fifth root".
(d) To undo taking a cube root, we cube the result. The inverse operation is "Cube" or "Raise to the 3rd power".

Undoing a Sequence of Operations

We have seen how to break down a function into a sequence of operations. The inverse function undoes that sequence of operations. How do we undo a sequence of operations? For example, how do we undo "Multiply by 2 and then add 5"? You can probably guess that we will be "dividing by 2" and "subtracting 5", but order matters in this sequence of operations and we need to undo the operations in the correct order.

Example 2   Find the sequence of operations to undo "multiply by 2 and then add 5".

Solution   We write this sequence of operations as the function $y = 2x+5$, and undo the operations by solving for x:

$$y = 2x + 5 \qquad \text{(Subtract 5 from both sides)}$$
$$y - 5 = 2x \qquad \text{(Divide both sides by 2)}$$
$$\frac{y - 5}{2} = x.$$

The resulting equation $x = (y-5)/2$ shows us that to undo "multiply by 2 and add 5", we subtract 5 and then divide by 2.

When one function undoes another , we say that it is the *inverse function* of the other. To test whether one function undoes another, we compose the two functions and see if we get back where we started. This leads to the following definition:

Inverse Functions

Given a function $f(x)$, we say that $g(x)$ is the inverse function to $f(x)$ if

$$f(g(x)) = x \quad \text{and} \quad g(f(x)) = x.$$

Example 3   Let

$$f(x) = 2x + 5 \qquad \text{and} \qquad g(y) = \frac{y-5}{2}.$$

Show that g is the inverse of f.

Solution   We test to see if composing the two functions in either order gets us back to the variable we started with. We have

$$f(g(y)) = 2g(y) + 5 = 2\left(\frac{y-5}{2}\right) + 5 \qquad \text{Multiplying by 2 undoes dividing by 2}$$

$$= (y-5) + 5 \qquad\qquad\qquad \text{Adding 5 undoes subtracting 5}$$

$$= y. \qquad\qquad\qquad\qquad \text{We are back to where we started.}$$

Also,

$$g(f(x)) = \frac{f(x) - 5}{2} = \frac{(2x+5) - 5}{2} \qquad \text{Subtracting 5 undoes adding 5}$$

$$= \frac{2x}{2} \qquad\qquad\qquad \text{Dividing by 2 undoes multiplying by 2}$$

$$= x. \qquad\qquad\qquad \text{We are back to where we started.}$$

Finding an Inverse

To see how to undo a sequence of operations in a function, we "unwrap" the function by solving for the input variable in terms of the output variable.

Example 4   (a) Write a function of x that says to raise x to the third power, multiply by 5, and then subtract 2.
(b) Find the inverse of the function in part (a).
(c) Describe in words the sequence of operations in the inverse.

Solution   (a) We have $y = f(x) = 5x^3 - 2$.
(b) We undo these operations by solving for x:

$$y = 5x^3 - 2 \qquad \text{Add 2 to both sides}$$

$$y + 2 = 5x^3 \qquad \text{Divide both sides by 5}$$

$$\frac{y+2}{5} = x^3 \qquad \text{Take the cube root of both sides}$$

$$x = \sqrt[3]{\frac{y+2}{5}}.$$

So the inverse function is

$$g(y) = \sqrt[3]{\frac{y+2}{5}}.$$

(c) The sequence of operations in the inverse is to add 2, divide by 5, and then take the cube root.

Warning! Not all functions have an inverse function, and not all operations have an inverse operation. For example, you might think that the inverse of squaring a number would be to take the square root, but these operations only undo each other if the number we start with is positive. If we start with a negative number, such as -2, we have

$$-2 \quad \rightarrow \quad \underset{\text{(square)}}{(-2)^2 = 4} \quad \rightarrow \quad \underset{\text{(take square root)}}{\sqrt{4}} \quad \rightarrow \quad 2$$

Notice that when we perform these operations in order, we do not get back to the number we started with. There is no inverse operation for squaring a number that will work for all numbers. Some operations do not have inverse operations.

Exercises and Problems for Section 4.3

Exercises

In Exercises 1–4, state in words the inverse operation.

1. Subtract 8

2. Divide by 10

3. Raise to the 5th power

4. Take the cube root

In Problems 5–8, find the sequence of operations to undo the sequence given.

5. Multiply by 5 and then subtract 2.

6. Add 10 and multiply the result by 3.

7. Raise to the 5th power and then multiply by 2.

8. Multiply by 6, add 10, then take the cube root.

In Problems 9–12, show that the equations describe inverse operations by showing that they undo each other, in the sense that composing the two in either order gets us back to the starting variable.

9. $y = 7x - 5$ and $x = \dfrac{y + 5}{7}$

10. $y = 8x^3$ and $x = \sqrt[3]{\dfrac{y}{8}}$

11. $y = x^5 + 1$ and $x = \sqrt[5]{y - 1}$

12. $y = \dfrac{10 + x}{3}$ and $x = 3y - 10$

In Problems 13–14,

(a) Write a function of x that performs the operations described.

(b) Find the inverse and describe in words the sequence of operations in the inverse.

13. Raise x to the fifth power, multiply by 8, and then add 4.

14. Subtract 5, divide by 2, and take the cube root.

Problems

In Exercises 15–16, check that the functions are inverses.

15. $f(x) = \dfrac{x}{4} - \dfrac{3}{2}$ and $g(t) = 4\left(t + \dfrac{3}{2}\right)$

16. $f(x) = 1 + 7x^3$ and $g(t) = \sqrt[3]{\dfrac{t - 1}{7}}$

Find the inverse, $g(y)$, of the functions in Problems 17–19.

17. $h(x) = 12x^3$

18. $h(x) = \dfrac{x}{2x + 1}$

19. $h(x) = \dfrac{\sqrt{x}}{\sqrt{x} + 1}$

Solve the equations in Problems 20–21 exactly. Use an inverse function when appropriate.

20. $\dfrac{2x + 3}{x + 3} = 8$

21. $\sqrt{x + \sqrt{x}} = 3$

REVIEW EXERCISES AND PROBLEMS FOR CHAPTER FOUR

Exercises

In Exercises 1–8, give the domain and range of the power function.

1. $y = x^4$ **2.** $y = x^{-3}$

3. $y = 2x^{3/2}$ **4.** $A = 5t^{-1}$

5. $P = 6x^{2/3}$ **6.** $Q = r^{5/3}$

7. $y = x^{-1/2}$ **8.** $M = 6n^3$

In Exercises 9–11, find the range.

9. $m(x) = 9 - x$ **10.** $f(x) = \dfrac{1}{\sqrt{x-4}}$

11. $f(x) = x - 3$

In Exercises 12–15, the graph of the function $g(x)$ is a horizontal and/or vertical shift of the graph of $f(x) = 2x^2 - 1$, shown in Figure 4.20. For each of the shifts described, sketch the graph of $g(x)$ and find a formula for $g(x)$.

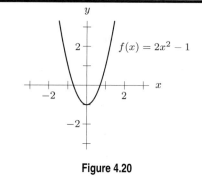

Figure 4.20

12. Shifted horizontally to the left 1 unit.

13. Shifted vertically down 3 units.

14. Shifted vertically up 2 units and horizontally to the right 3 units.

15. Shifted vertically down 1 unit and horizontally to the left 2 units.

Problems

16. A movie theater seats 200 people. For any particular show, the amount of money the theater makes is a function of the number of people, n, in attendance. If a ticket costs $4.00, find the domain and range of this function. Sketch its graph.

17. For what values of k does the equation

$$\frac{5x}{x-1} = k$$

have a solution? What does your answer say about the range of the function $f(x) = 5x/(x-1)$?

18. (a) Give the domain and range of the linear function $y = 100 - 25x$.
(b) If x and y represent quantities which cannot be negative, give the domain and range.

19. Find the range of the function $G = 0.75 - 0.3h$ in Example 5 giving the amount of gasoline, G gallons, in a portable electric generator h hours after it starts.

20. Tuition cost, T in dollars, for part-time students at a college is given by $T = 300 + 200C$, where C represents the number of credits taken. Part-time students cannot take more than 10 credit hours. Give a reasonable domain and range for this function.

21. (a) What is the domain the function $P = -100{,}000 + 50{,}000s$?
(b) If P represents the profit of a silver mine at price s dollars per ounce, and if the silver mine closes if profits fall below zero, what is the domain?

In Problems 22–23, the two functions share either an inside function or an outside function. Which is it? Describe the common function.

22. $y = 5(1 - 3x)^2$ and $y = 5(x^2 + 1)^2$

23. $y = \sqrt{x^2 + 1}$ and $y = (x^2 + 1)^3 + 5$

In Problems 24–26, decompose the function by defining a new variable u as a function of x so that when we write y as a function of u, composing these two functions gives the original function.

24. $y = \dfrac{5}{x^2 + 1}$

25. $y = 1 + 2(x - 1) + 5(x - 1)^2$

26. $y = 25(3x - 2)^5 + 100$

In Problems 27–28, find

(a) $f(g(x))$ (b) $g(f(x))$

27. $f(x) = 5x^2$ and $g(x) = 3x$

28. $f(x) = \dfrac{1}{x}$ and $g(x) = x^2 + 1$

29. Using $f(t) = 2 - 5t$ and $g(t) = t^2 + 1$, find

      (a) $f(g(t))$ (b) $g(f(t))$ (c) $f(f(t))$ (d) $g(g(t))$

30. If $f(g(x)) = \sqrt{5x + 1}$ and $f(x) = \sqrt{x}$, find $g(x)$.

31. A line has equation $y = b + mx$.

  (a) Find the new equation if the line is shifted vertically up by k units. What is the y-intercept of this line?
  (b) Find the new equation if the line is shifted horizontally to the right by k units. What is the y-intercept of this line?

32. Give three different composite functions with the property that the outside function takes the square root of the inside function.

33. Give a formula for a function with the property that the inside function raises the input to the 5th power, and the outside function multiplies by 2 and subtracts 1.

34. The amount A of pollution in a certain city is a function of the population P, with $A = 100P^{0.3}$. The population is growing over time, with $P = 10000 + 2000t$, with t in years since 2000. Express the amount of pollution A as a function of time t.

35. The cost, C in dollars, for an amusement park includes an entry fee and a certain amount per ride. We have $C = f(r)$ where r represents the number of rides. Give a formula in terms of $f(r)$ for the cost as a function of number of rides if the situation is modified as described:

  (a) The entry fee is increased by \$5.
  (b) The entry fee includes 3 free rides.

36. Give a formula for a function with the property that the outside function raises the input to the 5th power, and the inside function multiplies by 2 and subtracts 1.

37. If we compose the two functions $U = f(V)$ and $V = g(W)$ using substitution, what is the input variable of the resulting function? What is the output variable?

In Exercises 38–39, check that the functions are inverses.

38. $g(x) = 1 - \dfrac{1}{x - 1}$ and $f(t) = 1 + \dfrac{1}{1 - t}$

39. $h(x) = \sqrt{2x}$ and $k(t) = \dfrac{t^2}{2}$, for $x, t \geq 0$

Find the inverse, $g(y)$, of the functions in Problems 40–42.

40. $f(x) = \dfrac{x - 2}{2x + 3}$ **41.** $f(x) = \sqrt{\dfrac{4 - 7x}{4 - x}}$

42. $f(x) = \dfrac{\sqrt{x} + 3}{11 - \sqrt{x}}$

Chapter Five

Quadratic Functions, Expressions, and Equations

Contents

5.1 QUADRATIC FUNCTIONS

Example 1   The height in feet of a ball thrown upward from the top of a building after t seconds is given by

$$h(t) = -16t^2 + 32t + 128, \quad t \geq 0.$$

Find the ball's height after 0, 1, 2, 3, and 4 seconds and describe the path of the ball.

Solution   We have

$$h(0) = -16 \cdot 0^2 + 32 \cdot 0 + 128 = 128 \text{ ft.}$$
$$h(1) = -16 \cdot 1^2 + 32 \cdot 1 + 128 = 144 \text{ ft.}$$
$$h(2) = -16 \cdot 2^2 + 32 \cdot 2 + 128 = 128 \text{ ft.}$$

Continuing, we get Table 5.1. The ball rises from 128 ft when $t = 0$ to a height of 144 ft when $t = 1$, returns to 128 ft when $t = 2$ and reaches the ground, height 0 ft, when $t = 4$.

Table 5.1   *Height of a ball*

t (seconds)	$h(t)$ (ft)
0	128
1	144
2	128
3	80
4	0

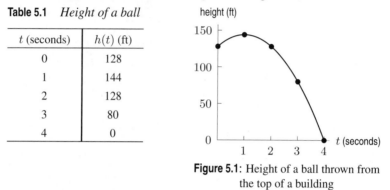

Figure 5.1: Height of a ball thrown from the top of a building

Quadratic Functions Expressed in Standard Form

The function in Example 1 has a term, $-16t^2$, involving the square of the independent variable. This term is called the *quadratic term* and the function is called a *quadratic function*. In general,

> A quantity y is a **quadratic function** of x if it can be written in the form
>
> $$y = f(x) = ax^2 + bx + c, \quad a, b, c \text{ constants}, a \neq 0.$$
>
> - The expression $ax^2 + bx + c$ is a **quadratic expression** in **standard form**.
> - The term ax^2 is called the **squared term** or **quadratic term**.
> - The term bx is called the **linear term**.
> - The term c is called the **constant term**.

Example 2   For the function $h(t) = -16t^2 + 32t + 128$ in Example 1, interpret in terms of the ball's motion
            (a)   the constant term 128   (b)   the sign of the quadratic term $-16t^2$.

Solution   (a) The constant term is the value of the function when $t = 0$, and so represents the height of the
            building.

(b) The negative quadratic term $-16t^2$ counteracts the positive terms $32t$ and 128 for $t > 0$, and eventually causes the values of $h(t)$ to decrease, which makes sense since the ball eventually starts to fall to the ground.

Figure 5.1 shows an important feature of quadratic functions. Unlike linear functions, quadratic functions have graphs that bend. This is the result of the presence of the quadratic term. See Figure 5.2, which illustrates this for the power functions $f(x) = x^2$ and $g(x) = -x^2$. The shape of the graph of a quadratic function is called a *parabola*. If $a > 0$ then the parabola opens upward, and if $a < 0$ it opens downward.

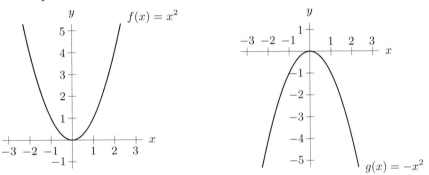

Figure 5.2: Graphs of $f(x) = x^2$ and $g(x) = -x^2$

Quadratic Functions Expressed in Factored Form

Example 3   The function $h(t) = -16t^2 + 32t + 128$ in Example 1 can be expressed in factored form as[1]

$$h(t) = -16(t - 4)(t + 2)   t \geq 0.$$

What is the practical interpretation of the factors $(t - 4)$ and $(t + 2)$?

Solution   When $t = 4$, the factor $(t - 4)$ has the value $4 - 4 = 0$. So

$$h(4) = -16(0)(4 + 2) = 0.$$

In practical terms, this means that the ball hits the ground 4 seconds after it is thrown. When $t = -2$, the factor $(t + 2)$ has the value $-2 + 2 = 0$. However, there is no practical interpretation for this, since the domain of h is $t \geq 0$. See Figure 5.3.

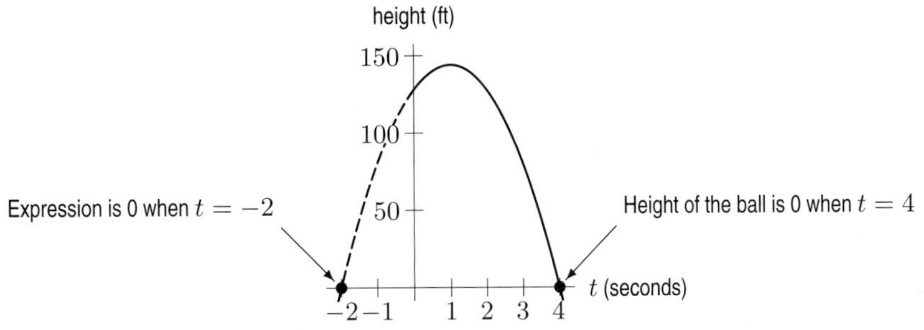

height (ft)

150

100

Expression is 0 when $t = -2$ 50   Height of the ball is 0 when $t = 4$

t (seconds)

$-2\ -1$ 1 2 3 4

Figure 5.3: Interpretation of the factored form $h(t) = -16(t - 4)(t + 2)$

---

[1] See the Tools sections beginning on page 223 for a review of factoring.

Example 4   When a company charges a price p dollars for one of its products, its revenue is given by

$$\text{Revenue} = f(p) = 500p(30 - p).$$

For what price(s) does the company have no revenue?

Solution   If the company has no revenue then

$$500p(30 - p) = 0.$$

The expression on the left is zero when one of its factors is zero, that is, when $p = 0$ or $p = 30$. The fact that the revenue is zero when $p = 0$ tells us that if the company charges \$0, it makes no money. The fact that it is zero when $p = 30$ tells us that if the company charges \$30 per item, it makes no sales because no one will pay that much.

Values of the independent variable where a function has the value zero, such as the $p = 0$ and $p = 30$ in the previous example, are called **zeros** of the function.

A quadratic function in x is expressed in **factored form** if it is written as

$$y = f(x) = a(x - r)(x - s), \quad \text{where } a, r, \text{ and } s \text{ are constants and } a \neq 0.$$

- The constants r and s are zeros of the function $f(x) = a(x - r)(x - s)$.
- The constant a is the leading coefficient, the same as the constant a in the standard form.

Quadratic Functions Expressed in Vertex Form

The next example shows a form for expressing a quadratic function which shows conveniently where the function reaches its maximum value.

Example 5   Show that the function $h(t) = -16t^2 + 32t + 128$ in Example 1 can be expressed in the form

$$h(t) = -16(t - 1)^2 + 144,$$

and use this form to show that the ball reaches its maximum height $h = 144$ when $t = 1$.

Solution   To check that $-16(t - 1)^2 + 144$ and $-16t^2 + 32t + 128$ are equivalent expressions, we expand the first one:

$$-16(t - 1)^2 + 144 = -16(t^2 - 2t + 1) + 144$$
$$= -16t^2 + 32t - 16 + 144$$
$$= -16t^2 + 32t + 128.$$

Looking at the left-hand side, we see that the term $-16(t - 1)^2$ is a negative number times a square, so it is always negative or zero, and it is zero when $t = 1$. Therefore $h(t)$ is always less than or equal to 144 and is equal to 144 when $t = 1$. This means the maximum height the ball reaches is 144 feet, and it reaches that height after 1 second. See Table 5.2 and Figure 5.3.

Table 5.2 *Values of $h(t) = -16(t - 1)^2 + 144$ are less than or equal to 144*

t (seconds)	$-16(t - 1)^2$	$-16(t - 1)^2 + 144$
0	-16	128
1	0	144
2	-16	128
3	-64	80
4	-144	0

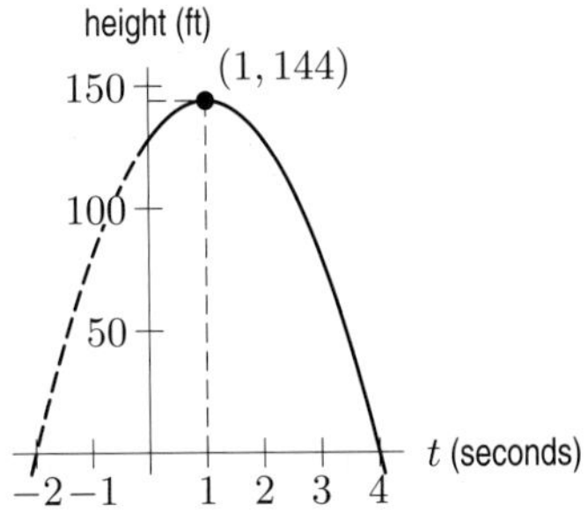

height (ft)

(1, 144)

Figure 5.4: Ball reaches its greatest height of 144 ft at $t = 1$

The point $(1, 144)$ in Figure 5.4 is called the *vertex* of the graph. For quadratic functions the vertex shows where the function reaches either its largest value, called the *maximum*, or its smallest value, called the *minimum*. In Example 5 the function reaches its maximum value at the vertex because the coefficient is negative in the term $-16(t - 1)^2$.

In general:

A quadratic function in x is expressed in **vertex form** if it is written as

$$y = f(x) = a(x - h)^2 + k, \quad \text{where } a, h, \text{ and } k \text{ are constants and } a \neq 0.$$

For the function $f(x) = a(x - h)^2 + k$,
- $f(h) = k$, and the point (h, k) is the vertex of the graph.
- The coefficient a is the leading coefficient, the same a as in the standard form.
    - If $a > 0$ then k is the minimum value of the function, and the graph opens upward.
    - If $a < 0$ then k is the maximum value of the function, and the graph opens downward.

Example 6 For each function, find the maximum or minimum and sketch the graph, indicating the vertex.

(a) $g(x) = (x - 3)^2 + 2$ (b) $A(t) = 5 - (t + 2)^2$ (c) $h(x) = x^2 - 4x + 4$

Solution (a) The expression for g is in vertex form. We have

$$g(x) = (x - 3)^2 + 2 = \text{Positive number (or zero)} + 2.$$

Thus $g(x) \geq 2$ for all values of x except $x = 3$, where it equals 2. The minimum value is 2, and the vertex is at $(3, 2)$ where the graph reaches its lowest point. See Figure 5.5 (a).

(b) The expression for A is also in vertex form. We have

$$A(t) = 5 - (t + 2)^2 = 5 - \text{Positive number (or zero)}.$$

Thus $A(t) \leq 5$ for all t except $t = -2$, where it equals 5. The maximum value is 5, and the vertex is at $(-2, 5)$, where the graph reaches its highest point. See Figure 5.5 (b).

(c) The expression for h is not in vertex form. However, recognizing that it is a perfect square,[2] we can write it as

$$h(x) = x^2 - 4x + 4 = (x - 2)^2,$$

which is in vertex form with $k = 0$. So $h(2) = 0$ and $h(x)$ is positive for all other values of x. Thus the minimum value is 0, and it occurs at $x = 2$. The vertex is at $(2, 0)$. See Figure 5.5 (c).

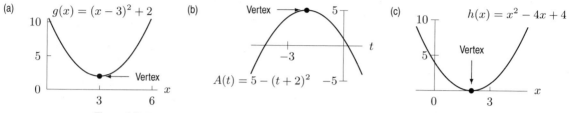

Figure 5.5: Interpretation of the vertex form of a quadratic expression

Using the Different Forms of a Quadratic Expression

Expressing a quadratic function in factored form allows us to see where it is positive and where it is negative.

Example 7   A college bookstore finds that if it charges p dollars for a T-shirt, it sells $1000 - 20p$ T-shirts. For what prices, p is the revenue $R = f(p) = p(1000 - 20p)$ positive?

Solution   We know the price, p is positive, so to make R positive we need to make the factor $1000 - 20p$ positive as well. Writing it in the form

$$1000 - 20p = 20(50 - p)$$

we see that R is positive only then $p < 50$. Therefore the revenue is positive when $0 < p < 50$.

Which form we express a quadratic function in depends on what information we have about it.

Example 8   Find a quadratic function whose graph could be

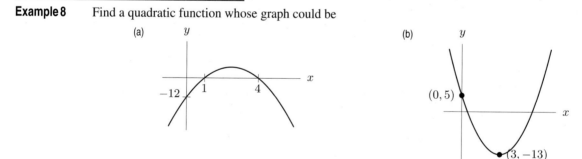

Figure 5.6: Find possible expressions for these functions

Solution   (a) Since we know the zeros, we start with a function in factored form:

$$f(x) = a(x - r)(x - s).$$

The graph has x-intercepts at $x = 1$ and $x = 4$, so the function has zeros at those values, so we choose $r = 1$ and $s = 4$, which gives

$$f(x) = a(x - 1)(x - 4).$$

---

[2]See page 227 for a review of perfect squares.

Since the y-intercept is -12, we know that $y = -12$ when $x = 0$. So

$$-12 = a(0-1)(0-4)$$
$$-12 = 4a$$
$$-3 = a.$$

So the function

$$f(x) = -3(x-1)(x-4)$$

has the right graph. Notice that the value of a is negative, which we expect because the graph opens downward.

(b) We are given the vertex of the parabola, so we try to write its equation using the vertex form $y = f(x) = a(x-h)^2 + k$. Since the coordinates of the vertex are $(3, -13)$, we let $h = 3$ and $k = -13$. This gives

$$f(x) = a(x-3)^2 - 13.$$

The y-intercept is $(0, 5)$, so we know that $y = 5$ when $x = 0$. Substituting, we get

$$5 = a(0-3)^2 - 13$$
$$5 = 9a - 13$$
$$18 = 9a$$
$$2 = a.$$

Therefore, the function

$$f(x) = 2(x-3)^2 - 13$$

has the correct graph. Notice that the value of a in positive, which we expect because the graph opens upward.

Example 9   Can each function graphed in Figure 5.7 be expressed in factored form $f(x) = a(x-r)(x-s)$? If so, is each of the parameters r and s positive, negative, or zero? (Assume $r \le s$.)

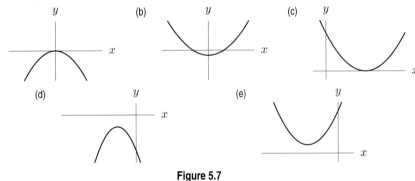

Figure 5.7

Solution   Since r and s are the x-intercepts, we look at whether the x-intercepts are positive, negative or zero.

(a) There is only one x-intercept at $x = 0$, so $r = s = 0$.
(b) There are two x-intercepts, one positive and one negative, so $r < 0$ and $s > 0$.
(c) There is only one x-intercept. It is positive and $r = s$.
(d) There are no x-intercepts, so it is not possible to write the quadratic in factored form.
(e) There are no x-intercepts, so it is not possible to write the quadratic in factored form.

Exercises and Problems for Section 5.1

Exercises

In Exercises 1–10, express the quadratic function in standard form $f(x) = ax^2 + bx + c$, and identify a, b, and c.

1. $f(x) = x(x - 3)$

2. $g(p) = 1 - \sqrt{2}p^2$

3. $q(m) = (m - 7)^2$

4. $p(x) = a(x - h)^2 + k$

5. $h(x) = (x - r)(x - s)$

6. $f(n) = (n - 4)(n + 7)$

7. $m(t) = 2(t - 1)^2 + 12$

8. $p(q) = (q + 2)(3q - 4)$

9. $q(p) = (p - 1)(p - 6) + p(3p + 2)$

10. $h(t) = 3(2t - 1)(t + 5)$

11. A rectangle is 6 feet narrower than it is long. Express its area as a function of its length l in feet.

12. The height of a triangle is 3 feet more than twice the length of its base. Express its area as a function of the length of its base, x, in feet.

In Exercises 13–20, is the expression quadratic in the given variable?

13. $y(xy + b)$, in y

14. $y(x^2 + 3)$, in y

15. $y(x^2 + 3)$, in x

16. $\dfrac{k}{r^2}$, in r

17. πr^2, in r

18. $2\pi r$, in r

19. $\frac{4}{3}\pi r^3$, in r

20. $\dfrac{x^2 + y^2}{2}$, in x

In Exercises 21–24, determine which values of x will make the function positive and which values will make the function negative.

21. $f(x) = (x - 4)(x + 5)$

22. $g(x) = x^2 - x - 56$

23. $h(x) = x^2 - 12x + 36$

24. $k(x) = -(x - 1)(x - 2)$

25.

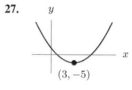
$(6, 15)$

26.

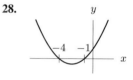
-2 5

27.

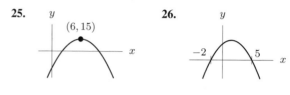
$(3, -5)$

28.

-4 -1

Problems

Find possible formulas for the quadratic functions described in Problems 29–32.

29. This function's graph has a vertex of $(2, 3)$ and a y-intercept of -4.

30. This function has zeros at $x = 3$ and $x = -5$, and its graph has a y-intercept of 12.

31. This function's graph has a vertex at $(3, 9)$ and passes through the origin.

32. This function's graph has x-intercepts 8 and 12 and y-intercept 50.

For each function $f(x) = a(x - h)^2 + k$ graphed in 33–37, is each of the constants h and k positive, negative, or zero?

33.

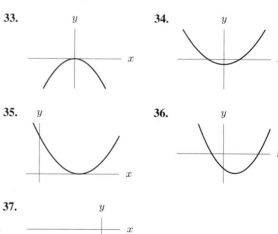

34.

35.

36.

37.

Each of the graphs in Problems 38–43 is the graph of a quadratic function.

(a) If the function is expressed in the form $y = ax^2 + bx + c$, say whether a and c are positive, negative, zero.

(b) If the function is expressed in the form $y = a(x-h)^2 + k$, say whether h and k are positive, negative, zero.

(c) Can the expression for the function be factored as $y = a(x-r)(x-s)$? If it can, are r and s equal to each other? Say whether they are positive, negative, or zero (assume $r \leq s$).

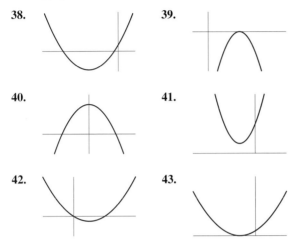

38.

39.

40.

41.

42.

43.

Match the graphs in Exercises 44–51 to the following equations, or state that there is no match.

(a) $y = (x-2)^2 + 3$ (b) $y = -(x-2)^2 + 3$
(c) $y = (x+2)^2 + 3$ (d) $y = (x+2)^2 - 3$
(e) $y = -(x+2)^2 + 3$ (f) $y = 2(x-2)^2 + 3$
(g) $y = (x-3)^2 + 2$ (h) $y = (x+3)^2 - 2$

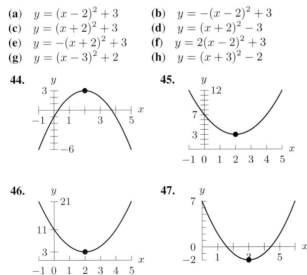

44.

45.

46.

47.

48.

49.

50.

51.

52. A peanut, dropped at time $t = 0$ from an upper floor of the Empire State Building, is at a height, h, in feet above the ground t seconds later given by

$$h(t) = -16t^2 + 1024.$$

What does the factored form

$$h(t) = -16(t-8)(t+8)$$

tell us about when the peanut hits the ground?

53. A coin thrown upward from an office in the Empire State Building at time $t = 0$ seconds has height, h, in feet above the ground, at time t given by

$$h(t) = -16t^2 + 64t + 960 = -16(t-10)(t+6).$$

(a) From what height is the coin thrown?
(b) At what time does the coin reach the ground?

54. A ball is dropped from the top of a tower. Its height above the ground in feet t seconds after it is dropped is given by $100 - 16t^2$.

(a) Explain why the 16 tells you something about how fast the speed is changing.
(b) When dropped from the top of a tree, the height of the ball at time t is $120 - 16t^2$. Which is taller, the tower or the tree?
(c) When dropped from a building on another planet, the height of the ball is given by $100 - 20t^2$. How does the height of the building compare to the height of the tower? How does the motion of the ball on the other planet compare to its motion on the earth?

55. The average weight of a baby during the first year of life is roughly a quadratic function of time. At month m, its average weight in pounds, is approximated by[3]

$$w(m) = -0.042m^2 + 1.75m + 8.$$

(a) What is the practical interpretation of the 8?
(b) What is the average weight of a one-year-old?

56. A company finds that if it charges x dollars for a widget it can sell $1500 - 3x$ of them. It costs \$5 to produce a widget.

(a) Express the revenue, $R(x)$, as a function of price.
(b) Express the cost, $C(x)$, as a function of price.
(c) Express the profit, $P(x)$, which is revenue minus cost, as a function of price.

57. A farmer has a square plot of size x feet on each side. He wants to put hedges around the plot which cost \$10 for every foot and on the inside of the square he wants to sow seed which costs \$0.01 for every square foot. Write an expression for the total cost.

In Problems 58–59, write an expression $f(x)$ for the result of the given operations on x, and put it in standard form.

58. Add 5, multiply by x, subtract 2.

59. Subtract 3, multiply by x, add 2, multiply by 5.

In Problems 60–61, find a possible quadratic function with the given zeros and write it in standard form.

60. 3 and 4

61. $a + 1$ and $3a$, where a is a constant

62. Explain how you can determine the coefficient of x^2 in the standard form without expanding out:

$$x(2x + 3) - 5(x^2 + 2x + 1) - 5(10x + 2) + 3x + 25$$

What is the coefficient?

What changes to the parameters a, h, k in the equation

$$y = a(x - h)^2 + k$$

produce the effects described in Problems 63–64?

63. The vertex of the graph is shifted down and to the right.

64. The graph changes from upward-opening to downward-opening, but the zeros (x-intercepts) do not change.

5.2 QUADRATIC EXPRESSIONS

Different aspects of the behavior of a quadratic function can be seen by expressing it in different forms. The following example illustrates a use of the factored form.

Example 1   Suppose that in Example 7 on page 196, each T-shirt costs the bookstore \$3, so its profit is given by

$$P = g(p) = p(1000 - 20p) - 3(1000 - 20p).$$

For what values of p is the profit positive?

Solution   Taking out a common factor of $(1000 - 20p)$, we get

$$g(p) = p(1000 - 20p) - 3(1000 - 20p) = (p - 3)(1000 - 20p) = 20(p - 3)(50 - p).$$

This first factor is $p - 3$, which is positive when $p > 3$ and negative when $p < 3$. So

$$
\begin{aligned}
p > 50 : \quad & g(p) = 20(p - 3)(50 - p) = \text{positive} \times \text{negative} = \text{negative} \\
3 < p < 50 : \quad & g(p) = 20(p - 3)(50 - p) = \text{positive} \times \text{positive} = \text{positive} \\
p < 3 : \quad & g(p) = 20(p - 3)(50 - p) = \text{negative} \times \text{positive} = \text{negative.}
\end{aligned}
$$

So the profit is positive if the price is greater than \$3 but less than \$50.

[3]http://www.cdc.gov/growthcharts/, accessed June 6, 2003

Transforming Quadratic Expressions

Example 2 Write each of the following expressions in the indicated form.

(a) $2 - x^2 + 3x(2 - x)$ (standard)

(b) $\dfrac{(n-1)(2-n)}{2}$ (factored)

(c) $(z + 3)(z - 2) + z - 2$ (factored)

(d) $5(x^2 - 2x + 1) - 9$ (vertex)

Solution (a) We have
$$2 - x^2 + 3x(2 - x) = 2 - x^2 + 6x - 3x^2 = -4x^2 + 6x + 2.$$
This is a quadratic expression in standard form with $a = -4$, $b = 6$, $c = 2$.

(b) We have
$$\frac{(n-1)(2-n)}{2} = \frac{1}{2}(n-1)(2-n) = \frac{1}{2}(n-1)(-(-2+n)) = -\frac{1}{2}(n-1)(n-2).$$
This is a quadratic expression in factored form with $a = -1/2$, $r = 1$, $s = 2$.

(c) We have
$$\begin{aligned}
(z+3)(z-2) + z - 2 &= (z-2)((z+3)+1) \quad \text{factor out } (z-2)\\
&= (z-2)(z+4)\\
&= (z-2)(z-(-4)).
\end{aligned}$$
This is a quadratic expression in factored form with $a = 1$, $r = 2$, $s = -4$.

(d) The key to putting this in vertex form is to recognize that the expression in parenthesis is a perfect square:
$$5\underbrace{\left(x^2 - 2x + 1\right)}_{(x-1)^2} - 9 = 5(x-1)^2 - 9.$$
This is a quadratic expression in vertex form with $a = 5$, $h = 1$, $k = -9$.

In general, we can express a quadratic function in standard form by expanding and collecting like terms, and we can express it in factored form using the factoring techniques reviewed in the the Tools section at the end of this chapter.

How Do We Put an Expression in Vertex Form?

In Example 2 (d) we had to rely on recognizing a perfect square to put the expression in vertex form. Now we look for a systematic method.

Example 3 Find the vertex of the parabolas (a) $y = x^2 + 6x + 9$ (b) $y = x^2 + 6x + 8$.

Solution (a) We recognize the expression on the right-hand side of the equal sign as a perfect square: $y = (x+3)^2$, so the vertex is at $(-3, 0)$.

(b) Unlike part (a), the expression on the right-hand side of the equal sign is not a perfect square. In order to have a perfect square, the $6x$ term should be followed by a 9. We can make this happen by adding a 9 and then subtracting it in order to keep the expression equivalent:
$$x^2 + 6x + 8 = \underbrace{x^2 + 6x + 9}_{\text{Perfect square}} - 9 + 8 \quad \text{add and subtract 9}$$
$$= (x+3)^2 - 1 \quad\quad \text{since } -9 + 8 = -1$$
Therefore, $y = x^2 + 6x + 8 = (x+3)^2 - 1$. This means the vertex is at the point $(-3, -1)$.

In the last example, we put the equation $y = x^2 + 6x + 8$ into a form where the right-hand side contains a perfect square, in a process that is called **completing the square**. To complete the square, we use the form of a perfect square

$$(x + p)^2 = x^2 + 2px + p^2.$$

Example 4   Put each expression in vertex form by completing the square.

(a) $x^2 - 8x$ (b) $x^2 + 4x - 7$

Solution   (a) We compare the expression with the form of a perfect square

$$x^2 - 8x$$
$$x^2 - 2px + p^2$$

To match the pattern, we must have $2p = -8$, so $p = -4$. Thus, if we add $(-4)^2 = 16$ to $x^2 - 8x$ we obtain a perfect square. Of course, adding a constant changes the value of the expression, so we must subtract the constant as well.

$$x^2 - 8x = \underbrace{x^2 - 8x + 16}_{\text{Perfect square}} - 16 \quad \text{add and subtract 16}$$
$$x^2 - 8x = (x - 4)^2 - 16$$

Notice that the constant, 16, that was added and subtracted could have been obtained by taking half the coefficient of the x-term, $(-8/2)$, and squaring this result. This gives $(-8/2)^2 = 16$.

(b) We compare with the form of a perfect square:

$$x^2 + 4x - 7$$
$$x^2 + 2px + p^2.$$

Note that in each of the previous examples, we chose p to be half the coefficient of x. In this case, half the coefficient of x is 2, so we add and subtract $2^2 = 4$ to each side in order to complete the square:

$$x^2 + 4x - 7 = \underbrace{x^2 + 4x + 4}_{\text{Perfect square}} - 4 - 7 \quad \text{add and subtract 4}$$
$$x^2 + 4x - 7 = (x + 2)^2 - 11.$$

In Examples 3 and 4 we transformed the quadratic expression $f(x)$. The following example illustrates a variation on the method of completing the square that uses the equation $y = f(x)$. Since there are more operations available for transforming equations than for transforming expressions, this method is more flexible.

Example 5   A bookstore finds that if it charges $\$p$ for a T-shirt then its revenue from T-shirt sales is given by

$$R = f(p) = p(1000 - 20p)$$

What price should it charge in order to maximize the revenue?

Solution   We first expand the revenue function into standard form:

$$R = f(p) = p(1000 - 20p) = 1000p - 20p^2.$$

Since the coefficient of p^2 is -20, we know the graph of the quadratic opens downward. So the vertex form of the equation gives us its maximum value. The expression on the right is more complicated than the ones we have dealt with so far, because it has a coefficient -20 on the quadratic terms. In order to deal with this we work with the equation $R = 1000p - 20p^2$ rather than the expression $1000 - 20p^2$ directly:

$$R = 1000p - 20p^2$$

$$-\frac{R}{20} = p^2 - 50p \qquad \text{dividing by } -20$$

$$-\frac{R}{20} + 25^2 = p^2 - 50p + 25^2 \qquad \text{add } 25^2 \text{ to both sides}$$

$$-\frac{R}{20} + 25^2 = (p - 25)^2 \qquad \text{rewrite the right side as a square}$$

$$R = -20(p - 25)^2 + 20 \cdot 25^2 \qquad \text{subtracting } 25^2 \text{ and multiplying by } -20$$

$$R = -20(p - 25)^2 + 12{,}500.$$

So the vertex is $(25, 12{,}500)$, and the price that maximizes the revenue is $p = \$25$. Figure 5.8 shows the graph of f. The graph reveals that revenue initially rises as the price increases, but eventually starts to fall again when the high price begins to deter customers.

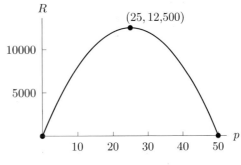

Figure 5.8: Revenue from the sale of T-shirts

Notice two differences between the method in Example 5 and the method in Examples 3 and 4: we can move the -20 out of the way temporarily by dividing both sides, and instead of adding and subtracting 25^2 to an expression, we add to both sides of an equation. Since in the end we subtract the 25^2 from both sides and multiply both sides by -20, we end up with an equivalent expression on the right hand side.

Visualizing The Process of Completing The Square

We can visualize how to find the constant that needs to be added to $x^2 + bx$ in order to obtain a perfect square by thinking of $x^2 + bx$ as the area of a rectangle. For example, the rectangle in Figure 5.9 has area $x(x + 8) = x^2 + 8x$. Now imagine cutting the rectangle into pieces as in Figure 5.10 and trying to rearrange them to make a square, as in Figure 5.11. The corner piece, whose area is $4^2 = 16$, is missing. By adding this piece to our expression, we "complete" the square: $x^2 + 8x + 16 = (x + 4)^2$.

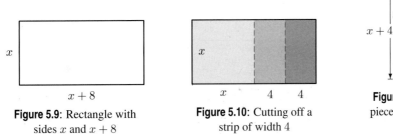

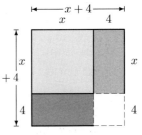

Figure 5.9: Rectangle with sides x and $x + 8$

Figure 5.10: Cutting off a strip of width 4

Figure 5.11: Rearranging the piece to make a square with a missing corner

Exercises and Problems for Section 5.2

Exercises

In Exercises 1–4, write the expression in factored form.

1. $x^2 + 8x + 15$

2. $x^2 + 2x - 24$

3. $(x-4)^2 - 4$

4. $(x+2)(x-3) + 2(x-3)$

In Exercises 5–8, find the x- and y-intercepts.

5. $y = 3x^2 + 15x + 12$

6. $y = x^2 - 3x - 10$

7. $y = x^2 + 5x + 2$

8. $y = 3x^2 + 18x - 30$

In Exercises 9–12, find the vertex of the parabola.

9. $y = x(x-1)$

10. $y = x^2 - 2x + 3$

11. $y = 2 - x^2$

12. $y = 2x + 2 - x^2$

In Exercises 13–17, write the expression in vertex form.

13. $x^2 - 12x + 36$

14. $-x^2 + 2bx - b^2$

15. $6(x^2 - 8x + 16) + 2$

16. $\dfrac{(t-6)^2 - 3}{4}$

17. $b + c(x^2 - 4dx + 4d^2) + 5$

In Exercises 18–21, find the minimum value of the function.

18. $f(x) = (x-3)^2 + 10$

19. $g(x) = x^2 - 2x - 8$

20. $h(x) = (x-5)(x-1)$

21. $j(x) = -(x-4)^2 + 7$

In Exercises 22–27, put the equation in the form $y = a(x - h)^2 + k$ by completing the square. State the values of $a, h,$ and k.

22. $y - 5 = x^2 + 2x + 1$

23. $y - 12 = 2x^2 + 8x + 8$

24. $y = x^2 + 6x + 4$

25. $y = 2x^2 + 20x + 12$

26. $y + 2 = x^2 - 3x$

27. $y = 3x^2 + 7x + 5$

In Exercises 28–29, a quadratic expression is written in vertex form.

(a) Write the expression in standard form and in factored form.

(b) Evaluate the expression at $x = 0$ and $x = 3$ using each of the three forms and compare the results.

28. $(x+3)^2 - 1$

29. $(x-2)^2 - 25$

Write the quadratic functions in Exercises 30–31 in the following forms and state the values of all constants.

(a) Standard form $y = ax^2 + bx + c$.

(b) Factored form $y = a(x-r)(x-s)$.

(c) Vertex form $y = a(x-h)^2 + k$.

30. $y = x(x-3) - 7(x-3)$

31. $y = 21 - 23x + 6x^2$.

Problems

32. Group expressions (a)–(f) together so that expressions in each group are equivalent. Note that some groups may contain only one expression.

(a) $(t+3)^2$

(b) $t^2 + 9$

(c) $t^2 + 6t + 9$

(d) $t(t+9) - 3(t+8)$

(e) $\sqrt{t^4 + 81}$

(f) $\dfrac{2t^2 + 18}{2}$

Express the quadratic functions in Problems 33–38 in standard form and state the values of a, b, c.

33. $f(x) = -5x^2 - 2x + 3$

34. $g(t) = 7 - \dfrac{t^2}{2} - \dfrac{t}{3}$

35. $h(z) = \dfrac{z^2 + 4z + 7}{5}$

36. $u(r) = 2(r-2)(3-2r)$

37. $v(s) = 2\left(s - s(4-s) - 1\right)$

38. $w(z) = (z^2 + 3)(z^2 + 2) - (z^2 + 4)(z^2 - 1)$

Write the quadratic expressions in Problems 39–42 in factored form.

39. $x^2 + 7x + 10$

40. $4z^2 - 49$

41. $9t^2 + 60t + 100$

42. $6w^2 + 31w + 40$

In Problems 43–48, put the quadratic functions in vertex form and state the values of a, h, k.

43. $y = -2(x-5)^2 + 5$ **44.** $y = \dfrac{(x+4)^2}{5} - 7$

45. $y = (2x-4)^2 + 6$ **46.** $y - 5 = (2-x)^2$

47. $y = x^2 + 12x + 20$ **48.** $y = 4x^2 + 24x + 17$

In Problems 49–50, put the quadratic function in factored form, and use the factored form to sketch a graph of the function without a calculator.

49. $y = x^2 + 8x + 12$ **50.** $y = x^2 - 6x - 7$

Write the quadratic function

$$y = 8x^2 - 2x - 15$$

in the forms indicated in Problems 51–53. Give the values of all constants.

51. $y = a(x-r)(x-s)$ **52.** $y = a(x-h)^2 + k$

53. $y = kx(vx+1) + w$

Problems 54–53 show the graph of a quadratic function. Find a possible formula for the function.

54. The three different forms for a quadratic expression are:

$$\text{Standard form: } x^2 - 10x + 16$$
$$\text{Factored form: } (x-2)(x-8)$$
$$\text{Vertex form: } (x-5)^2 - 9.$$

  (a) Show that the three forms are equivalent.
  (b) Which form is most useful for finding the
     (i) Smallest value of the expression? Use it to find that value.
     (ii) Values of x when the expression is 0? Use it to find those values of x.
     (iii) Value of the expression when $x = 0$? Use it to find that value.

55. The profit (in thousands of dollars) a company makes from selling a certain item depends on the price of the item. The three different forms for the profit at a price of p dollars are:

$$\text{Standard form: } -2p^2 + 24p - 54$$
$$\text{Factored form: } -2(p-3)(p-9)$$
$$\text{Vertex form: } -2(p-6)^2 + 18.$$

  (a) Show that the three forms are equivalent.
  (b) Which form is most useful for finding the prices that give a profit of zero dollars? (These are called the break-even prices.) Use it to find these prices.
  (c) Which form is most useful for finding the profit when the price is zero? Use it to find that profit.
  (d) The company would like to maximize profits. Which form is most useful for finding the price that gives the maximum profit? Use it to find the optimal price and the maximum profit.

56. Write a possible quadratic function $F(x)$ that takes its largest value of 100 at $x = 3$, and then write it in standard form.

57. A farmer encloses a rectangular paddock with 200 feet of fencing.

  (a) Express the area A in square feet of the paddock as a function of the width w in feet.
  (b) Find the maximum area that can be enclosed.

58. A farmer makes two adjacent rectangular paddocks with 200 feet of fencing.

  (a) Express the total area A in square feet as a function of the length y in feet of the shared side.
  (b) Find the maximum total area that can be enclosed.

5.3 QUADRATIC EQUATIONS

In Example 3 on page 193 we solved the equation $h(t) = 0$ to find when the ball hit the ground. When we find the zeros of a quadratic function $f(x) = ax^2 + bx + c$, we are solving the equation

$$ax^2 + bx + c = 0.$$

A **quadratic equation** in x is one which can be put into the standard form

$$ax^2 + bx + c = 0, \qquad \text{where } a, b, c \text{ are constants, with } a \neq 0.$$

Some quadratic equations can be solved by taking square roots.

Example 1   Solve (a) $x^2 - 4 = 0$ (b) $x^2 - 5 = 0$

Solution   (a) Rewriting the equation as

$$x^2 = 4,$$

we see the solutions are

$$x = \sqrt{4} = 2 \quad \text{and} \quad x = -\sqrt{4} = -2.$$

(b) By a similar method, we see that the solutions to $x^2 - 5 = 0$ are $x = \sqrt{5}$ and $x = -\sqrt{5}$. Both solutions can be written at once as $x = \pm\sqrt{5}$. Using a calculator, we see these solutions are approximately $x = 2.236$ and $x = -2.236$.

Example 2   Use the result of Example 1 to solve $(x - 2)^2 = 5$.

Solution   Since the equation

$$x^2 = 5 \qquad \text{has solutions} \qquad x = \pm\sqrt{5},$$

the equation

$$(x - 2)^2 = 5 \qquad \text{has solutions} \qquad x - 2 = \pm\sqrt{5}.$$

Thus the solutions to $(x - 2)^2 = 5$ are

$$x = 2 + \sqrt{5} \qquad \text{and} \qquad x = 2 - \sqrt{5},$$

which can be combined as $x = 2 \pm \sqrt{5}$. Using a calculator, these solutions are approximately

$$x = 2 + 2.236 = 4.236 \qquad \text{and} \qquad x = 2 - 2.236 = -0.236.$$

Example 3   Solve (a) $2(y + 1)^2 = 0$ (b) $2(y - 3)^2 + 4 = 0$

Solution   (a) Since $2(y + 1)^2 = 0$, dividing by 2 gives

$$(y + 1)^2 = 0 \qquad \text{so} \qquad y + 1 = 0.$$

Thus the only solution is $y = -1$. It is possible for a quadratic equation to have only one solution.

(b) Since $2(y - 3)^2 + 4 = 0$, we have

$$2(y - 3)^2 = -4,$$

so dividing by 2 gives

$$(y - 3)^2 = -2.$$

But since no number squared is -2, this equation has no real number solutions. Solutions to these types of quadratic equations will be discussed in Section 5.4.

In general, if we can put a quadratic equation in the form

$$(x - h)^2 = \text{Constant},$$

then we can solve it by taking square roots of both sides.

Example 4   For the function

$$h(t) = -16(t - 1)^2 + 144$$

giving the height of a ball after t seconds, find the times where the ball reaches a height of 135 feet.

Solution We want to find the values of t such that $h(t) = 135$, so we want to solve the equation

$$-16(t-1)^2 + 144 = 135.$$

Isolating the $(t-1)^2$ term we get

$$(t-1)^2 = \frac{135 - 144}{-16} = \frac{9}{16}$$

$$t - 1 = \pm\frac{3}{4}$$

$$t = 1 \pm \frac{3}{4}.$$

Therefore, the solutions are $t = 0.25$ and $t = 1.75$. So the ball reaches a height of 135 ft on its way up very soon after being thrown and again on its way down about 2 seconds after being thrown.

Next we develop a systematic method for solving quadratic equations by taking square roots. The solutions to quadratic equations are sometimes called *roots* of the equation.

Solving Quadratic Equations by Completing the Square

In Section 5.2 we saw how to find the vertex of a parabola by completing the square. A similar method can be used to solve quadratic equations.

Example 5 Solve $x^2 + 6x + 8 = 1$.

Solution First, we move the constant to the right by adding -8 to both sides:

$$x^2 + 6x + 8 - 8 = 1 - 8 \qquad \text{add } -8 \text{ to each side}$$

$$x^2 + 6x = -7$$

$$x^2 + 6x + 9 = -7 + 9 \qquad \begin{array}{l}\text{complete the square by adding}\\ (6/2)^2 = 9 \text{ to both sides}\end{array}$$

$$(x+3)^2 = 2$$

$$x + 3 = \pm\sqrt{2} \qquad \begin{array}{l}\text{take the square root of both}\\ \text{sides}\end{array}$$

$$x = -3 \pm \sqrt{2}.$$

If the coefficient of x^2 is not 1 we can divide through by it before completing the square.

Example 6 Solve $3x^2 + 6x - 2 = 0$ for x.

Solution First, divide both sides of the equation by 3:

$$x^2 + 2x - \frac{2}{3} = 0.$$

Next, move the constant to the right by adding 2/3 to both sides:

$$x^2 + 2x = \frac{2}{3}.$$

Now complete the square. The coefficient of x is 2, and half this is 1, so we add $1^2 = 1$ to each side:

$$x^2 + 2x + 1 = \frac{2}{3} + 1 \qquad \text{completing the square}$$
$$(x+1)^2 = \frac{5}{3}$$
$$x + 1 = \pm\sqrt{\frac{5}{3}}$$
$$x = -1 \pm \sqrt{\frac{5}{3}}.$$

The Quadratic Formula

Completing the square on the equation $ax^2 + bx + c = 0$ gives a formula for the solution of any quadratic equation. First we divide by a, getting

$$x^2 + \frac{b}{a}x + \frac{c}{a} = 0$$

Now subtract the constant c/a from both sides:

$$x^2 + \frac{b}{a}x = -\frac{c}{a}.$$

We complete the square by adding a constant to both sides of the equation. The coefficient of x is b/a, and half this is

$$\frac{1}{2} \cdot \frac{b}{a} = \frac{b}{2a}.$$

We square this and add the result, $(b/2a)^2$, to each side:

$$x^2 + \frac{b}{a}x + \left(\frac{b}{2a}\right)^2 = -\frac{c}{a} + \left(\frac{b}{2a}\right)^2 \qquad \text{complete the square}$$
$$x^2 + \frac{b}{a}x + \frac{b^2}{4a^2} = -\frac{c}{a} + \frac{b^2}{4a^2} \qquad \text{expand parentheses}$$
$$= -\frac{c}{a} \cdot \frac{4a}{4a} + \frac{b^2}{4a^2} \qquad \text{find a common denominator}$$
$$= \frac{b^2 - 4ac}{4a^2} \qquad \text{simplify right-hand side.}$$

We now rewrite the left-hand side as a perfect square:

$$\left(x + \frac{b}{2a}\right)^2 = \frac{b^2 - 4ac}{4a^2}.$$

Taking square roots gives

$$x + \frac{b}{2a} = \frac{\sqrt{b^2 - 4ac}}{2a} \quad \text{or} \quad x + \frac{b}{2a} = \frac{-\sqrt{b^2 - 4ac}}{2a},$$

so

$$x = \frac{-b}{2a} + \frac{\sqrt{b^2 - 4ac}}{2a} \quad \text{or} \quad x = \frac{-b}{2a} - \frac{\sqrt{b^2 - 4ac}}{2a},$$

so the solutions are

$$x = \frac{-b + \sqrt{b^2 - 4ac}}{2a} \quad \text{and} \quad x = \frac{-b - \sqrt{b^2 - 4ac}}{2a}.$$

The **quadratic formula** combines both solutions of $ax^2 + bx + c = 0$:

$$x = \frac{-b \pm \sqrt{b^2 - 4ac}}{2a}.$$

Example 7   Use the quadratic formula to solve for x: (a) $2x^2 - 2x - 7 = 0$ (b) $3x^2 + 3x = 10$

Solution   (a) We have $a = 2$, $b = -2$, $c = -7$, so

$$x = \frac{-(-2) \pm \sqrt{(-2)^2 - 4 \cdot 2(-7)}}{2 \cdot 2} = \frac{2 \pm \sqrt{4 + 56}}{4} = \frac{2 \pm \sqrt{60}}{4}.$$

If we write $\sqrt{60} = \sqrt{4 \cdot 15} = 2\sqrt{15}$, we get

$$x = \frac{2 \pm 2\sqrt{15}}{4} = \frac{1}{2} \pm \frac{\sqrt{15}}{2}.$$

(b) Before reading the values of a, b, c, we must write the equation in standard form, so we subtract 10 from both sides:

$$3x^2 + 3x - 10 = 0.$$

Thus $a = 3$, $b = 3$, and $c = -10$, so

$$x = \frac{-3 \pm \sqrt{3^2 - 4 \cdot 3(-10)}}{2 \cdot 3} = \frac{-3 \pm \sqrt{9 + 120}}{6} = \frac{-3 \pm \sqrt{129}}{6} = -\frac{1}{2} \pm \frac{\sqrt{129}}{6}.$$

Example 8   A ball is thrown into the air and its height above the ground t second afterward is given in feet by $y = -16t^2 + 32t + 8$. How long is the ball in the air?

Solution   The ball is in the air until it hits the ground, which is at a height of $y = 0$, so we want to know when

$$-16t^2 + 32t + 8 = 0.$$

Using the quadratic formula gives

$$t = \frac{-32 \pm \sqrt{32^2 - 4(-16)(8)}}{2(-16)} = \frac{-32 \pm \sqrt{1536}}{-32} = 1 \pm \frac{\sqrt{6}}{2} = 2.22, -0.22.$$

The negative root does not make sense in this context, so the time in the air is $t = 2.22$ seconds, a little more than 2 seconds.

The Discriminant

We can tell how many solutions a quadratic equation has without actually solving the equation. Look at the expression, $b^2 - 4ac$, under the square root sign in the quadratic formula.

> For the equation $ax^2 + bx + c = 0$, we define the **discriminant** $D = b^2 - 4ac$.
> - If $D = b^2 - 4ac$ is positive, then $\pm\sqrt{b^2 - 4ac}$ has two different values, so the quadratic equation has two distinct solutions (roots).
> - If $D = b^2 - 4ac$ is negative, then $\sqrt{b^2 - 4ac}$ is the square root of a negative number, so the quadratic equation has no real solutions.
> - If $D = b^2 - 4ac = 0$, then $\sqrt{b^2 - 4ac} = 0$, so the quadratic equation has only one solution. This is sometimes referred to as a repeated root.

[handwritten margin note: pos = 2 sol'n; neg = no real sol'n; 0 = 1 sol'n]

Example 9   How many solutions does each equation have?

(a) $4x^2 - 10x + 7 = 0$ (b) $0.3w^2 + 1.5w + 1.8 = 0$ (c) $3t^2 - 18t + 27 = 0$

Solution   (a) We use the discriminant. In this case, $a = 4$, $b = -10$, and $c = 7$. The value of the discriminant is thus

$$b^2 - 4ac = (-10)^2 - 4 \cdot 4 \cdot 7 = 100 - 112 = -12.$$

Since the discriminant is negative, we know that the quadratic equation has no real solutions.

(b) We again use the discriminant. In this case, $a = 0.3$, $b = 1.5$, and $c = 1.8$. The value of the discriminant is thus

$$b^2 - 4ac = (1.5)^2 - 4 \cdot 0.3 \cdot 1.8 = 0.09.$$

Since the discriminant is positive, we know that the quadratic equation has two distinct solutions.

(c) Checking the discriminant,

$$b^2 - 4ac = (-18)^2 - 4 \cdot 3 \cdot 27 = 0.$$

Since the discriminant is zero, we know that the quadratic equation has only one solution.

Exercises and Problems for Section 5.3

Exercises

Write the equations in Exercises 1–9 in the standard form $ax^2 + bx + c = 0$ and give possible values of a, b, c. Note that there may be more than one possible answer.

1. $2x^2 - 0.3x = 9$

2. $3x - 2x^2 = -7$

3. $4 - x^2 = 0$

4. $A = \pi \left(\dfrac{x}{2}\right)^2$

5. $-2(2x - 3)(x - 1) = 0$

6. $\dfrac{1}{1 - x} - 4 = \dfrac{4 - 3x}{2 - x}$

7. $7x(7 - x - 5(x - 7)) = (2x - 3)(3x - 2)$

8. $t^2 x - x^2 t^3 + tx^2 - t^3 - 4x^2 - 3x = 5$

9. $5x\left((x + 1)^2 - 2\right) = 5x\left((x + 1)^2 - x\right)$

Solve the quadratic equations in Exercises 10–21 by taking square roots.

10. $x^2 = 9$

11. $x^2 - 7 = 0$

12. $x^2 + 3 = 17$

13. $(x + 2)^2 - 4 = 0$

14. $(x - 3)^2 - 6 = 10$

15. $(x - 1)^2 + 5 = 0$

16. $(x - 5)^2 = 6$

17. $7(x - 3)^2 = 21$

18. $1 - 4(9 - x)^2 = 13$

19. $2(x - 1)^2 = 5$

20. $\left((x - 3)^2 + 1\right)^2 = 16$

21. $\left(x^2 - 5\right)^2 - 5 = 0$

Solve the quadratic equations in Exercises 22–27 or state that there are no solutions.

22. $x^2 + 6x + 9 = 4$ **23.** $x^2 - 12x - 5 = 0$

24. $2x^2 + 3x - 1 = 0$ **25.** $5x - 2x^2 - 5 = 0$

26. $x^2 + 5x - 7 = 0$ **27.** $(2x + 5)(x - 3) = 7$

In Exercises 28–44, solve for x by

(a) Completing the square   **(b)** Using the quadratic formula

28. $x^2 + 8x + 12 = 0$ **29.** $x^2 - 10x - 15 = 0$

30. $2x^2 + 16x - 24 = 0$ **31.** $x^2 + 7x + 5 = 0$

32. $x^2 - 9x + 2 = 0$ **33.** $x^2 + 17x - 8 = 0$

34. $x^2 - 22x + 10 = 0$ **35.** $2x^2 - 32x + 7 = 0$

36. $3x^2 + 18x + 2 = 0$ **37.** $5x^2 + 17x + 1 = 0$

38. $6x^2 + 11x - 10 = 0$ **39.** $2x^2 = -3 - 7x$

40. $7x^2 = x + 8$ **41.** $4x^2 + 4x + 1 = 0$

42. $9x^2 - 6x + 1 = 0$ **43.** $4x^2 + 4x + 3 = 0$

44. $9x^2 - 6x + 2 = 0$

Find all zeros (if any) of the quadratic functions in Exercises 45–46.

45. $y = 3x^2 - 2x - 4$ **46.** $y = 5x^2 - 2x + 2$

Write the quadratic functions in Exercises 47–52 in vertex form. Identify the values of a, h, k.

47. $y = -2(x + 3)^2 - 4$ **48.** $y = 5 - 3(3x - 6)^2$

49. $y = x^2 + 6x + 12$ **50.** $y = x^2 - 8x - 20$

51. $y = 2x^2 + 12x + 40$ **52.** $y = 5x^2 - 7x + 3$

In Exercises 53–58, use the discriminant to say whether the equation has two, one, or no solutions.

53. $2x^2 + 7x + 3 = 0$ **54.** $7x^2 - x - 8 = 0$

55. $9x^2 - 6x + 1 = 0$ **56.** $4x^2 + 4x + 1 = 0$

57. $9x^2 - 6x + 2 = 0$ **58.** $4x^2 + 4x + 3 = 0$

Problems

59. At time $t = 0$, in seconds, a pair of sunglasses is dropped from the Eiffel Tower in Paris. At time t, its height in feet above the ground is given by

$$\text{Height} = -16t^2 + 900.$$

(a) What does this expression tell us about the height from which the sunglasses were dropped?

(b) When do the sunglasses hit the ground?

60. You wish to fence a circular garden of area 80 square meters. How much fence do you need? [Hint: Remember that the circumference of a circle of radius r is $2\pi r$, and the area is πr^2.]

61. A ball, dropped from the top of a building, is h feet above the ground t seconds after it is dropped, where h is given by

$$h = -16t^2 + 100.$$

How long does it take the ball to fall k feet, where $0 \le k \le 100$?

In Problems 62–65, decide for what values of the constant A (if any) the equation has no solution. Give a reason for your answer.

62. $3(x - 2)^2 = A$ **63.** $(x - A)^2 = 10$

64. $A(x - 2)^2 + 5 = 0$ **65.** $5(x - 3)^2 + A = 10$

66. If a and c have opposite signs, the equation $ax^2 + bx + c = 0$ has two solutions. Explain why this is true in two different ways:

(a) Using what you know about the graph of $y = ax^2 + bx + c$.

(b) Using what you know about the quadratic formula.

67. Use what you know about the discriminant $b^2 - 4ac$ to decide what must be true about c in order for the quadratic equation $3x^2 + 2x + c = 0$ to have two different solutions.

68. Use what you know about the discriminant $b^2 - 4ac$ to decide what must be true about b in order for the quadratic equation $2x^2 + bx + 8 = 0$ to have two different solutions.

69. Show that $2ax^2 - 2(a - 1)x - 1 = 0$ has two solutions for all values of the constant a, except for $a = 0$. What happens if $a = 0$?

70. Under what conditions on the constants b and c do the line $y = -x + b$ and the curve $y = c/x$ intersect in

(a) No points?
(b) Exactly one point?
(c) Exactly two points?
(d) Is it possible for the two graphs to intersect in more than two points?

71. A Norman window is composed of a rectangle surmounted by a semicircle whose diameter is equal to the width of the rectangle.

(a) Find a formula for the area in square feet of a Norman window in which the rectangle has length l feet and width w feet.
(b) A Norman window has area 20 square feet, and the rectangle is twice as long as it is wide, what are the dimensions of the window?

72. The New River Gorge Bridge in West Virginia is the second longest steel arch bridge in the world.[4] (The longest is the Lupu Bridge in Shanghai, China.) The height, in feet, of the arch of the bridge is approximated by $h(x) = -0.00121246x^2 + 876$, where x represents the horizontal distance, in feet, from the center of the arch.

(a) What is the height of the arch?

(b) To the nearest foot, what is the span of the arch of the New River Gorge Bridge at a height of 575 feet above the ground?

73. The length of a rectangular piece of paper is 2 inches more than the width. A one inch square is cut from each corner of the paper, and the paper is folded up to make a box. The volume of the box is 80 cubic inches. ($V = lwh$.) Use the method of completing the square to find the dimensions of the rectangular piece of paper.

74. The stopping distance, in feet, of a car that was traveling with speed, v, in miles per hour, when it braked, is given by[5]

$$d = 2.2v + \frac{v^2}{20}.$$

(a) What is the stopping distance of a car going 30 mph? 60 mph? 90 mph?
(b) If the stopping distance of a car is 500 feet, use a graph to determine how fast it was going when it braked.

75. Graph each function on the same set of axes and find the x-intercepts.

(a) $f(x) = x^2 - 4x + 5$ (b) $f(x) = x^2 - 4x + 4$
(c) $f(x) = x^2 - 4x + 3$ (d) $f(x) = x^2 - 4x + 2$

5.4 SOLVING QUADRATIC EQUATIONS BY FACTORING

In Section 5.1 we saw that r and s are solutions of the equation

$$a(x - r)(x - s) = 0.$$

The following principle shows that they are the only solutions.

The **zero-factor principle** states that

If $A \cdot B = 0$ then either $A = 0$ or $B = 0$ (or both).

The factors A and B can be any numbers, including those represented by algebraic expressions.

Example 1 Find the zeros of the quadratic function $f(x) = x^2 - 4x + 3$ by expressing it in factored form.

Solution We have

$$f(x) = x^2 - 4x + 3 = (x - 1)(x - 3),$$

---
[4] www.nps.gov/neri/bridge.htm, accessed on 2/19/05.
[5] http://www.arachnoid.com/lutusp/auto.html, accessed June 6, 2003

so $x = 1$ and $x = 3$ are zeros. To see that they are the only zeros, we let $A = x - 1$ and $B = x - 3$, and we apply the zero-factor principle. If x is a zero, then

$$\underbrace{(x - 1)}_{A}\underbrace{(x - 3)}_{B} = 0,$$

so either $A = 0$, which implies

$$x - 1 = 0$$
$$x = 1,$$

or $B = 0$, which implies

$$x - 3 = 0$$
$$x = 3.$$

In general, the zero-factor principle tells us that for a quadratic equation in the form

$$a(x - r)(x - s) = 0, \qquad \text{where } a, r, s \text{ are constants, } a \neq 0,$$

the only solutions are $x = r$ and $x = s$.

Example 2   Solve for x.

(a) $(x - 2)(x - 3) = 0$ (b) $(x - q)(x + 5) = 0$ (c) $x^2 + 4x = 21$
(d) $(2x - 2)(x - 4) = 20$

Solution
(a) This is a quadratic equation in factored form, with solutions $x = 2$ and $x = 3$.
(b) This is a quadratic equation in factored form, with solutions $x = q$ and $x = -5$.
(c) We put the equation into factored form:

$$x^2 + 4x = 21$$
$$x^2 + 4x - 21 = 0$$
$$(x + 7)(x - 3) = 0.$$

Thus the solutions are $x = -7$ and $x = 3$.

(d) The left-hand side of this equation is in factored form, but the right-hand side does not equal zero, so we cannot apply the zero-factor principle. To make the right hand side zero, we rewrite the equation as follows:

$$(2x - 2)(x - 4) = 20$$
$$2(x - 1)(x - 4) = 20 \quad \text{factoring out a 2}$$
$$(x - 1)(x - 4) = 10 \quad \text{dividing both sides by 2}$$
$$x^2 - 5x + 4 = 10 \quad \text{expand left-hand side}$$
$$x^2 - 5x - 6 = 0$$
$$(x + 1)(x - 6) = 0.$$

Applying the zero-factor principle we have either

$$x + 1 = 0 \quad \text{or} \quad x - 6 = 0.$$

Solving these equations for x gives $x = -1$ and $x = 6$.

Example 3   Find the horizontal and vertical intercepts of the graph of the quadratic function $h(x) = 2x^2 - 16x + 30$.

Solution   The vertical intercept is the the constant term, $h(0) = 30$. The horizontal intercepts are the solutions to the equation $h(x) = 0$, which we find by factoring the expression for $h(x)$:

$$h(x) = 2(x^2 - 8x + 15)$$
$$0 = 2(x - 3)(x - 5) \quad \text{letting } h(x) = 0$$

The x-intercepts are at $x = 3, 5$. See Figure 5.12.

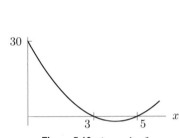

Figure 5.12: A graph of
$h(x) = 2x^2 - 16x + 30$

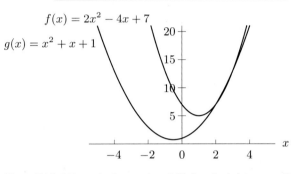

Figure 5.13: Where do the graphs of $f(x)$ and $g(x)$ intersect?

Example 4   Find the points where the graphs of

$$f(x) = 2x^2 - 4x + 7 \quad \text{and} \quad g(x) = x^2 + x + 1$$

intersect.

Solution   See Figure 5.13. At the points where the graphs intersect,

$$2x^2 - 4x + 7 = x^2 + x + 1$$
$$2x^2 - x^2 - 4x - x + 7 - 1 = 0 \qquad \text{combine like terms}$$
$$x^2 - 5x + 6 = 0 \qquad \text{simplify.}$$

To apply the zero factor principle, we factor the left-hand side:

$$(x - 2)(x - 3) = 0.$$

If x is a solution to this equation then either $x - 2 = 0$ or $x - 3 = 0$, by the zero-factor principle. So the possible solutions are $x = 2$ and $x = 3$. We check these values in the original equation:

$$f(2) = 2 \cdot 2^2 - 4 \cdot 2 + 7 = 7$$
$$g(2) = 2^2 + 2 + 1 \qquad = 7,$$

so the first intersection point is $(2, 7)$, and

$$f(3) = 2 \cdot 3^2 - 4 \cdot 3 + 7 = 13$$
$$g(3) = 3^2 + 3 + 1 \qquad = 13,$$

so the second intersection point is $(3, 13)$.

Example 5   Write a quadratic equation in t that has the given solutions.

(a) -1 and 2 (b) $2 + \sqrt{3}$ and $2 - \sqrt{3}$ (c) a and b, constants

Solution   (a) Using the factored form we get

$$(t - (-1))(t - 2) = (t + 1)(t - 2) = 0.$$

(b) If $t = 2 \pm \sqrt{3}$, then $t - 2 = \pm\sqrt{3}$, so

$$(t - 2)^2 = 3$$

is a quadratic equation with the required solutions.

(c) Again we use the factored form $(t - a)(t - b) = 0$.

Solving Other Equations Using Quadratic Equations

Factoring can be used to solve some equations which are not quadratic, as in the following examples.

Example 6   Solve $2x^3 - 5x^2 = 3x$.

Solution   We arrange for a 0 on the right by moving the $3x$ to the left:

$$2x^3 - 5x^2 - 3x = 0,$$

and then we factor out an x:

$$x(2x^2 - 5x - 3) = 0.$$

Now we factor the quadratic $2x^2 - 5x - 3$, giving

$$x(2x + 1)(x - 3) = 0.$$

If the product of three numbers is 0, then at least one of them must be 0. Thus

$$x = 0 \qquad \text{or} \qquad 2x + 1 = 0 \qquad \text{or} \qquad x - 3 = 0,$$

so

$$x = 0 \qquad \text{or} \qquad x = -\frac{1}{2} \qquad \text{or} \qquad x = 3.$$

We call $2x^3 - 5x^2 - 3x = 0$ a *cubic equation*, because the highest power of x is 3. This cubic equation has 3 solutions. In Chapter 8 we discuss the possible number of solutions of a cubic equation.

You might be tempted to divide through by x in the original equation, $2x^3 - 5x^2 = 3x$, giving $2x^2 - 5x = 3$. However this is not allowed, because x could be 0. In fact, the result of dividing by x is to lose the solution $x = 0$.

Example 7   Solve $y^4 - 10y^2 + 9 = 0$.

Solution   Since $y^4 = (y^2)^2$, the equation can be written as

$$y^4 - 10y^2 + 9 = (y^2)^2 - 10y^2 + 9 = 0.$$

If we replace y^2 with z, we can write

$$(y^2)^2 - 10y^2 + 9 = 0$$
$$z^2 - 10z + 9 = 0 \quad \text{replacing } y^2 \text{ with } z$$
$$(z - 1)(z - 9) = 0 \quad \text{factoring the left side}$$
$$(y^2 - 1)(y^2 - 9) = 0 \quad \text{replacing } z \text{ with } y^2.$$

Thus, the solutions are given by

$$y^2 - 1 = 0 \quad \text{or} \quad y^2 - 9 = 0.$$

Solving for y gives

$$y = \pm 1 \quad \text{or} \quad y = \pm 3.$$

The equation $y^4 - 10y^2 + 9 = 0$ is called a *quartic equation* because the highest power of y is 4.

Sometimes an equation that does not look like a quadratic equation can be transformed into one. We must be careful to check our answers after making a transformation, because sometimes the transformed equation has solutions that are not solutions to the original equation.

Example 8   Solve (a) $\dfrac{a}{a-1} = \dfrac{4}{a}$ (b) $\dfrac{2}{y} - \dfrac{2}{y-3} - 3 = 0$

Solution   (a) Multiplying both sides by $a(a-1)$ gives

$$\frac{a}{a-1} \cdot a(a-1) = \frac{4}{a} \cdot a(a-1).$$

Canceling $(a-1)$ on the left and a on the right, we get

$$a^2 = 4(a - 1),$$

so we get the quadratic equation

$$a^2 - 4a + 4 = 0.$$

Factoring gives

$$(a - 2)^2 = 0,$$

which has solution $a = 2$. Since the denominators a and $(a-1)$ in the original equation are not zero when $a = 2$, the operations that led to this equation are valid, so $a = 2$ is also a solution to the original equation. We can also check this directly:

$$\frac{2}{2-1} = \frac{4}{2} = 2.$$

(b) Multiplying both sides by $y(y-3)$ means multiplying all three terms on the left

$$\frac{2}{y} \cdot y(y-3) - \frac{2}{y-3} \cdot y(y-3) - 3y(y-3) = 0.$$

Canceling gives

$$2(y-3) - 2y - 3y(y-3) = 0.$$

To solve, we first expand and collect terms, giving

$$2y - 6 - 2y - 3y^2 + 9y = 0$$
$$-3y^2 + 9y - 6 = 0.$$

Factoring this quadratic, we see

$$-3(y^2 - 3y + 2) = 0$$
$$-3(y-1)(y-2) = 0,$$

so the solutions are $y = 1$ and $y = 2$. Since these solutions do not make the denominator of the original equation zero, these are the solutions to the original equation.

Example 9   Solve (a) $\dfrac{x^2 - 1}{x+3} = \dfrac{8}{x+3}$ (b) $\dfrac{3}{z-2} - \dfrac{12}{z^2 - 4} = 1$

Solution

(a) Notice that the two equal fractions have the same denominator. Therefore, it will suffice to find the values of x that make the numerators equal, *provided that such values do not make the denominator zero.*

$$x^2 - 1 = 8$$

Rewriting this equation as

$$x^2 = 9$$

and taking the square root of both sides, we see that

$$x = 3 \quad \text{and} \quad x = -3.$$

The denominator of original equation, $x+3$, is not zero when $x = 3$. Therefore, it is a solution to the original equation. However, the denominator in the original equation is zero when $x = -3$. This means that $x = -3$ is not a solution to the original equation. Answers such as $x = -3$, that are not solutions to the original equation, are called *extraneous solutions*.

(b) Multiplying both sides by $z^2 - 4$ gives

$$\frac{3}{z-2} \cdot (z^2 - 4) - \frac{12}{z^2 - 4} \cdot (z^2 - 4) = 1 \cdot (z^2 - 4).$$

Canceling produces

$$3(z+2) - 12 = z^2 - 4.$$

To solve, we set the equation equal to zero, expand, and collect terms.

$$3(z+2) - 12 - z^2 + 4 = 0$$
$$3z + 6 - 12 - z^2 + 4 = 0$$
$$-z^2 + 3z - 2 = 0$$

Factoring this quadratic, we get

$$-1(z^2 - 3z + 2) = 0$$
$$-1(z - 1)(z - 2) = 0.$$

This means

$$z = 1 \quad \text{and} \quad z = 2.$$

However, $z = 1$ is the only solution to the original equation. Since the denominator of the original equation is zero when $z = 2$, it is an extraneous solution.

Exercises and Problems for Section 5.4

Exercises

Solve the equations in Exercises 1–10 by factoring.

1. $(x - 2)(x - 3) = 0$ **2.** $x(5 - x) = 0$

3. $x^2 + 5x + 6 = 0$ **4.** $6x^2 + 13x + 6 = 0$

5. $(x - 1)(x - 3) = 8$ **6.** $(2x - 5)(x - 2)^2 = 0$

7. $x^4 + 2x^2 + 1 = 0$ **8.** $x^4 - 1 = 0$

9. $(x - 3)(x + 2)(x + 7) = 0$

10. $x(x^2 - 4)(x^2 + 1) = 0$

In Exercises 11–22, solve for x.

11. $2(x - 3)(x + 5) = 0$ **12.** $x^2 - 4 = 0$

13. $5x(x + 2) = 0$ **14.** $x(x + 3) = 10$

15. $x^2 + 2x = 5x + 4$ **16.** $2x^2 + 5x = 0$

17. $x^2 - 8x + 12 = 0$ **18.** $x^2 + 3x + 7 = 0$

19. $x^2 + 6x - 4 = 0$ **20.** $2x^2 - 5x - 12 = 0$

21. $x^2 - 3x + 12 = 5x + 5$ **22.** $2x(x + 1) = 5(x - 4)$

Find the zeros (if any) of the quadratic functions in Exercises 23–25.

23. $y = 3x^2 - 2x - 11$ **24.** $y = 5x^2 + 3x + 3$

25. $y = (2x - 3)(3x - 1)$

In Exercises 26–34, write a quadratic equation in x with the given solutions.

26. $\sqrt{5}$ and $-\sqrt{5}$ **27.** 2 and -3

28. 0 and $3/2$ **29.** $2 + \sqrt{3}$ and $2 - \sqrt{3}$

30. $-p$ and 0   **31.** $p + \sqrt{q}$ and $p - \sqrt{q}$

32. a and b **33.** a, no other solutions

34. With no solutions

In Exercises 35–37, solve for x.

35. $x^2 - x^3 + 2x = 0$ **36.** $-2x^2 - 3 + x^4 = 0$

37. $x + \dfrac{1}{x} = 2.$

Problems

38. If a diver jumps off a diving board that is 6 ft above the water at a velocity of 20 ft/sec, his height, s, in feet, above the water can be modeled by $s(t) = -16t^2 + 20t + 6$, where $t \geq 0$ is in seconds.

    (a) How long is the diver in the air before he hits the water?

    (b) What is the maximum height achieved and when did it occur?

39. A ball is thrown straight upward from the ground. Its height above the ground in meters after t seconds is given by $-4.9t^2 + 30t + c$.

    (a) Find the constant c

    (b) Find the values of t that make the height zero and give a practical interpretation of each value.

Find the zeros (if any) of the quadratic functions in Problems 40–41.

40. $y = x(6x - 10) - 7(3x - 5)$

41. $y = 3x^2 - 5x - 1$

Without solving them, say whether the equations in Problems 42–49 have two solutions, one solution, or no solution. Give a reason for your answer.

42. $3(x - 3)(x + 2) = 0$ **43.** $(x - 2)(x - 2) = 0$

44. $(x + 5)(x + 5) = -10$ **45.** $(x + 2)^2 = 17$

46. $(x - 3)^2 = 0$ **47.** $3(x + 2)^2 + 5 = 1$

48. $-2(x - 1)^2 + 7 = 5$ **49.** $2(x - 3)^2 + 10 = 10$

Solve the equations in Problems 50–53 using quadratic equations.

50. $t^4 - 13t^2 + 36 = 0$.

51. $t^4 - 3t^2 - 10 = 0$.

52. $(t - 3)^6 - 5(t - 3)^3 + 6 = 0$.

53. $t + 2\sqrt{t} - 15 = 0$.

54. A gardener wishes to double the area of her 4 feet by 6 feet rectangular garden. She wishes to add a strip of uniform width to all of the sides of her garden. How wide should the strip be?

55. Does $x^{-1} + 2^{-1} = (x + 2)^{-1}$ have solutions? If so, find them.

In Problems 56–59, solve **(a)** For p **(b)** For q. In each case, assume that the other quantity is nonzero and restricted so that solutions exist.

56. $p^2 + 2pq + 5q = 0$ **57.** $q^2 + 3pq = 10$

58. $p^2q^2 - p + 2 = 0$ **59.** $pq^2 + 2p^2q = 0$

60. Consider the equation $(x - 3)(x + 2) = 0$.

    (a) What are the solutions?

    (b) Use the quadratic formula as an alternative way to find the solutions. Compare your answers.

61. Consider the equation $x^2 + 7x + 12 = 0$.

    (a) Solve the equation by factoring.

    (b) Solve the equation using the quadratic formula. Compare your answers.

62. Consider the equation $ax^2 + bx = 0$ with $a \neq 0$.

    (a) Use the discriminant to show that this equation has solutions.

    (b) Use factoring to find the solutions.

    (c) Use the quadratic formula to find the solutions.

63. We know that if $A \cdot B = 0$, that either $A = 0$ or $B = 0$. If $A \cdot B = 6$, does that imply that either $A = 6$ or $B = 6$? Explain your answer.

64. In response to the problem "Solve $x(x + 1) = 2 \cdot 6$," a student writes "We must have $x = 2$ or $x + 1 = 6$, which leads to $x = 2$ or $x = 5$ as the solutions." Is the student correct?

REVIEW EXERCISES AND PROBLEMS FOR CHAPTER FIVE

Exercises

In Exercises 1–8, sketch the graph without using a calculator.

1. $y = (x - 2)^2$ **2.** $y = 5 + 2(x + 1)^2$

3. $y = (3 - x)(x + 2)$ **4.** $y = (2x + 3)(x + 3)$

5. $y = (x - 1)(x - 5)$ **6.** $y = -2(x + 3)(x - 4)$

7. $y = 2(x - 3)^2 + 5$ **8.** $y = -(x + 1)^2 + 25$

In Exercises 9–15, is the expression quadratic in the given variable? Assume all constants are non-zero.

9. $ax^2 + bx + c^3$, a.

10. $ax^2 + bx + c^3$, x.

11. $mx + b + c^3x^2$, x.

12. $2ax + bx + c$, a.

13. $2ax + bx + c$, x.

14. $(a+b)^2 - (a-b)^2$, a

15. $aP(P-b)(c-P)$, P.

In Exercises 16–19, find the minimum or maximum value of the quadratic expression in x.

16. $(x+7)^2 - 8$

17. $a - (x+2)^2$

18. $q - 7(x+a)^2$

19. $2(x^2 + 6x + 9) + 2$

In Exercises 20–22, write the expressions in vertex form and identify the constants $a, h,$ and k.

20. $2(x^2 - 6x + 9) + 4$

21. $11 - 7(3-x)^2$

22. $(2x+4)^2 - 7$

In Exercises 23–25, rewrite the equation in a form that clearly shows its solutions and give the solutions.

23. $x^2 + 3x + 2 = 0$

24. $1 + x^2 + 2x = 0$

25. $5z + 6z^2 + 1 = 0$

Write the quadratic functions in Exercises 26–27 in the following forms and state the values of all constants.

(a) Standard form $y = ax^2 + bx + c$.

(b) Factored form $y = a(x-r)(x-s)$.

(c) Vertex form $y = a(x-h)^2 + k$.

26. $y - 8 = -2(x+3)^2$

27. $y = 2x(3x-7) + 5(7-3x)$

In Exercises 28–47, solve for x.

28. $x^2 - 16 = 0$

29. $x^2 + 5 = 9$

30. $(x+4)^2 = 25$

31. $(x-6)^2 + 7 = 7$

32. $(x-5)^2 + 15 = 17$

33. $(x-1)(2-x) = 0$

34. $(x-93)(x+115) = 0$

35. $(x+47)(x+59) = 0$

36. $x^2 + 14x + 45 = 0$

37. $x^2 + 21x + 98 = 0$

38. $2x^2 + x - 55 = 0$

39. $x^2 + 16x + 64 = 0$

40. $x^2 + 26x + 169 = 0$

41. $x^2 + 12x + 7 = 0$

42. $x^2 + 24x - 56 = 0$

43. $3x^2 + 30x + 9 = 0$

44. $2x^3 - 4x^2 + 2x = 0$

45. $x^5 + 3x^3 + 2x = 0$

46. $x + \sqrt{x} = 6$

47. Solve $x^{-1} + 2^{-1} = -x - 2$.

Solve Exercises 48–53 with the quadratic formula.

48. $x^2 - 4x - 12 = 0$

49. $2x^2 - 5x - 12 = 0$

50. $y^2 + 3y + 4 = 6$

51. $3y^2 + y - 2 = 7$

52. $7t^2 - 15t + 5 = 3$

53. $-2t^2 + 12t + 5 = 2t + 8$

Solve Exercises 54–59 by completing the square.

54. $x^2 - 8x + 8 = 0$

55. $y^2 + 10y - 2 = 0$

56. $s^2 + 3s - 1 = 2$

57. $r^2 - r + 2 = 7$

58. $2t^2 - 4t + 4 = 6$

59. $3v^2 + 9v = 12$

60. Explain why the equation $(x-3)^2 = -4$ has no real solution.

61. Explain why the equation $x^2 + x + 7 = 0$ has no positive solution.

Find possible quadratic equations in standard form that have the solutions given in Exercises 62–65.

62. $x = 2, -6$

63. $t = \frac{2}{3}, t = -\frac{1}{3}$

64. $x = 2 \pm \sqrt{5}$

65. $x = \sqrt{5}, -\sqrt{3}$

Problems

66. The height in feet of a stone, t seconds after it is dropped from the top of the Petronas Towers (one of the highest buildings in the world), is given by

$$h = 1483 - 16t^2.$$

   (a) How tall are the Petronas Towers?
   (b) When does the stone hit the ground?

67. The height, in feet, of a rocket t seconds after it is launched is given by $h = -16t^2 + 160t$. How long does the rocket stay in the air?

68. Write the expression $x^2 - 6x + 9$ in a form that demonstrates that the value of the expression is always greater than or equal to zero, no matter what the value of x.

69. Explain why the smallest value of $f(x) = 3(x-6)^2 + 10$ occurs when $x = 6$ and give the value.

70. Explain why the largest value of $g(t) = -4(t+1)^2 + 21$ occurs when $t = -1$ and give the value.

71. **(a)** Use the method of completing the square to write $y = x^2 - 6x + 20$ in vertex form.
   (b) Use the vertex form to identify the smallest value of the function, and the x-value at which it occurs.

In Problems 72–75, find the number of x-intercepts, and give an explanation of your answer that does not involve changing the form of the right-hand side.

72. $y = -3(x+3)(x-5)$ **73.** $y = 1.5(x-10)^2$

74. $y = -(x-1)^2 + 5$ **75.** $y = 2(x+3)^2 + 4$

Problems 76–77 show the graph of a quadratic function.

   (a) If the function is in standard form $y = ax^2 + bx + c$, is a positive or negative? What is c?
   (b) If the function is in factored form $y = a(x-r)(x-s)$ with $r \le s$, what is r? What is s?
   (c) If the function is in vertex form $y = a(x-h)^2 + k$, what is h? What is k?

76.

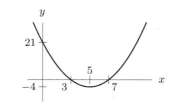

77.

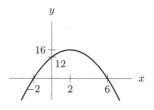

Write the quadratic function

$$y = 4x - 30 + 2x^2$$

in the forms indicated in Problems 78–81. Give the values of all constants.

78. $y = ax^2 + bx + c$ **79.** $y = a(x-r)(x-s)$

80. $y = a(x-h)^2 + k$ **81.** $y = ax(x-v) + w$

82. A rectangle has one corner at $(0,0)$ and the opposite corner on the line $y = -x + b$, where b is a positive constant.

   (a) Express the rectangle's area, A, in terms of x.
   (b) What value of x gives the largest area?
   (c) With x as in part (b), what is the shape of the rectangle?

83. A carpenter finds that if she charges p dollars for a chair, she sells $1200 - 3p$ of them each year.

   (a) At what price will she price herself out of the market, that is, have no customers at all?
   (b) How much should she charge to maximize her annual revenue?

84. When P dollars is invested at an annual interest rate r compounded once a year, the balance, A, after 2 years is given by $A = P(1+r)^2$.

   (a) Evaluate A when $r = 0$, and interpret the answer in practical terms.
   (b) If r is between 5% and 6%, what can you conclude about the percentage growth after 2 years?
   (c) Express r as a function of A. Under what circumstances might this function be useful?
   (d) What interest rate is necessary to obtain an increase of 25% in 2 years?

In Problems 85–87, use a calculator or computer to sketch the graphs of the equations. What do you observe? Use algebra to confirm your observation.

85. $y = (x+2)(x+1) + (x-2)(x-3) - 2x(x-1)$

86. $y = (x-1)(x-2) + 3x$ and $y = x^2$

87. $y = (x - a)^2 + 2a(x - a) + a^2$ for $a = 0$, $a = 1$, and $a = 2$

In Problems 88–91, decide for what values of the constant A (if any) the equation has a positive solution. Give a reason for your answer.

88. $x^2 + x + A = 0$

89. $3(x + 4)(x + A) = 0$

90. $A(x - 3)^2 - 6 = 0$

91. $5(x - 3)^2 - A = 0$

In Problems 92–94, decide what conditions on the parameters ensure that the equation has the solution $x = 0$. Assume $a \neq 0$. Give a reason for your answer.

92. $ax^2 + bx + c = 0$

93. $a(x - r)(x - s) = 0$

94. $a(x - h)^2 + k = 0$

Problems 95–98 refer to the equation $(x - a)(x - 1) = 0$.

95. Solve the equation for x, assuming a is a constant.

96. Solve the equation for a, assuming x is a constant not equal to 1.

97. For what value(s) of the constant a does the equation have only one solution in x?

98. For what value(s) of x does the equation have infinitely many solutions in a?

In Problems 99–102, give the values of a for which the equation has a solution in x. You do not need to solve the equation.

99. $(3x - 2)^{10} + a = 0$

100. $ax - 5 + 6x = 0$

101. $6x - 5 = 8x + a - 2x$

102. $4x^2 - (1 - 2x)^2 + a = 0$

103. For what values of a does the equation $ax^2 = x^2 - x$ have exactly one solution in x?

104. (a) Show that if $(x - h)^2 + k = 0$ has two distinct solutions in x, then k must be negative.
(b) Use your answer to part (a) to explain why k must be negative if $(x - h)^2 + k = (x - r)(x - s)$, with $r \neq s$.
(c) What can you conclude about k if $-2(x - h)^2 + k = -2(x - r)(x - s)$, with $r \neq s$?

105. (a) Explain why $2(x - h)^2 + k$ is greater than or equal to k for all values of x.
(b) If $2(x - h)^2 + k = ax^2 + bx + c$, explain why it must be true that $c \geq k$.
(c) What can you conclude about c and k if $-2(x - h)^2 + k = ax^2 + bx + c$?

106. If a is a constant, rewrite $x^2 + (a - 1)x - a = 0$ in a form that clearly shows its solutions. What are the solutions?

107. Rewrite $2a + 2 + a^2 = 0$ in a form that clearly shows it has no solutions.

108. The length of a rectangular swimming pool is twice its width. The pool is surrounded by a walk that is 2 feet wide. The area of the region consisting of the pool and the walk is 1056 square feet.

(a) Use the method of completing the square to determine the dimensions of the swimming pool.
(b) If the material for the walk costs $10 per square foot, how much would the material cost for the entire walk?

109. When the square of a certain number is added to the number, the result is the same as when 48 is added to three times the number. Use the method of completing the square to determine the number.

TOOLS FOR CHAPTER 5

FACTORING TECHNIQUES

In Section 0.2 we used the distributive law to expand products like $a(b+c)$, and to take common factors out of expressions like $ab+ac$. Now we consider products in which both factors are sums of more than one term. Such products can be multiplied out or *expanded* by repeated use of the distributive law.

Expanding Using the Distributive Law

$$(x+r)(y+s) = (x+r)y + (x+r)s \quad \text{distribute } x+r \text{ over } y+s$$
$$= xy + ry + xs + rs \quad \text{use the distributive law twice more}$$

Notice that in the final result, each term in the first factor, $x+r$, has been multiplied by each term in the second factor, $y+s$.

Factoring by Grouping

When an expression has a common factor in every term, that factor can be pulled out in front. For example, $8a^3 - 4a^2 + 6a = 2a(4a^2 - 2a + 3)$. An expression of the form $xy + ry + xs + rs$ has no common factors. However, the four terms in the expression can be grouped in pairs as

$$xy + ry + xs + rs = (xy + ry) + (xs + rs).$$

Notice that the first two terms have a common factor of y and the last two terms have a common factor of s. Factoring out the common factors in each group gives

$$xy + ry + xs + rs = (xy + ry) + (xs + rs) = y(x+r) + s(x+r).$$

Notice that we are lucky—there is now another factor which is common to both groups, namely $(x+r)$. Factoring out $(x+r)$ enables us to factor the entire expression

$$xy + ry + xs + rs = y(x+r) + s(x+r) = (x+r)(y+s).$$

Factoring Quadratic Expressions

When we expand a product of expressions that are linear in the same variable, the result can be simplified to an expression with fewer than four terms.

Example 1 Expand the product $(x-4)(x+6)$.

Solution When we expand, we can combine the like terms, $-4x$ and $6x$, giving

$$(x-4)(x+6) = x^2 - 4x + 6x - 24 = x^2 + 2x - 24.$$

An expression of the form

$$ax^2 + bx + c, \quad \text{where } a, b, \text{ and } c \text{ are constants,}$$

is called a *quadratic expression* in x. Expanding a product of the form $(x + r)(x + s)$ always gives a quadratic expression:

$$(x + r)(x + s) = (x + r)x + (x + r)s$$
$$= x^2 + rx + sx + rs$$
$$= x^2 + (r + s)x + rs.$$

Notice that the sum $r + s$ is the coefficient of the x term, and the product rs is the constant term. This gives a quick way to expand $(x+r)(x+s)$. For instance, in Example 1, we could have obtained the answer by noticing that $-4+6 = 2$, giving a term $2x$, and $-4 \cdot 6 = -24$, giving a constant term -24.

Example 2   Expand and collect like terms in $(s - 3)(s + 5) + s(s - 2)$.

Solution   Multiplying and collecting like terms gives

$$(s - 3)(s + 5) + s(s - 2) = s^2 + 2s - 15 + s^2 - 2s$$
$$= 2s^2 - 15.$$

Factoring Quadratic Expressions by Guess and Check

We start by considering expressions of the form x^2+bx+c, where the coefficient of x^2 is 1. Suppose in Example 1 that we were given the answer, $x^2 + 2x - 24$, and had to factor it into $(x - 4)(x + 6)$. We have just seen how to factor by grouping, but in this case we don't know which terms to group, since the x terms have been collected into the one term $2x$. One approach to factoring $x^2 + bx + c$ as $(x + r)(x + s)$ is to find r and s by trying various values and seeing which ones fit, if any.

Example 3   If possible, factor $x^2 - 9x + 18$ into $(x + r)(x + s)$, where r and s are integers.

Solution   If this is possible, then $r+s$ must equal -9 and rs must equal 18. The numbers $r = -3$ and $s = -6$ satisfy both conditions. Therefore, $x^2 - 9x + 18 = (x - 3)(x - 6)$.

Example 4   If possible, factor $x^2 + 3x + 4$ into $(x + r)(x + s)$, where r and s are integers.

Solution   If this is possible, then $r + s = 3$ and $rs = 4$. There are no integers that satisfy both conditions. Therefore, $x^2 + 3x + 4$ cannot be factored in this way.

Example 5   Factor $x^2 + 10xy + 24y^2$.

Solution   If the expression $x^2 + 10xy + 24y^2$ can be factored, it can be written as $(x + ry)(x + sy)$, where $r + s = 10$ and $rs = 24$. Since 4 and 6 satisfy these conditions, we have

$$x^2 + 10xy + 24y^2 = (x + 4y)(x + 6y).$$

What if the Coefficient of x^2 is Not 1?

Expanding products of the form $(px+r)(qx+s)$ gives a quadratic expression in which the coefficient of x^2 is not necessarily 1.

Example 6   Expand $(2x + 1)(x + 3)$.

Solution   Expanding using the distributive law gives

$$(2x + 1)(x + 3) = x(2x + 1) + 3(2x + 1)$$
$$= 2x^2 + x + 6x + 3$$
$$= 2x^2 + 7x + 3.$$

Given a quadratic expression $ax^2 + bx + c$, we can try to factor it by reversing the process of multiplying out. We start by factoring $2x^2 + 7x + 3$ by reversing the multiplication of Example 6.

Example 7   Factor $2x^2 + 7x + 3$.

Solution   The strategy for factoring this type of expression is to break the middle term, $7x$, into two terms, in such a way that we can factor by grouping. Following Example 6, we break $7x$ into $x + 6x$, so

$$2x^2 + 7x + 3 = 2x^2 + x + 6x + 3.$$

Then we factor by grouping

$$2x^2 + 7x + 3 = 2x^2 + x + 6x + 3$$
$$= (2x^2 + x) + (6x + 3)$$
$$= x(2x + 1) + 3(2x + 1)$$
$$= (2x + 1)(x + 3).$$

Notice that this method works because we get a common factor $(2x + 1)$ at the last stage. If we had split the $7x$ up differently, this might not have happened.

How would we know to write $7x = x + 6x$ in Example 7, without Example 6 to guide us? There is a method which always lead to the right way of breaking up the x term, if it can be done at all. In the expression $ax^2 + bx + c$ that we want to factor, we form the product of the constant term and the coefficient of x^2. In Example 7, this is $2 \cdot 3 = 6$. Then we try to write the x term as a sum of two terms whose coefficients multiply to this product. In Example 7, we find $7x = x + 6x$ works, since $1 \cdot 6 = 6$.

Example 8   Factor $8x^2 + 14x - 15$.

Solution   We multiply the coefficient of the x^2 term by the constant term: $8 \cdot (-15) = -120$. Now we try to write $14x$ as a sum of two terms whose coefficients multiply to -120. Writing $14x = -6x + 20x$ works, since $-6 \cdot 20 = -120$. Breaking up the $14x$ in this way enables us to factor by grouping:

$$8x^2 + 14x - 15 = 8x^2 - 6x + 20x - 15$$
$$= (8x^2 - 6x) + (20x - 15)$$
$$= 2x(4x - 3) + 5(4x - 3)$$
$$= (4x - 3)(2x + 5).$$

Here is the general process.

Factoring Quadratic Expressions

If the expression $ax^2 + bx + c$ is factorable, the following steps work:

- Factor out all common constant factors.
- On the remaining expression, multiply the coefficient of the x^2 term by the constant term, giving ac.
- Find two numbers that multiply to ac and sum to b, the coefficient of the x term.
- Break the middle term, bx, into two terms using the result of the previous step.
- Factor the four terms by grouping.

Example 9   Factor $12x^2 - 44x + 24$.

Solution   First, take out the common factor of 4. This gives

$$12x^2 - 44x + 24 = 4(3x^2 - 11x + 6).$$

To factor $3x^2 - 11x + 6$, we multiply 3 and 6 to get 18, and then we look for two numbers that multiply to 18 and sum to -11. We find that -9 and -2 work, so we write $-11x = -9x - 2x$. This gives

$$
\begin{aligned}
3x^2 - 11x + 6 &= 3x^2 - 9x - 2x + 6 \\
&= (3x^2 - 9x) + (-2x + 6) \\
&= 3x(x - 3) - 2(x - 3) \\
&= (x - 3)(3x - 2).
\end{aligned}
$$

Therefore,

$$12x^2 - 44x + 24 = 4(x - 3)(3x - 2).$$

In Example 9, how can we find the numbers -9 and -2 if they don't jump out at us? A systematic method is to list all the ways of factoring 18 into two integer factors and to pick out the pair with the correct sum. The factorizations are

$$18 = 18 \times 1 = 6 \times 3 = 9 \times 2 = -18 \times -1 = -6 \times -3 = -9 \times -2.$$

Only the last pair, -9 and -2, sums to -11.

All the methods of factorization in this section are aimed at finding factors whose coefficients are integers. If we allow roots, further factorizations might be possible.

Perfect Squares

There are some special quadratic expressions that can be easily factored if we recognize their form.

Example 10   Expand $(2z + 5)^2$.

Solution   The expression $(2z + 5)^2$ can be rewritten as $(2z + 5)(2z + 5)$. Using the distributive property and combining like terms, we get

$$
\begin{aligned}
(2z + 5)^2 &= (2z + 5)(2z + 5) \\
&= 2z(2z + 5) + 5(2z + 5) \\
&= 4z^2 + 10z + 10z + 25 \\
&= 4z^2 + 20z + 25.
\end{aligned}
$$

Notice that the first and last term of the final expression are squares of the first and last term of the expression $2z + 5$. The middle term is twice the product of the first and last term. This pattern works in general:

$$
\begin{aligned}
(x + r)^2 &= (x + r)(x + r) \\
&= x(x + r) + r(x + r) \\
&= x^2 + rx + rx + r^2 \\
&= x^2 + 2rx + r^2.
\end{aligned}
$$

We also have

$$(x - r)^2 = (x + (-r))^2 = x^2 + 2(-r)x + (-r)^2 = x^2 - 2rx + r^2.$$

Since the expression $x^2 + 2rx + r^2$ is the square of $(x + r)$ and the expression $x^2 - 2rx + r^2$ is the square of $(x - r)$, we call them *perfect squares*.

Factoring Perfect Squares

We can sometimes recognize an expression as a perfect square and thus factor it. The clue is to recognize that two terms of a perfect square are squares of other terms, and the third term is twice the product of those two other terms.

Perfect Squares

$$x^2 + 2rx + r^2 = (x + r)^2$$

$$x^2 - 2rx + r^2 = (x - r)^2$$

Example 11   If possible, factor as perfect squares.

(a) $4r^2 + 10r + 25$ (b) $9p^2 + 60p + 100q^2$ (c) $25y^2 + 30yz + 9z^2$

Solution   (a) In the expression $4r^2 + 10r + 25$ we see that the first and last terms are squares:

$$4r^2 = (2r)^2 \quad \text{and} \quad 25 = 5^2.$$

Twice the product of $2r$ and 5 is $2(2r)(5) = 20r$. Since $20r$ is not the middle term of the expression to be factored, $4r^2 + 10r + 25$ does not appear to be a perfect square.

(b) We have
$$9p^2 = (3p)^2 \quad \text{and} \quad 100q^2 = (10q)^2 \quad \text{and} \quad 2(3p)(10q) = 60pq.$$

Since the middle term of does not contain a q, it appears that $9p^2 + 60p + 100q^2$ is not perfect square.

(c) In the expression $25y^2 + 30yz + 9z^2$ we see that the first and last terms are squares:

$$25y^2 = (5y)^2 \quad \text{and} \quad 9z^2 = (3z)^2.$$

Twice the product of $5y$ and $3z$ is $2(5y)(3z) = 30yz$, which is the middle term of the expression to be factored. Thus, $25y^2 + 30yz + 9z^2$ is a perfect square, and $25y^2 + 30yz + 9z^2 = (5y+3z)^2$.

In summary:

Determining if an Expression is a Perfect Square

An expression with three terms is a perfect square if
- Two terms are squares, and
- The third term is twice the product of the expressions whose squares are the other terms.

Difference of Squares

We have looked at perfect squares $(x+r)(x+r)$ and $(x-r)(x-r)$. We now consider products of the form $(x-r)(x+r)$, in which the sign in one of the terms has been changed.

Factoring the Difference of Squares

Difference of Squares

If an expression is in the form $x^2 - r^2$, it can be factored as

$$x^2 - r^2 = (x - r)(x + r).$$

Example 12   If possible, factor the following expressions as the difference of squares.

(a) $x^2 - 100$

(b) $8a^2 - 2b^2$

(c) $49y^2 + 25$

(d) $48s^{1/3} - 27s^{7/3}$

Solution   (a) We see that $100 = 10^2$, so $x^2 - 100$ is a difference of squares and can be factored as

$$x^2 - 100 = (x - 10)(x + 10).$$

(b) First we take out a common factor of 2:

$$8a^2 - 2b^2 = 2(4a^2 - b^2).$$

Since $4a^2$ is the square of $2a$, we see that $4a^2 - b^2$ is a difference of squares. Therefore,

$$8a^2 - 2b^2 = 2(2a + b)(2a - b).$$

(c) We have $49y^2 = (7y)^2$ and $25 = 5^2$. However, $49y^2 + 25$ is a sum of squares, not a difference. Therefore it cannot be factored as the difference of squares.

(d) We take out a common factor of $3s^{1/3}$:

$$48s^{1/3} - 27s^{7/3} = 3s^{1/3}(16 - 9s^2).$$

Since $16 = 4^2$ and $9s^2 = (3s)^2$, we see that $(16 - 9s^2)$ is difference of squares. Hence,

$$48s^{1/3} - 27s^{7/3} = 3s^{1/3}(4 - 3s)(4 + 3s).$$

Example 13   Factor $4a^2 + 4ab + b^2 - 4$.

Solution   Since there are four terms with no common factors, we might try to group any two terms that have a common factor and hope that a common expression will be in each term. One possibility is

$$4a^2 + 4ab + b^2 - 4 = (4a^2 + 4ab) + (b^2 - 4) = 4a(a + b) + (b^2 - 4).$$

Another possibility is

$$4a^2 + 4ab + b^2 - 4 = (4a^2 - 4) + (4ab + b^2) = 4(a^2 - 1) + b(4a + b).$$

The third possibility is

$$4a^2 + 4ab + b^2 - 4 = (4a^2 + b^2) + (4ab - 4) = (4a^2 + b^2) + 4(ab - 1).$$

Unfortunately, none of these groupings leads to a common factor. Another approach is to notice that the first three terms form a perfect square. Therefore,

$$\begin{aligned}
4a^2 + 4ab + b^2 - 4 &= (4a^2 + 4ab + b^2) - 4 && \text{grouping the first three terms} \\
&= (2a + b)^2 - 4 && \text{factoring the perfect square} \\
&= ((2a + b) - 2)((2a + b) + 2) && \text{factoring the difference of squares} \\
&= (2a + b - 2)(2a + b + 2).
\end{aligned}$$

The previous example shows that an expression which cannot be factored by one method may still be factored by another method. Thus it is important to try as many different methods as possible before concluding that an expression cannot be factored.

Exercises and Problems on Factoring Techniques

Exercises

Factor out the common factors in Exercises 1–3.

1. $b(b+3) - 6(b+3)$

2. $6r(s-2) - 12(s-2)$

3. $4ax(x+4) - 2x(x+4)$

For what values of t in Exercises 4–11 is the expression zero?

4. $t^2 - 4t$

5. $t^3 - 25t$

6. $t^3 + 16t$

7. $t(t-3) + 2(t-3)$

8. $4t^{1/2} - 4t$

9. $t^2(t+5) + 9(t+5)$

10. $t^2(t+8) - 36(t+8)$

11. $4t^2(2t-3) - 36(2t-3)$

In Exercises 12–17, determine the values of x for which the expression is equal to zero.

12. $x^3 - 4x^2 + 25x - 100$

13. $x^3 - 4x^2 - 25x + 100$

14. $x^4 + 6x^3 + 8x + 48$

15. $4x^4 + 6x^3 + 10x + 15$

16. $15x^3 - 12x^2 - 10x + 8$

17. $x^3 + ax^2 + 4x + 4a$

In Exercises 18–57, expand and combine terms.

18. $(x+5)(x+2)$

19. $(y+3)(y-1)$

20. $(z-5)(z-6)$

21. $(2a+3)(3a-2)$

22. $(3b+c)(b+2c)$

23. $(r-2)(r^2-r)$

24. $(t+2)(t^2-3t+6)$

25. $(r^2-r+9)(r-4)$

26. $(n^2+3n+2)(n^2+n-4)$

27. $(a+b+c)(a-b-c)$

28. $3(x-4)^2 + 8x - 48$

29. $(5-(x+h)^2 - (5-x^2))/h$

30. $(x+6)^2$

31. $(x-8)^2$

32. $(x+11)^2$

33. $(x-13)^2$

34. $(x+7)(x-7)$

35. $(x+y)^2$

36. $(2a+3b)^2$

37. $(5p^2-q)^2$

38. $(3a^{1/2} - 2b^{1/3})^2$

39. $(x-y)^3$

40. $(2a-3b)^3$

41. $(q+r)^4$

42. $(c-2d)^4$

43. $(x+9)(x-9)$

44. $(x-8)(x+8)$

45. $(x-12)(x+12)$

46. $(2x+6)(x-8)$

47. $(x-11)(3x+2)$

48. $x(4x-7)(2x+2)$

49. $x(3x+7)(5x-8)$

50. $2x(5x+8)(7x+2)$

51. $(x+2)(x^2+7x-2)$

52. $(x-1)(2x^2+15x-8)$

53. $(x+1)(6x-7)(8x+2)$

54. $(x-2)(5x-12)(2x+3)$

55. $(x^2-5)(2x+2)(3x+3)$

56. $(x^2+8)(x^2-1)(x^2+2)$

57. $(x^3+1)(x^2+8)(x-2)$

Factor the expressions in Exercises 58–123.

58. $x^2 + x + 3x + 3$

59. $ax + bx - ay - by$

60. $\pi r + \pi r^2 + 6 + 6r$

61. $8x^2 - 4xy - 6x + 3y$

62. $12a^2 + 2b + 24a + ab$

63. $6v^3 - 4v^2w + 9w^3v - 6w^4$

64. $x^2 + 5x + 6$

65. $y^2 - 5y - 6$

66. $z^2 - 5z + 6$

67. $n^2 - n - 30$

68. $a^2 + 8a - 48$

69. $g^2 - 12g + 20$

70. $v^2 - 4v - 32$

71. $t^2 - 27t + 50$

72. $b^2 - 23b - 50$

73. $q^2 + 15q + 50$

74. $w^2 + 2w + 24$

75. $b^2 + 2b - 24$

76. $x^2 + x - 72$

77. $x^2 + 11xy + 24y^2$

78. $y^3 + 7y^2 - 18y$

79. $2z^2 + 12z - 14$

80. $6x^2 + 5x - 6$

81. $4z^2 + 19z + 12$

82. $30t^2 + 26t + 4$

83. $y^2 - 6y + 7$

84. $q^2 - 4qz + 3z^2$

85. $3w^2 + 12w - 36$

86. $12s^2 + 17s - 5$

87. $t^4 + 3t^2 - 54$

88. $2n^2 - 12n - 54$

89. $12w^2 - 10w - 8$

90. $x^6 - 2x^3 - 63$

91. $a^2 - a - 16$

92. $10z^2 - 21z - 10$

93. $x^2 - 16$

94. $r^2 + 4$

95. $s^2 - 12st + 36t^2$

96. $r^3 - 14r^2s^3 + 49rs^6$

97. $n^2 + 10n + 25q^2$

98. $7y^5 - 28yz^6$

99. $18x^7 + 48x^4z^2 + 32xz^4$

100. $8a^3 + 50ab^2$

101. $r^{2p} + 18r^p + 81$

102. $(c + 3)^2 - d^4$

103. $-3t^7 + 24t^5v^2 - 48t^3v^4$

104. $x^2 + 7x$

105. $x^3 + 4x$

106. $x^2 - 36$

107. $x^2 - 81$

108. $x^2 - 169$

109. $x^2 - 144$

110. $x^2 + 10x + 25$

111. $x^2 - 14x + 49$

112. $x^2 + 26x + 169$

113. $x^2 - 22x + 121$

114. $x^3 - 16x^2 + 64x$

115. $x^4 - 18x^3 + 81x^2$

116. $x^2 + 15x + 56$

117. $x^2 - 3x - 54$

118. $x^2 - 19x + 90$

119. $x^3 + 23x^2 + 132x$

120. $3x^2 + 22x + 35$

121. $5x^2 - 37x - 72$

122. $x^2 - 5x + dx - 5d$

123. $2qx^2 + pqx - 14x - 7p$

Problems

124. For what values of x is $\dfrac{x^3 + 6x^2 + 6x + 36}{4x^2 + 24}$ positive?

125. For what values of x is $\dfrac{3x^3 - 5x^2 + 6x - 10}{3x^3 - 5x^2}$ negative?

In Problems 126–127, which of the expressions I, II, and III, if any, are equivalent to each other?

126. I. $x^2 + 4x^2$ II. $5x^2$ III. $4x^4$

127. I. $(x + 2)^2$ II. $x^2 + 4$ III. $\sqrt{x^4 + 16}$

128. For what value(s) of the constant a does the equation $\dfrac{(x + a)(x + 2)}{x + 1} = 0$ have only one solution?

129. Verify the identity

$$x^2 + (2a - 1)x + (a^2 - a) = (x + a)(x + a - 1).$$

In Problems 130–133, give the value of a that matches the given expression with the left-hand side of the identity

$$x^2 + (2a - 1)x + (a^2 - a) = (x + a)(x + a - 1),$$

then use the identity to factor the expression.

130. $x^2 + 5x + 6$

131. $x^2 + 15x + 56$

132. $x^2 - 3x + 2$

133. $x^2 + (4b - 1)x + (4b^2 - 2b)$

In Problems 134–135, perform the calculations, state an identity that generalizes them, then prove the identity.

134. (a) $(5 - 1)(5 + 1) + 1$
     (b) $(6 - 1)(6 + 1) + 1$
     (c) $(8 - 1)(8 + 1) + 1$
     (d) $(100 - 1)(100 + 1) + 1$

135. (a) $\left(\frac{5+3}{2}\right)^2 - \left(\frac{5-3}{2}\right)^2$
     (b) $\left(\frac{15+13}{2}\right)^2 - \left(\frac{15-13}{2}\right)^2$
     (c) $\left(\frac{8+4}{2}\right)^2 - \left(\frac{8-4}{2}\right)^2$
     (d) $\left(\frac{7+2}{2}\right)^2 - \left(\frac{7-2}{2}\right)^2$

In Problems 136–140, decide if the statement is true or false. If it is true, use an algebraic identity to explain why. If it is false, provide one case to show that it is false.

136. The product of an integer n with the integer $n + 2$ is 1 less than the square of an integer.

137. The sum of two squares, a^2 and b^2, is another square.

138. Take two numbers a and b. The square of their sum minus the square of their difference is 4 times their product.

139. The sum of 4 consecutive integers, n, $n + 1$, $n + 2$, and $n + 3$, is divisible by 4.

140. Any odd integer, $2n + 1$, is the difference of two squares of integers.

141. Establish the identity $(x+y)^3 = x^3 + 3x^2y + 3xy^2 + y^3$.

142. Establish the following identities:
     (a) $x^3 - y^3 = (x - y)(x^2 + xy + y^2)$
     (b) $x^3 + y^3 = (x + y)(x^2 - xy + y^2)$

143. Show that $(a + b)^3 - 3ab(a + b) = a^3 + b^3$.

144. For what values of a does $(ax + b)^2 - x^2$ simplify to an expression which is linear in x?

145. Explain how you can tell that the expression $(ax + by + c)^2 - (ax - by - c)^2$ must have x as a factor, without actually expanding it out.

In Problems 146–148 carry out the calculations (using a calculator or computer if necessary) and describe the algebraic pattern in the answers.

146. (a) $(1 - x)(1 + x)$
     (b) $(1 - x)(1 + x + x^2)$
     (c) $(1 - x)(1 + x + x^2 + x^3)$
     (d) $(1 - x)(1 + x + x^2 + x^3 + x^4)$

147. (a) $(1 + x)(1 + x^2)(1 + x^4)$
     (b) $(1 + x)(1 + x^2)(1 + x^4)(1 + x^8)$
     (c) $(1 + x)(1 + x^2)(1 + x^4)(1 + x^8)(1 + x^{16})$

148. (a) $(a + b)(a^2 + b^2) - ab(a + b)$
     (b) $(a + b)(a^3 + b^3) - ab(a^2 + b^2)$
     (c) $(a + b)(a^4 + b^4) - ab(a^3 + b^3)$

Chapter Six

Exponential Functions

Contents

6.1 WHAT IS AN EXPONENTIAL FUNCTION?

We often describe the change in a quantity in terms of factors instead of absolute amounts. For instance, if ticket prices rise from $7 to $14, instead of saying "Ticket prices went up by $7" we might say "Ticket prices doubled." Likewise, we might say theater attendance has dropped by a third, or that online ordering of tickets has increased tenfold.

In Chapter 2 we saw that linear functions describe quantities that grow at a constant rate. Now we consider functions that describe quantities that grow by a constant factor.

Example 1   In a petri dish, there are initially 3 million bacteria, and the number doubles every hour. Find a formula for $f(t)$, the number of bacteria (in millions) after t hours.

Solution   At $t = 0$, we have
$$f(0) = 3.$$
One hour later, at $t = 1$, the population has doubled, so we have
$$f(1) = 3 \cdot 2 = 6.$$
After 2 hours, at $t = 2$, the population has doubled again, so we have
$$f(2) = 3 \cdot 2 \cdot 2 = 3 \cdot 2^2 = 12.$$
After each hour, the population grows by another factor of 2, so after 3 hours the population is $3 \cdot 2^3$, after 4 hours the population is $3 \cdot 2^4$, and so on. Thus, after t hours the population is
$$f(t) = 3 \cdot 2^t.$$

Exponents Are Used to Group Repeated Growth Factors

In the case of linear functions, we express repeated addition as a multiplication. For example, in an investment that starts with $500 and grows by $50 per year for three years, the final balance is given by
$$\text{Final balance} = 500 + \underbrace{50 + 50 + 50}_{\text{3 terms of 50}} = 500 + 50 \cdot 3.$$
Now we express repeated multiplication using exponents. For example, after 3 hours, the bacteria population has doubled 3 times:
$$f(3) = 3 \cdot \underbrace{2 \cdot 2 \cdot 2}_{\text{3 factors of 2}} = 3 \cdot 2^3.$$
Notice that to use exponential notation, we must multiply by the same quantity each time.

Example 2   Write an expression that represents the final value of a property that is initially worth $180,000 and

(a) Increases tenfold   (b) Quadruples twice
(c) Increases seven times by a factor of 1.12   (d) Increases by 25% twice

Solution   (a) The final value is $180,000 \cdot 10$ dollars.
(b) Since the initial value quadruples two times, the final value is $180,000 \cdot 4 \cdot 4$ dollars, which can be written $180,000 \cdot 4^2$ dollars.

(c) Since the initial value increases by a factor of 1.12 seven times, the final value is

$$180{,}000 \underbrace{(1.12)(1.12)\cdots(1.12)}_{7 \text{ times}},$$

which can be written $180{,}000(1.12)^7$ dollars.

(d) If something increases by 25%, its value becomes 1.25 times the original. Since this happens twice in a row, the final value is

$$180{,}000 \underbrace{(1.25)(1.25)}_{2 \text{ times}},$$

which can be written $180{,}000(1.25)^2$ dollars.

Exponential Functions Change by a Constant Factor

Functions describing quantities that grow by a constant positive factor, called the *growth factor*, are called *exponential functions*. For instance, the function $Q = 3 \cdot 2^t$ in Example 1 has the form

$$Q = (\text{Initial value}) \cdot (\text{Growth factor})^t,$$

where the initial value is 3 and the growth factor is 2. In general:

Exponential Functions

A quantity Q is an **exponential function** of t if it can be written as

$$Q = f(t) = a \cdot b^t, \quad a \text{ and } b \text{ constants}, b > 0.$$

Here, a is the **initial value** and b is the **growth factor**. We sometimes call b the **base**.

The base b is restricted to positive values because if $b \leq 0$, then b^t is not defined for all exponents t. For example, $(-2)^{1/2}$ and 0^{-1} are not defined.

Example 3   Are the following functions exponential? If so, identify the initial value and growth factor.

(a) $Q = 275 \cdot 3^t$ (b) $Q = 80(1.05)^t$ (c) $Q = 120 \cdot t^3$

Solution

(a) This is exponential with an initial value of 275 and a growth factor of 3.

(b) This is exponential with an initial value of 80 and a growth factor of 1.05.

(c) This is not exponential because the variable t appears in the base, not the exponent. It is a power function.

Example 4   Model each situation with an exponential function.

(a) The number of animals N in a population is initially 880, and the size of the population increases by a factor of 1.12 each year.

(b) The price P of an item is initially $55, and it increases by a factor of 1.031 each year.

Solution

(a) The initial size is 880 and the growth factor is 1.12, so $N = 880(1.12)^t$ after t years.

(b) The initial price is $55 and the growth factor is 1.031, so $P = 55(1.031)^t$ after t years.

Example 5   The following exponential functions describe different quantities. What do the functions tell you about the quantities?

(a) The number of bacteria (in millions) in a sample after t hours is given by $N = 25 \cdot 2^t$.
(b) The balance (in dollars) of a bank account after t years is given by $B = 1200(1.033)^t$.
(c) The value of a commercial property (in millions of dollars) after t years is given by $V = 22.1(1.041)^t$.

Solution   (a) Initially there are 25 million bacteria, and the number doubles every hour.
(b) The initial balance is $1200, and the balance increases by a factor of 1.033 every year.
(c) The initial value is $22.1 million, and the value increases by a factor of 1.041 every year.

In addition to describing growth by a constant factor, exponential functions also describe growth by a constant percent, as we discuss in Section 6.2.

Decreasing by a Constant Factor

So far, we have only considered increasing exponential functions. However, A function that describes a quantity which decreases by a constant factor is also exponential.

Example 6   After being treated with bleach, a population of bacteria begins to decrease. The number, N, of bacteria remaining after t hours is given by

$$N = g(t) = 800 \cdot \left(\frac{3}{4}\right)^t.$$

Find the number of bacteria at $t = 0, 1, 2$ hours.

Solution   We have

$$g(0) = 800 \cdot \left(\frac{3}{4}\right)^0 = 800$$

$$g(1) = 800 \cdot \left(\frac{3}{4}\right)^1 = 600$$

$$g(2) = 800 \cdot \left(\frac{3}{4}\right)^2 = 450.$$

Here the constant 800 tells us the initial size of the bacteria population, and the constant $3/4$ tells us that each hour the population is three-fourths as large as it was an hour earlier.

Graphs of Exponential Functions

Figure 6.1 shows a graph of the bacteria population from Example 1, and Figure 6.2 shows a graph of the population from Example 6. These graphs are typical of exponential functions. Notice in particular that:

- Both graphs cross the vertical axis at the initial value, $f(0) = 3$ in the case of Figure 6.1, and $g(0) = 800$ in the case of Figure 6.2.

- Neither graph crosses the horizontal axis.

- The graph of $N = f(t) = 3 \cdot 2^t$ rises as we move from left to right, indicating that the number of bacteria increases over time. This is because the growth factor, 2, is greater than 1.

- The graph of $N = g(t) = 800(3/4)^t$ falls as we move from left to right, indicating that the number of bacteria decreases over time. This is because the growth factor, $3/4$, is less than 1.

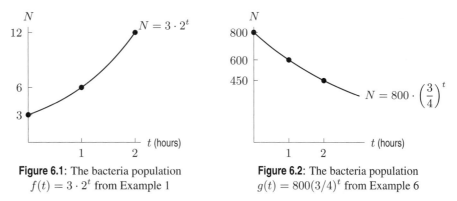

Figure 6.1: The bacteria population
$f(t) = 3 \cdot 2^t$ from Example 1

Figure 6.2: The bacteria population
$g(t) = 800(3/4)^t$ from Example 6

Notice that exponential functions for which $b \neq 1$ are either increasing or decreasing—in other words, their graphs do not change directions. This is because an exponential function represents a quantity that changes by a constant factor. If the factor is greater than 1, the function's value increases and the graph rises (when read from left to right); if the factor is between 0 and 1, the function's value decreases and the graph falls.

The Domain of an Exponential Function

We can use any input value in an exponential function.

Example 7   Let $P = f(t) = 1500(1.025)^t$ give the population of a town t years after 2005. What do the following expressions tell you about the town?

(a) $f(0)$ (b) $f(4)$ (c) $f(-4)$

Solution   (a) We have
$$f(0) = 1500(1.025)^0 = 1500.$$
This tells us that in year 2005, the population was 1500. Here we are thinking of 1500 as the initial size of the population in 2005, the starting year.

(b) We have
$$f(4) = 1500(1.025)^4 = 1656.$$
This tells us that 4 years after 2005, or in year 2009, the population will be 1656.

(c) We have
$$f(-4) = 1500(1.025)^{-4} = 1359.$$
This tells us that 4 years *before* 2005, or in year 2001, the population was 1359.

In Example 7, negative input values are interpreted as times in the past, and positive input values are interpreted as times in the future. As the example illustrates, any number can serve as an input for an exponential function. Thus, the domain for an exponential function is the set of all real numbers.

Recognizing Exponential Functions

Sometimes we need to use the exponent laws to find out whether an expression can be put in a form that defines an exponential function.

Example 8   Are the following functions exponential? If so, identify the values of a and b.

(a) $Q = \dfrac{2^{-t}}{3}$ (b) $Q = \sqrt{5 \cdot 4^t}$ (c) $Q = \sqrt{5t^4}$

Solution   (a) Yes. Writing this as

$$\frac{2^{-t}}{3} = \frac{1}{3}\left(2^{-1}\right)^t = \frac{1}{3}\left(\frac{1}{2}\right)^t,$$

we have $a = 1/3$ and $b = 1/2$.

(b) Yes. Writing this as

$$\sqrt{5 \cdot 4^t} = (5 \cdot 4^t)^{1/2} = 5^{1/2}(4^{1/2})^t = \sqrt{5} \cdot 2^t,$$

we have $a = \sqrt{5}$ and $b = 2$.

(c) No. Writing this as

$$\sqrt{5t^4} = \sqrt{5}t^2,$$

we see that it is a power function with exponent 2.

Exercises and Problems for Section 6.1

Exercises

Are the functions in Exercises 1–4 exponential? If so, identify the initial value and the growth factor

1. $Q = 12t^4$

2. $Q = t \cdot 12^4$

3. $Q = 0.75(0.2)^t$

4. $Q = 0.2(3)^{0.75t}$

Write the exponential functions in Exercises 5–8 in the form $Q = ab^t$ and identify the initial value and growth factor.

5. $Q = 300 \cdot 3^t$

6. $Q = \dfrac{190}{3^t}$

7. $Q = 200 \cdot 3^{2t}$

8. $Q = 50 \cdot 2^{-t}$

Can the quantities in Exercises 9–12 be represented by exponential functions? Explain.

9. The price of gas if it grows by $0.02 a week.

10. The quantity of a prescribed drug in the bloodstream if it shrinks by a factor of 0.915 every 4 hours.

11. The speed of personal computers if it doubles every 3 years.

12. The height of a baseball thrown straight up into the air and then caught.

13. The population of a town is M in 2005 and is growing with a yearly growth factor of Z. Write a formula that gives the population, P, at time t years after 2005.

Write expressions representing the quantities described in Exercises 14–19.

14. The balance B increases n times in a row by a factor of 1.11.

15. The population is initially 800 and grows by a factor of d a total of k times in a row.

16. The population starts at 2000 and increases by a factor of k five times in a row.

17. The investment value V_0 drops by a third n times in a row.

18. The area covered by marsh starts at R acres and decreases by a factor of s nine times in a row.

19. The number infected N increases by a factor of z for t times in a row.

20. An investment initially worth $3000 grows by 6.2% per year. Complete Table 6.1, which shows the value of the investment over time.

Table 6.1

t	1	3	5	10	25	50
V						

21. A commercial property initially worth $200,000 decreases in value by 8.3% per year. Complete Table 6.2, which shows the value of the property (in $1000s) over time.

Table 6.2

t	0	3	5	10	25	50
V, $1000s	200					

22. A lump of Uranium is decaying exponentially with time, t. Which of graphs (I)–(IV) could represent this?

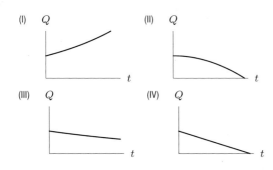

23. Which of graphs (I)–(IV) might be exponential?

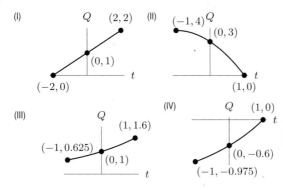

24. Match the exponential functions (a)–(d) with their graphs (I)–(IV).

   (a) $Q = 4(1.2)^t$ **(b)** $Q = 4(0.7)^t$

   (c) $Q = 8(1.2)^t$ **(d)** $Q = 8(1.4)^t$

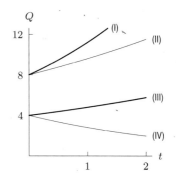

25. With time t, in years, on the horizontal axis and the balance on the vertical axis, plot the balance at $t = 0, 1, 2, 3$ of an account that starts in year 0 with \$5000 and increases by 12% per year.

Sketch graphs of the exponential functions in Exercises 26–27. Label your axes.

26. $y = 100(1.04)^t$ **27.** $y = 800(0.9)^t$

Problems

28. In 2005, Bhutan had a population[1] of about 2,200,000 and an annual growth factor of 1.0211. Let $f(t)$ be the population t years after 2005 assuming growth continues at this rate.

   (a) Evaluate the following expressions and explain what they tell you about Bhutan.

      (i) $f(1)$ (ii) $f(2)$

      (iii) $f(3)$ (iv) $f(5)$

   (b) Write a formula for $f(t)$.

29. In 2005, the Czech Republic had a population of about 10,200,000 and a growth factor of 0.9995 (that is, the population was shrinking).[2] Let $f(t)$ be the population t years after 2005 assuming growth continues at this rate.

   (a) Evaluate the following expressions and explain what they tell you about the Czech Republic.

      (i) $f(1)$ (ii) $f(2)$

      (iii) $f(3)$ (iv) $f(5)$

   (b) Write a formula for $f(t)$.

30. Write an expression that represents the population of a bacteria colony that starts with 10,000 members and

   (a) Halves twice

   (b) Is multiplied 8 times by 0.7

   (c) Is multiplied 4 times by 1.5

31. The population of a city after t years is given by $220,000(1.016)^t$. Identify the initial value and the growth factor and explain what they mean in terms of the city.

[1] www.odci.gov/cia/publications/factbook/geos/bt.html#People, accessed August 9, 2005.

[2] www.odci.gov/cia/publications/factbook/geos/ez.html, accessed August 9, 2005.

32. The number of grams of Palladium-98 remaining after t minutes is given by

$$Q = 600(0.962)^t.$$

Find the quantity remaining after

(a) 1 minute   **(b)** 2 minutes   **(c)** 1 hour

33. An investment grows by 1.2 times in the first year, by 1.3 times in the second year, then drops by a factor of 0.88 in the third year. By what overall factor does the value change over the three-year period?

34. The value V, in dollars, of an investment after t years is given by the function

$$V = g(t) = 3000(1.08)^t.$$

Plot the value of the investment at 5-year intervals over a 30-year period beginning with $t = 0$. By how much does the investment grow during the first ten years? The second ten years? The third ten years?

35. A mill in a town with a population of 400 closes and people begin to move away. Find a possible formula for the number of people P in year t if each year one-fifth of the remaining population leaves.

Write the exponential functions in Problems 36–41 in the form $Q = ab^t$, and identify the initial value and the growth factor.

36. $Q = \dfrac{50}{2^{t/12}}$

37. $Q = 250 \cdot 5^{-2t-1}$

38. $Q = 2^t\sqrt{300}$

39. $Q = 40 \cdot 2^{2t}$

40. $Q = 45 \cdot 3^{t-2}$

41. $Q = \dfrac{\sqrt{100^t}}{5}$

42. Do the following expressions define exponential functions? If so, identify the values of a and b.

(a) $3 \cdot 0.5^x$ **(b)** 8^t **(c)** 8^{1-2x}

6.2   INTERPRETING THE BASE

Growth Factors and Percentage Growth Rates

It is sometimes convenient to describe growth in terms of a percentage rate. The next example compares growth at a constant percent rate with linear growth.

Example 1   A person invests $500 in year $t = 0$. Find the annual balance over a five-year period assuming:
(a)   It grows by $50 each year.   (b)   It grows by 10% each year.

Solution   (a) See Table 6.3, which shows that the balance is a linear function of the year.

Table 6.3   *An investment earning $50 per year*

Year	0	1	2	3	4	5
Balance	500.00	550.00	600.00	650.00	700.00	750.00

(b) See Table 6.4. Notice that the balance goes up by a larger amount each year, because the 10% is taken of an ever-larger balance.

$$\text{Balance in year 1 is } \$500 + 10\%(500) = \$550.00$$
$$\text{Balance in year 2 is } \$550 + 10\%(550) = \$605.00$$
$$\text{Balance in year 3 is } \$605 + 10\%(605) = \$665.50,$$

and so on. So the balance is *not* a linear function of the year since it does not grow by the same dollar amount each year.

Table 6.4   *An investment growing by 10% each year*

Year	0	1	2	3	4	5
Balance	500.00	550.00	605.00	665.50	732.05	805.26

Exponential Functions Describe Growth at a Constant Percent Rate

In Example 1, we saw that growth at a constant percent rate is not the same as linear growth. In Example 2, we see that growth at a constant percent rate *is* the same as growth by a constant factor: In other words, a quantity that grows at a constant percent rate defines an exponential function.

Example 2

Example 1 examines the balance of a bank account that begins with $500 and that earns 10% interest each year. Find the balance after:

(a)   10 years   (b) t years.

Solution

We would like to avoid the tedious process of calculating the balance every year. Instead, we use what we know about growth factors and exponential growth.

(a) Let's see how to use growth factors to rework our original calculations. After 1 year, the balance in dollars is given by

$$\text{Balance} = \text{Original deposit} + \text{Interest}$$
$$= 500 + 10\% \cdot 500$$
$$= 500 + 0.10 \cdot 500.$$

At this step, instead of evaluating the right-hand side directly, we rewrite it using a growth factor:

$$\text{Balance} = 500(1 + 0.10) \quad \text{factor out 500}$$
$$= 500 \cdot 1.10 \quad\quad \text{growth factor is } 1.10.$$

Note that $500(1.10) = 550$, which is the same answer we get by evaluating $500 + 0.10(500)$ directly. Using the same reasoning, after two years the balance is

$$\text{Balance} = \text{Balance at end of first year} + \text{Interest}$$
$$= 550 + 0.10 \cdot 550$$
$$= 550(1 + 0.10) \quad\quad \text{factor out 550}$$
$$= 550 \cdot 1.10 \quad\quad\quad \text{growth factor is } 1.10$$
$$= 605 \text{ dollars.}$$

Since $550 = 500(1.10)$, notice that we have multiplied by 1.10 two times in a row:

$$\text{Balance at end of 2 years} = \underbrace{(500 \cdot 1.10)}_{550} 1.10 = 500 \cdot 1.10^2 = 605 \text{ dollars.}$$

Likewise, after three years the balance is

$$\text{Balance} = \text{Balance at end of second year} + \text{Interest}$$
$$= 605 + 0.10 \cdot 605$$
$$= 605(1 + 0.10) \quad\quad \text{factor out 605}$$
$$= 605 \cdot 1.10 \quad\quad\quad \text{growth factor is } 1.10$$
$$= 665.50 \text{ dollars.}$$

Since we have again multiplied by 1.10, we have

$$\text{Balance at the end of 3 years} = \underbrace{(500 \cdot 1.10^2)}_{605} 1.10 = 500(1.10)^3 = 665.50 \text{ dollars.}$$

Extending the pattern, we see

$$\text{Balance at end of 4 years} = 500(1.10)^4 = 732.05 \text{ dollars}$$
$$\text{Balance at end of 5 years} = 500(1.10)^5 = 805.26 \text{ dollars}$$

$$\vdots$$

$$\text{Balance at end of 10 years} = 500(1.10)^{10} = 1296.87 \text{ dollars.}$$

(b) Extending the pattern from part (a), we see

$$\text{Balance at end of } t \text{ years} = 500(1.10)^t \text{ dollars.}$$

Converting Between Growth Rates and Growth Factors

Example 2 shows that if a balance grows by 10% a year, the growth factor is

$$b = 1 + 0.10 = 1.10.$$

In general,

In the exponential function $f(t) = ab^t$, the growth factor b is given by

$$b = 1 + r,$$

where r is the percent rate of change, called the **growth rate**. For example, if $r = 9\%$, $b = 1 + r = 1 + 9\% = 1 + 0.09 = 1.09$.

Example 3   State the growth rate r and the growth factor b for the following quantities.

(a) The size of a population grows by 0.22% each year.
(b) The value of an investment grows by 11.3% each year.
(c) The area of a lake decreases by 7.4% each year.

Solution   (a) We have $r = 0.22\% = 0.0022$ and $b = 1 + r = 1.0022$.
(b) We have $r = 11.3\% = 0.113$ and $b = 1 + r = 1.113$.
(c) We have $r = -7.4\% = -0.074$ and $b = 1 + r = 1 + (-0.074) = 0.926$.

When the growth rate is negative, we can also express it as a *decay rate*. For example, the lake in part (c) of Example 3 has a decay rate of 7.4%.

Example 4   What is the growth or decay rate r of the following exponential functions?

(a) $f(t) = 200(1.041)^t$ (b) $g(t) = 50(0.992)^t$ (c) $h(t) = 75\left(\sqrt{5}\right)^t$

Solution   (a) Here, $b = 1.041$, so the growth rate is

$$1 + r = 1.041$$
$$r = 0.041 = 4.1\%.$$

(b) Here, $b = 0.992$, so the growth rate is

$$1 + r = 0.992$$
$$r = -0.008 = -0.8\%.$$

We could also say that the decay rate is 0.8% per year.

(c) Here, $b = \sqrt{5} = 2.2361$, so the growth rate is

$$1 + r = 2.2361$$
$$r = 1.2361 = 123.61\%.$$

It is easy to confuse growth rates and growth factors. Keep in mind that:

$$\text{Change in value} = \text{Starting value} \times \text{Growth rate}$$
$$\text{Ending value} = \text{Starting value} \times \text{Growth factor}.$$

Example 5   The starting value of a bank account is $1000. What does the annual growth rate of 4% tell us? What does the growth factor of 1.04 tell us?

Solution   The growth rate of 4% tells us that over the course of the year,

$$\text{Change in value} = \text{Starting value} \times \text{Growth rate}$$
$$= 1000(4\%) = \$40.$$

The growth factor of 1.04 tells us that after 1 year,

$$\text{Ending value} = \text{Starting value} \times \text{Growth factor}$$
$$= 1000(1.04) = \$1040.$$

Of course, both statements have the same consequences for the bank balance.

Example 6   Find the growth rate r and annual growth factor b for a quantity that
(a)   Doubles each year   (b)   Grows by 200% each year (c)   Halves each year

Solution   (a) The quantity doubles each year, so $b = 2$. We have

$$b = 1 + r$$
$$r = 2 - 1 = 1 = 100\%.$$

In other words, doubling each year is the same as growing by 100% each year.
(b) We have $r = 200\% = 2$, so

$$b = 1 + 200\%$$
$$= 1 + 2 = 3.$$

In other words, growing by 200% each year is the same as tripling each year.
(c) The quantity halves each year, so $b = 1/2$. We have

$$b = 1 + r$$
$$r = 1 - \frac{1}{2} = \frac{1}{2} = 50\%.$$

In other words, halving each year is the same as decreasing by 50% each year.

Recognizing Exponential Growth from a Table of Data

We can sometimes use the fact that exponential functions describe growth by a constant factor to determine if a table of data describes exponential growth.

Example 7   Could the table give points on the graph of an exponential function $y = ab^x$? If so, find a and b.

Table 6.5

x	0	1	2	3
y	4	4.4	4.84	5.324

Solution   Yes. Taking ratios, we see that

$$\frac{4.4}{4} = 1.1, \quad \frac{4.84}{4.4} = 1.1, \quad \frac{5.324}{4.84} = 1.1.$$

Thus, the value of y increases by a factor of 1.1 each time x increases by 1. This means the growth factor is $b = 1.1$. From the table, we see that the starting value is $a = 4$, so $y = 4(1.1)^x$.

Exercises and Problems for Section 6.2

Exercises

Do the exponential expressions in Exercises 1–6 represent growth or decay?

1. $203(1.03)^t$

2. $7.04(1.372)^t$

3. $42.7(0.92)^t$

4. $0.98(1.003)^t$

5. $109(0.81)^t$

6. $0.22(0.04)^t$

In Exercises 7–10 give the growth factor that corresponds to the given growth rate.

7. 8.5% growth

8. 46% decay

9. 215% growth

10. 99.99% decay

In Exercises 11–14 give the growth rate that corresponds to the given growth factor.

11. 1.7   **12.** 0.27   **13.** 5   **14.** 0.639

State the starting value a, the growth factor b, and the percentage growth rate r correct to 2 decimals for the exponential functions in Exercises 15–19.

15. $Q = 200(1.031)^t$

16. $Q = 700(0.988)^t$

17. $Q = \sqrt{3}\left(\sqrt{2}\right)^t$

18. $Q = 50\left(\frac{3}{4}\right)^t$

19. $Q = 5 \cdot 2^t$

The functions in Exercises 20–23 describe the value of different investments in year t. What do the functions tell you about the investments?

20. $V = 800(1.073)^t$

21. $V = 2200(1.211)^t$

22. $V = 4000 + 100t$

23. $V = 8800(0.954)^t$

Write an expression representing the quantities in Exercises 24–27.

24. A population at time t years if it is initially 2 million and growing at 3% per year.

25. The value of an investment which starts at $5 million and grows at 30% per year for t years.

26. The quantity of pollutant remaining in a lake if it is removed at 2% a year for 5 years.

27. The cost of doing a project, initially priced at $2 million, if the cost increases exponentially for 10 years.

In words, give a possible interpretation in terms of percentage growth of the expressions in Exercises 28–31.

28. $10,000(1.06)^t$

29. $(47 \text{ grams})(0.97)^x$

30. $(400 \text{ people})(1.006)^y$

31. $100,000(0.98)^a$

Problems

32. After t years, an initial population P_0 has grown to $P_0(1+r)^t$. If the population doubles during the first year, which of the following are possible values of r?

   (a) $r = 2\%$ **(b)** $r = 50\%$

   (c) $r = 100\%$ **(d)** $r = 200\%$

State the starting value a, the growth factor b, and the percentage growth rate r correct to 2 decimals for the exponential functions in Problems 33–37.

33. $Q = 90 \cdot 10^{-t}$ **34.** $Q = \dfrac{10^t}{40}$

35. $Q = \dfrac{210}{3 \cdot 2^t}$ **36.** $Q = 50\left(\dfrac{1}{2}\right)^{t/25}$

37. $Q = 2000\left(1 + \dfrac{0.06}{12}\right)^{12t}$

38. In Example 6 on page 236, describe the hourly change in population as a percentage.

39. With t in years since 2000, the population[3] of Plano, Texas, in thousands is given by $222(1.056)^t$.

   (a) What was the population in 2000? What is its growth rate?

   (b) What population is expected in 2010?

40. A quinine tablet is taken to prevent malaria; t days later, $50(0.23)^t$ mg remain in the body.

   (a) How much quinine is in the tablet? At what rate is it decaying?

   (b) How much quinine remains 12 hours after a tablet is taken?

41. In 2005, the population of Burkina Faso was about 13,900,000 and growing at 2.53% per year.[4] Estimate the population in

   (a) 2006   **(b)** 2007

   (c) 2008   **(d)** 2010

42. The mass of tritium in a 2000 mg sample decreases by 5.47% per year. Find the amount remaining after

   (a) 1 year   **(b)** 2 years

The population of Austin, Texas, is increasing by 2% a year, that of Bismark, North Dakota, is shrinking by 1% a year, and that of Phoenix, Arizona, is increasing by 3% a year. Interpret the quantities in Problems 43–45 in terms of one of these cities.

43. $(1.02)^3$ **44.** $(0.99)^1$ **45.** $(1.03)^2 - 1$

In Problems 46–48, could the table give points on the graph of a function $y = ab^x$, for constants a and b? If so, find the function.

46.

x	0	1	2	3
y	1	7	49	343

47.

x	0	2	4	6
y	0	1	2	4

48.

x	4	9	14	24
y	5	4.5	4.05	3.2805

49. The percent change (increase or decrease) in the value of an investment each year over a five-year period is shown in Table 6.6. By what percent does the investment's value change over this five-year period?

Table 6.6

t	1	2	3	4	5
r	12.9%	9.2%	11.3%	−4.5%	−13.6%

6.3 INTERPRETING THE EXPONENT

In the previous section we considered different ways of expressing the base of an exponential expression. In this section we consider different forms for the exponent.

Offsetting the Exponent

Sometimes it is useful to express an exponential function in a form that has a constant subtracted from the exponent. We call this offsetting the exponent.

[3]E. Glaeser, J. Shapiro, "City Growth and the 2000 Census: Which Places Grew, and Why," at www.brookings.edu, accessed on October 24, 2004.

[4]www.odci.gov/cia/publications/factbook/geos/uv.html#People, accessed August 9, 2005.

Example 1   The population of a town was 1500 people in 2005 and increasing at rate of 2.5% per year. Express the population in terms of (a)   the number of years, t, since 2005   (b)   the year, y.
Use each expression to calculate the population in the year 2014.

Solution   (a) Here the starting value, when $t = 0$, is 1500, so

$$P = f(t) = 1500 \cdot 1.025^t$$

Since 2014 is 9 years since 2005, we have $t = 9$. Substituting this into our equation, we get:

$$P = f(9) = 1500 \cdot 1.025^9$$
$$= 1873.294.$$

Thus, in the year 2014, we would have approximately 1873 people.
(b) If y is the year, then the relationship between y and t is given by

$$y = 2005 + t.$$

For instance, at $t = 9$ we have $y = 2005 + 9 = 2014$. Solving for t,

$$t = y - 2005,$$

we can rewrite our equation for P by substituting:

$$P = g(y) = 1500 \cdot 1.025^{y-2005} \quad \text{because } t = y - 2005.$$

So, in 2014, we would have

$$P = g(2014) = 1500 \cdot 1.025^{2014-2005}$$
$$= 1500 \cdot 1.025^9$$
$$= 1873.294,$$

which is the same answer that we got before.

In Example 1, we offset the exponent by 2005. We can convert the resulting formula to standard form by using exponent rules:

$$P = 1500 \cdot 1.025^{y-2005}$$
$$= 1500 \cdot 1.025^y \cdot 1.025^{-2005}$$
$$= \frac{1500}{1.025^{2005}} \cdot 1.205^y.$$

However, in this form of the expression, the starting value, $a = 1500/1.025^{2005}$, does not have a reasonable interpretation as the size of the population in the year 0, because it is a very small number, and the town probably did not exist in the year 0. It is more convenient to use the form $P = 1500 \cdot 1.025^{y-2005}$, because we can see at a glance that the population is 1500 in the year 2005.
In general, an offset of t_0 years gives us the formula

$$P = P_0 b^{t-t_0}.$$

Here, P_0 is the value of P in year t_0. This form is like the point-slope form $y = y_0 + m(x - x_0)$ for linear functions, where y_0 is the value of y at $x = x_0$.

Example 2   Let $B = 5000(1.071)^{t-1998}$ be the value of an investment in year t. Here, we have $B_0 = 5000$ and $t_0 = 1998$, which tells us that the investment's value in 1998 was $5000.

Doubling Time and Half-Life

Another useful form of an exponential expression is one in which the base is an easily interpreted growth factor, like 2, and the exponent is scaled to express the time period over which that growth factor applies.

Example 3   The value V, in dollars, of an investment t years after 2004 is given by

$$V = 4000 \cdot 2^{t/12}.$$

(a) How much is it worth in 2016? 2028? 2040?
(b) Use the form of the expression for V to explain why the value doubles every 12 years.
(c) Use exponent laws to put the expression for V in a form that shows it is an exponential function.

Solution   (a) In 2016, twelve years have passed, and $t = 12$. In 2028, twelve more years have passed, so $t = 24$, and in 2040 we have $t = 36$.

$$\text{At } t = 12: \quad V = 4000 \cdot 2^{12/12} = 4000 \cdot 2^1 = 8000$$
$$\text{At } t = 24: \quad V = 4000 \cdot 2^{24/12} = 4000 \cdot 2^2 = 16{,}000$$
$$\text{At } t = 36: \quad V = 4000 \cdot 2^{36/12} = 4000 \cdot 2^3 = 32{,}000,$$

Notice that the value seems to double every 12 years.

(b) The 12 in the denominator of the exponent means that every time we add 12 to t, we add 1 to the exponent:

$$4000 \cdot 2^{(t+12)/12} = 4000 \cdot 2^{(t/12)+1}.$$

Adding 1 to the exponent means multiplying by the base, 2:

$$4000 \cdot 2^{(t/12)+1} = 4000 \cdot 2^{t/12} \cdot 2^1 = 2 \cdot 4000 \cdot 2^{t/12}.$$

Thus when 12 is added to t, the value is multiplied by 2. So the value doubles every 12 years.

(c) Using exponent laws, we write

$$V = 4000 \cdot 2^{t/12} = 4000 \cdot (2^{1/12})^t = 4000 \cdot 1.0595^t.$$

Thus, the value is growing exponentially, at a rate of 5.95% per year.

The *doubling time* of an exponentially growing quantity is the amount of time it takes for the quantity to double. In the previous example, if you start with an investment worth $4000, then after 12 years you have $8000, and after another 12 years you have $16,000, and so on. The doubling time is 12 years.

For exponentially decaying quantities, we consider the half-life, the time it takes for the quantity to be halved.

Example 4   Let

$$P(t) = 1000 \cdot \left(\frac{1}{2}\right)^{t/7}$$

be the population of a town t years after it was founded.

(a) What is the initial population? What is the population after 7 years? After 14 years? When is the population 125?
(b) Rewrite $P(t)$ in the form $P(t) = a \cdot b^t$ for constants a and b, and give the annual growth rate of the population.

Solution   (a) Since $P(0) = 1000 \cdot (1/2)^{0/7} = 1000$, the initial population is 1000. After 7 years the population is

$$P(7) = 1000 \cdot \left(\frac{1}{2}\right)^{7/7} = 1000 \cdot \left(\frac{1}{2}\right)^{1} = 1000 \cdot \frac{1}{2} = 500,$$

and after 14 years it is

$$P(14) = 1000 \cdot \left(\frac{1}{2}\right)^{14/7} = 1000 \cdot \left(\frac{1}{2}\right)^{2} = 1000 \cdot \frac{1}{4} = 250.$$

Notice that the population halves in the first 7 years and then halves again in the next 7 years. To reach 125 it must halve yet again, which happens after the next 7 years, when $t = 21$:

$$P(21) = 1000 \cdot \left(\frac{1}{2}\right)^{21/7} = 1000 \cdot \left(\frac{1}{2}\right)^{3} = 1000 \cdot \frac{1}{8} = 125.$$

(b) We have

$$P(t) = 1000 \cdot \left(\frac{1}{2}\right)^{t/7} = 1000 \left(\left(\frac{1}{2}\right)^{1/7}\right)^{t} = 1000(0.906)^{t}.$$

So $a = 1000$ and $b = 0.906$. Since $b = 1 + r$, the growth rate is $r = -0.094 = -9.4\%$. Alternatively, we can say the population is decaying at a rate of 9.4% a year.

So far, we have considered quantities that double or halve in a fixed time period (12 years, 7 years, etc.). Example 5 uses the base 3 to represent a quantity that *triples* in a fixed time period.

Example 5   The number of people using the latest picture phone after t months is given by

$$N(t) = 450 \cdot 3^{t/6}.$$

How many people have the phone in 6 months? In 20 months?

Solution   In 6 months

$$N(6) = 450 \cdot 3^{6/6}$$
$$= 450 \cdot 3$$
$$= 1350 \text{ people.}$$

In 20 months, there are

$$N(20) = 450 \cdot 3^{20/6}$$
$$= 17{,}523.332,$$

or approximately 17,523 people.

Comparing the functions
$$V = 4000 \cdot 2^{t/12},$$
which describes a quantity that doubles every 12 years, and
$$N = 450 \cdot 3^{t/6},$$
which describes a quantity that triples every 6 months, we can generalize as follows: The function
$$Q = ab^{t/T}$$
describes a quantity that increases by a factor of b every T years.

Example 6   Describe what the following functions tell you about the value of the investments they describe.

(a) $V = 2000 \cdot 3^{t/15}$ (b) $V = 3000 \cdot 4^{t/40}$ (c) $V = 2500 \cdot 10^{t/100}$

Solution   (a) This investment begins with $2000 and triples (goes up by a factor of 3) every 15 years.
(b) This investment begins with $3000 and quadruples (goes up by a factor of 4) every 40 years.
(c) This investment begins with $2500 goes up by a factor of 10 every 100 years.

Negative Exponents

We have seen that exponential decay is represented by ab^t with $0 < b < 1$. An alternative way to represent it is by using a minus sign in the exponent.

Example 7   Let $Q(t) = 4^{-t}$. Show that Q represents an exponentially decaying quantity that decreases to $1/4$ of its previous amount for every unit increase in t.

Solution   We rewrite $Q(t)$ as follows:

$$Q(t) = 4^{-t} = (4^{-1})^t = \left(\frac{1}{4}\right)^t.$$

This represents exponential decay with a growth factor of $1/4$, so the quantity gets multiplied by $1/4$ every unit of time.

Example 8   Express the population in Example 4 using a negative exponent.

Solution   Since $1/2 = 2^{-1}$, we can write

$$P(t) = 1000 \cdot \left(\frac{1}{2}\right)^{t/7} = 1000 \cdot (2^{-1})^{t/7} = 1000 \cdot 2^{-t/7}.$$

Converting Between Different Time Scales

Sometimes, we know by how much a quantity changes over one period of time, and we want to know the change over another period. For instance, in Examples 9 and 10, we see how to convert a monthly growth rate to an annual growth rate.

Example 9   A bank account starts at $500 and grows by 1% each month, so its value after t months is given by $B(t) = 500 \cdot 1.01^t$. Explain the meaning of

(a) $500 \cdot 1.01^{12}$ (b) $500 \cdot 1.01^{24}$ (c) $500 \cdot 1.01^{12n}$

Solution   (a) The expression $500 \cdot 1.01^{12}$ represents the value after $t = 12$ months or one year.
(b) The expression $500 \cdot 1.01^{24}$ represents the value after $t = 24$ months or two years.
(c) In general, if n is the number of years, then

$$\underbrace{\text{Number of months}}_{t} = 12 \times \underbrace{\text{Number of years}}_{n}$$

$$t = 12n.$$

Thus the expression $500 \cdot 1.01^{12n}$ represents the value after $t = 12n$ months, or n years.

Example 10   For the bank account in Example 9, write a formula for the balance $C(n)$ after n years that shows the annual growth rate of the account. Which is better, an account earning 1% per month or one earning 12% per year?

Solution   Using the exponent rules, we get

$$C(n) = 500 \cdot 1.01^{12n} = 500\left(1.01^{12}\right)^n$$
$$= 500 \cdot 1.127^n,$$

because $1.01^{12} = 1.127$. This represents the account balance after n years, so the annual growth factor is 1.127. Since $1.127 = 1 + 0.127 = 1 + r$, the annual growth rate r is 0.127, or 12.7%. This is more than 12%, so the account earning 1% per month is better than one earning 12% per year.

Example 11   The population, in thousands, of a town after t years is given by

$$P = 50 \cdot 1.035^t.$$

(a) What is the annual growth rate?
(b) Write an expression that gives the population after m months.
(c) Use the expression in part (b) to find the town's monthly growth rate.

Solution   (a) From the given expression, we see that the growth factor is 1.035, so each year the population is 3.5% larger than the year before, and it is growing at 3.5% per year.
(b) We know that

$$\underbrace{\text{Number of months}}_{m} = 12 \times \underbrace{\text{Number of years}}_{t}$$

$$m = 12t,$$

so $t = m/12$. This means

$$50 \cdot 1.035^t = 50 \cdot 1.035^{m/12},$$

so the population after m months is given by

$$50 \cdot 1.035^{m/12}.$$

(c) Using the exponent laws, we rewrite this as

$$50 \cdot \left(1.035^{1/12}\right)^m = 50 \cdot 1.0029^m,$$

because $1.035^{1/12} = 1.0029$. So the monthly growth rate 0.0029, or 0.29%.

Exercises and Problems for Section 6.3

Exercises

Find values for the constants a, b, and T so that the quantities described in Exercises 1–4 are represented by the function

$$Q = ab^{t/T}.$$

1. A population begins with 1000 members and doubles every twelve years.

2. An investment is initially worth 400 and triples in value every four years.

3. A community that began with only fifty households having cable television service has seen the number increase fivefold every fifteen years.

4. The number of people in a village of 6000 susceptible to a certain strain of virus has gone down by one-third every two months since the virus first appeared.

Find values for the constants $a, b,$ and T so that the quantities described in Exercises 5–8 are represented by the function

$$Q = ab^{-t/T}.$$

5. A lake begins with 250 fish of a certain species, and one-half disappear every six years.

6. A commercial property initially worth $120,000 loses 25% of its value every two years.

7. The amount of a 200 mg injection of a therapeutic drug remaining in a patient's blood stream goes down by one-fifth every 90 minutes.

8. The number of people who haven't heard about a new movie is initially N and decreases by 10% every five days.

Find values for the constants $a, b,$ and t_0 so that the quantities described in Exercises 9–12 are represented by the function

$$Q = ab^{t-t_0}.$$

9. The size of a population in year $t = 1995$ is 800 and it grows by 2.1% per year.

10. After $t = 3$ hours there are 5 million bacteria in a culture and the number doubles every hour.

11. A house whose value increases by 8.2% per year is worth $220,000 in year $t = 7$.

12. A 200 mg sample of radioactive material is isolated at time $t = 8$ hours, and it decays at a rate of 17.5% per hour.

13. Find a formula for the value of an investment initially worth $12,000 that grows by 12% every 5 years.

14. Find the starting value a, the growth factor b, and the growth rate r for the exponential function $Q = 500 \cdot 2^{t/7}$.

15. Which is better, an account earning 2% per month or one earning 7% every 3 months?

16. A sunflower grows at a rate of 1% a day; another grows at a rate of 7% per week. Which is growing faster? Assuming they start at the same height, compare their heights at the end of 1 and 5 weeks.

Write the functions in Exercises 17–20 in the form $Q = ab^t$. Give the values of the constants a and b.

17. $Q = \dfrac{1}{3} \cdot 2^{t/3}$

18. $Q = -\dfrac{5}{3^t}$

19. $Q = 7 \cdot 2^t \cdot 4^t$

20. $Q = 4(2 \cdot 3^t)^3$

What do the functions in Exercises 21–22 tell you about the quantities they describe?

21. The size P of a population of animals in year t is $P = 1200(0.985)^{12t}$.

22. The value V of an investment in year t is $V = 3500 \cdot 2^{t/7}$.

Problems

23. The average rainfall in Hong Kong in January and February is about 1 inch each month. From March to June, however, average rainfall in each month is double the average rainfall of the previous month.

    (a) Make a table showing average rainfall for each month from January to June.
    (b) Write a formula for the average rainfall in month n, where $2 \le n \le 6$ and January is month 1.
    (c) What is the total average rainfall in the first six months of the year?

24. Formulas I–III all describe the growth of the same population, with time, t, in years: I. $P = 15(2)^{t/6}$
    II. $P = 15(4)^{t/12}$ III. $P = 15(16)^{t/24}$

    (a) Show that the three formulas are equivalent.
    (b) What does formula I tell you about the doubling time of the population?
    (c) What do formulas II and III tell you about the growth of the population? Give answers similar to the statement which is the answer to part (b).

For the investments described in Problems 25–30, assume that t is the elapsed number of years and that T is the elapsed number of months.

    (a) Describe in words how the value of the investment changes over time.
    (b) Given the annual growth rate.

25. $V = 3000(1.0088)^{12t}$

26. $V = 6000(1.021)^{4t}$

27. $V = 250(1.0011)^{365t}$

28. $V = 400(1.007)^T$

29. $V = 625(1.03)^{T/3}$

30. $V = 500(1.2)^{t/2}$

Prices are increasing at 5% per year. What is wrong with the statements in Problems 31–39? Correct the formula in the statement.

31. A $6 item costs $(6 \cdot 1.05)^7$ in 7 years time.

32. A $3 item costs $3(0.05)^{10}$ in ten years time.

33. The percent increase in prices over a 25-year period is $(1.05)^{25}$.

34. If time t is measured in months, then the price of a $100 item at the end of one year is $100(1.05)^{12t}$.

35. If the rate at which prices increase is doubled, then the price of a \$20 object in 7 years time is $\$20(2.10)^7$.

36. If time t is measured in decades (10 years), then the price of a \$45 item in t decades is $\$45(1.05)^{0.1t}$.

37. Prices change by $10 \cdot 5\% = 50\%$ over a decade.

38. Prices change by $(5/12)\%$ in one month.

39. A \$250 million town budget is trimmed by 1% but then increases with inflation as prices go up. Ten years later, the budget is $\$250(1.04)^{10}$ million.

40. Arrange the following expressions in order of increasing half-lives (from smallest to largest):

$$20(0.92)^t; \quad 120(0.98)^t; \quad 0.27(0.9)^t; \quad 90(0.09)^t.$$

41. Find the half-lives of the following quantities:

  (a) A decays by 50% in 1 week.
  (b) B decays to one quarter of the original in 10 days.
  (c) C decays by 7/8 in 6 weeks.

42. The population of bacteria m doubles every 24 hours. The population of bacteria q grows by 3% per hour. Which has the larger population in the long term?

43. The half-life of substance ϵ is 17 years, and substance s decays at a rate of 30% per decade. Of which substance is there less in the long term?

Exponential functions written in the form

$$Q = ab^{t/\tau}$$

can be used to describe growth in intervals of τ years. (τ is the Greek letter *tau*.) For the population functions in Problems 44–49:

(a) Give the values of the parameters a, b, and τ, and say what these parameters tell you about population growth.

(b) Give the annual growth rate.

44. $P = 400 \cdot 2^{t/4}$

45. $P = 800 \cdot 2^{t/15}$

46. $P = 80 \cdot 3^{t/5}$

47. $P = 75 \cdot 10^{t/30}$

48. $P = 50 \left(\frac{1}{2}\right)^{t/6}$

49. $P = 400 \left(\frac{2}{3}\right)^{t/14}$

State the starting value a, the growth factor b, and the growth rate r as a percent correct to 2 decimals for the exponential functions in Problems 50–53.

50. $V = 2000(1.0058)^t$

51. $V = 500 \left(3 \cdot 2^{t/5}\right)$

52. $V = 5000(0.95)^{t/4}$

53. $V = 3000 \cdot 2^{t/9} + 5000 \cdot 2^{t/9}$

54. The value of an investment grows by a factor of 1.011 each month. By what percent does it grow each year?

55. The value of an investment grows by 0.06% every day. By what percent does it increase in a year?

56. A property decreases in value by 0.5% each week. By what percent does it decrease after one year (52 weeks)?

57. The area of a wetland drops by a third every five years. What percent of its total area disappears after twenty years?

6.4 EXPONENTIAL EQUATIONS

Often we want to know when an exponential function attains a specific value. This question leads to an exponential equation.

Example 1   A biochemist needs a sample of 80 million bacteria. If she begins with 5 million bacteria and the population doubles every day, when is the sample ready?

Solution   The number of bacteria, P, in millions, on day t is given by

$$P = 5 \cdot 2^t.$$

We want to know when there are 80 million bacteria, so we solve for t in the equation

$$80 = 5 \cdot 2^t.$$

Dividing by 5 we get

$$\frac{80}{5} = \frac{5 \cdot 2^t}{5}$$
$$16 = 2^t.$$

Since $16 = 2^4$, we see that $t = 4$ is a solution to the equation, so she has 80 million bacteria after 4 days.

Example 2   A population is given by $P = 149 \cdot (2^{1/9})^t$, where t is in years. When does the population reach 596?

Solution   We solve the equation
$$149 \cdot (2^{1/9})^t = 596.$$

Dividing by 149 we get
$$(2^{1/9})^t = 4$$
$$2^{t/9} = 2^2.$$

Since the two exponents must be equal, we can see that
$$\frac{t}{9} = 2$$
$$t = 18 \text{ years.}$$

Using Tables and Graphs to Approximate Solutions to Equations

In Examples 1 and 2, we solved exponential equations by inspection. However, suppose the biochemist from Example 1 wants a sample of 70 million bacteria. To find when the sample is ready, we must solve the equation
$$5 \cdot 2^t = 70.$$

This equation cannot be solved so easily. Since there are 40 million bacteria after 3 days and 80 million bacteria after 4 days, we know that the solution must be between 3 and 4 days, but can we find a more precise value?

In Chapter 7 we use logarithms to solve exponential equations algebraically. For now, we consider numerical and graphical methods. We try values of t between 3 and 4. Table 6.7 shows that $t \approx 3.81$ days is approximately the solution to $5 \cdot 2^t = 70$. Alternatively, we can trace along a graph of $y = 5 \cdot 2^t$ until we find the point with y-coordinate 70, as shown in Figure 6.3.

Table 6.7   *Million bacteria after t days*

t	3	3.5	3.6	3.7	3.8	3.81	3.9	4
$y = 5 \cdot 2^t$	40	56.6	60.6	65.0	69.6	70.1	74.6	80

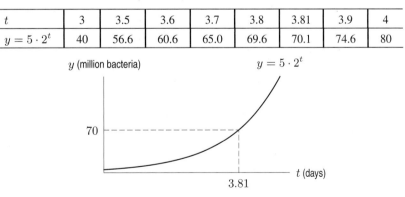

Figure 6.3: Graph of bacteria after t days

Example 3   A bank account begins at \$22,000 and earns 3% interest per year. When does the balance reach \$62,000?

Solution   After t years the balance is given by

$$B = 22{,}000(1.03)^t. \tag{6.1}$$

Figure 6.4 shows that when $B = 62{,}000$ we have $t \approx 35$. So it takes about 35 years for the balance to reach \$62,000.

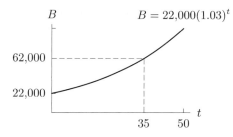

Figure 6.4: Graph of balance after t years

Describing Solutions to Exponential Equations

We can sometimes describe the solutions to exponential equations without actually solving them. Notice that for an exponential function with a positive starting value a and a growth factor b,

$$f(t) = ab^t, \quad a > 0,$$

the function increases for $b > 1$ and decreases for $0 < b < 1$. (See Figures 6.1 and 6.2 on page 237.) This means that:

- For $b > 1$, the value of $f(t)$ is greater than its starting value for positive input values, $t > 0$, and is less than its starting value for negative input values, $t < 0$.

- For $0 < b < 1$, the value of $f(t)$ is less than its starting value for positive input values, $t > 0$, and is greater than its starting value for negative input values, $t < 0$.

Example 4   Is the solution to each of the following equations a positive or negative number?

(a) $2^x = 12$ (b) $2^x = 0.3$ (c) $3(0.5)^x = 0.3$ (d) $3(0.5)^x = 12$

Solution   (a) Letting $f(x) = ab^x = 2^x$, we see that the starting value is $a = 1$. An output of 12 is greater than the starting value, so, since $b > 1$, this happens for a positive input value. Thus, the solution must be a positive number.

(b) Again, the starting value is $a = 1$. An output of 0.3 is less than the starting value, so, since $b > 1$, this happens for a negative input value. Thus, the solution must be a negative number.

(c) Letting $g(x) = ab^x = 3(0.5)^x$, we see that the starting value is $a = 3$. An output value of 0.3 is less than the starting value, so, since $b < 1$, this happens for a positive input value. Thus, the solution must be a positive number.

(d) Again, the starting value is $a = 3$. An output value of 12 is greater than the starting value, so, since $b < 1$, this happens for a negative input value. Thus, the solution must be a negative number.

Often we can say more about a solution than just whether it is positive or negative.

Example 5   Say whether the solution is greater than 1, between 0 and 1, between -1 and 0, or less than -1.

(a) $4^x = 3$ (b) $5 \cdot 4^x = 25$ (c) $\left(\dfrac{1}{4}\right)^x = 3$

Solution   (a) Since 3 is between $4^0 = 1$ and $4^1 = 4$, the value of x that makes $4^x = 3$ is between 0 and 1.
   (b) Since 25 is between $5 \cdot 4^1 = 20$ and $5 \cdot 4^2 = 80$, the value of x that makes $5 \cdot 4^x = 25$ is between 1 and 2 and thus greater than 1.
   (c) Here x must be negative, since $1/4$ raised to a positive power is less than $1/4$, and so can never equal 3. Moreover, since 3 is between $(1/4)^0 = 1$ and $(1/4)^{-1} = 4$, then value of x that makes $(1/4)^x = 3$ is between -1 and 0.

Example 6   Is the solution for x to the following equation positive or negative?

$$a \cdot 5^x = 4, \quad a > 4,$$

Solution   The starting value a is greater than the output value of 4. Since the base $b = 5$ is greater than 1, this happens for an input value $x < 0$. Therefore, the solution must be negative.

The Range of an Exponential Function

Not every exponential equation has a solution, as Example 7 illustrates.

Example 7   Describe the solution to
$$2^x = -4.$$

Solution   When 2 is raised to any power, positive or negative, the result is always a positive number. So, $2^x = -4$ has no solutions.

The observation in Example 7 that 2^x is positive for all values of x applies to any positive base: Provided $b > 0$,
$$b^x > 0 \quad \text{for all } x.$$

This means that the output value of the exponential function $f(x) = ab^x$ has the same sign as the starting value a, regardless of the value of x:

$$\text{Sign of output value} = (\text{Sign of } a) \cdot \underbrace{(\text{Sign of } b^x)}_{\text{Always positive}}$$

$$= \text{Sign of } a.$$

Thus, if a is positive, $f(x) = ab^x$ is positive for all x, and if a is negative, $f(x)$ is negative for all x. In fact, we have
   • The range of an exponential function with $a > 0$ is the set of all positive numbers.
   • The range of an exponential function with $a < 0$ is the set of all negative numbers.

Exercises and Problems for Section 6.4

Exercises

Solve the equations in Exercises 1–8 given that

$$f(t) = 2^t, \quad g(t) = 3^t, \quad h(t) = 4^t.$$

1. $f(t) = 4$ **2.** $h(t) = 4$

3. $g(t) = 1$ **4.** $2g(t) = 162$

5. $2 + h(t) = \dfrac{33}{16}$ **6.** $2(1 - f(t)) = 1$

7. $f(t) = h(t)$ **8.** $h(t) = f(6)$

Answer Exercises 9–12 based on Table 6.8, which gives values of the exponential function $Q = 12(1.32)^t$. Your answers may be approximate.

Table 6.8

t	2.0	2.2	2.4	2.6	2.8	3.0
Q	20.9	22.1	23.4	24.7	26.1	27.6

9. Solve $12(1.32)^t = 24.7$.

10. Solve $6(1.32)^t = 10.45$.

11. Solve $12(1.032)^t = 25$.

12. Solve $25 - 12(1.32)^t = 2.9$.

Without solving them, say whether the equations in Exercises 13–16 have a positive solution, a negative solution, a zero solution, or no solution. Give a reason for your answer.

13. $9^x = 250$ **14.** $2.5 = 5^t$

15. $7^x = 0.3$ **16.** $6^t = -1$

Problems

17. (a) What are the domain and range of $Q = 200(0.97)^t$?
   (b) If Q represents the quantity of a 200-gram sample of a radioactive substance remaining after t days in a lab, what are the domain and range?

18. Match the statement (a)–(b) with the solutions to one or more of the equations (I)–(VI).

   I. $10(1.2)^t = 5$ II. $10 = 5(1.2)^t$
   III. $10 + 5(1.2)^t = 0$ IV. $5 + 10(1.2)^t = 0$
   V. $10(0.8)^t = 5$ VI. $5(0.8)^t = 10$

   (a) The time an exponentially growing quantity takes to grow from 5 to 10 grams.
   (b) The time an exponentially decaying quantity takes to drop from 10 to 5 grams.

19. Match the statements (a)–(c) with one or more of the equations (I)–(VIII). The solution for t in the equation should be the quantity described in the statement. An equation may be used more than once. (Do not solve the equations.)

   I. $4(1.1)^t = 2$ II. $(1.1)^{2t} = 4$
   III. $2^t = 4$ IV. $2(1.1)^t = 4$
   V. $4(0.9)^t = 2$ VI. $(0.9)^{2t} = 4$
   VII. $2^{-t} = 4$ VIII. $2(0.9)^t = 4$

   (a) The doubling time for a bank balance.

   (b) The half-life of a radioactive compound.
   (c) The time for a quantity growing exponentially to quadruple if its doubling time is 1.

20. The balance in dollars in a bank account after t years is given by the function $f(t) = 4622(1.04)^t$ rounded to the nearest dollar. For which values of t in Table 6.9 is

   (a) $f(t) < 5199$?
   (b) $f(t) > 5199$?
   (c) $f(t) = 5199$?

   Say what your answers tell you about the bank account.

Table 6.9

t	0	1	2	3	4
$4622(1.04)^t$	4622	4807	4999	5199	5407

21. (a) Construct a table of values for the function $g(x) = 28(1.1)^x$ for $x = 0, 1, 2, 3, 4$.
   (b) For which values of x in the table is
   (i) $g(x) < 33.88$ (ii) $g(x) > 30.8$
   (iii) $g(x) = 37.268$

22. (a) Construct a table of $y = 526(0.87)^x$ for $x = -2, -1, 0, 1, 2$.
   (b) Use your table to solve $604.598 = 526(0.87)^x$ for x.

23. (a) Construct a table of $y = 253(2.65)^x$ for $x = -3.5, -1.5, 0.5, 2.5$.

(b) At which x-values in your table is $253(2.65)^x \geq 58.648$?

24. The value, V, of a $5000 investment at time t in years is $V = 5000(1.012)^t$.

(a) Complete the following table:

t	55	56	57	58	59	60
v						

(b) What does your answer to part (a) tell you about the doubling time of your investment?

25. The quantity, Q, of caffeine in the body t hours after drinking a cup of coffee containing 100 mg is $Q = 100(0.83)^t$.

(a) Complete the following table:

t	3	3.5	3.6	3.7	3.8	3.9	4	5
Q								

(b) What does your answer to part (a) tell you about the half-life of caffeine?

26. A lab receives a 1000 grams of an unknown radioactive substance that decays at a rate of 7% per day.

(a) Write an expression for Q, the quantity of substance remaining after t days.

(b) Make a table showing the quantity of the substance remaining at the end of $8, 9, 10, 11, 12$ days.

(c) For what values of t in the table is the quantity left

   (i) Less than 500 gm?

   (ii) More than 500 gm?

(d) A lab worker says that the half-life of the substance is between 11 and 12 days. Is this consistent with your table? If not, how would you correct the estimate?

27. In year t, the population, L, of a colony of large ants is $L = 2000(1.05)^t$, and the population of a colony of small ants is $S = 1000(1.1)^t$.

(a) Construct a table showing each colony's population in years $t = 5, 10, 15, 20, 25, 30, 35, 40$.

(b) The small ants go to war against the large ants; they destroy the large ant colony when there are twice as many small ants as large ants. Use your table to determine in which year this happens.

(c) As long as the large ant population is greater than the small ant population, the large ants harvest fruit that falls on the ground between the two colonies. In which years in your table do the large ants harvest the fruit?

28. From Figure 6.5, when is

(a) $100(1.05)^t > 121.55$

(b) $100(1.05)^t < 121.55$

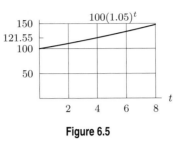

Figure 6.5

29. Figure 6.6 shows the resale value of a car given by $10{,}000(0.5)^t$, where t is in years.

(a) In which years in the figure is the car's value greater than $1250?

(b) What value of t is a solution to $5000 = 10{,}000(0.5)^t$?

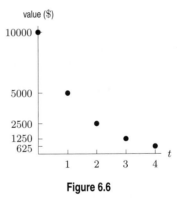

Figure 6.6

30. (a) Graph $y = 500(0.8)^x$ using the points $x = -2, -1.5, -1, -0.5, 0, 0.5, 1, 1.5, 2$.

(b) Use your graph to solve

   (i) $500(0.8)^x = 698.71$

   (ii) $500(0.8)^x \geq 400$

31. (a) Graph $y = 250(1.1)^x$ and $y = 200(1.2)^x$ using the points $x = 1, 2, 3, 4, 5$.

(b) Using the points in your graph, for what x-values is

   (i) $250(1.1)^x > 200(1.2)^x$

   (ii) $250(1.1)^x < 200(1.2)^x$

(c) How might you make your answers to part (b) more precise?

32. An employee signs a contract for a salary t years in the future given in thousands of dollars by $45(1.041)^t$.

    (a) What do the numbers 45 and 1.041 represent in terms of the salary?

    (b) What is the salary in 15 years? 20 years?

    (c) By trial and error, find to the nearest year how long it takes for the salary to double.

Without solving them, say whether the equations in Problems 33–44 have a positive solution, a negative solution, a zero solution, or no solution. Give a reason for your answer.

33. $7 + 2^y = 5$

34. $25 \cdot 3^z = 15$

35. $13 \cdot 5^{t+1} = 5^{2t}$

36. $(0.1)^x = 2$

37. $5(0.5)^y = 1$

38. $5 = -(0.7)^t$

39. $28 = 7(0.4)^z$

40. $7 = 28(0.4)^z$

41. $0.01(0.3)^t = 0.1$

42. $10^t = 7 \cdot 5^t$

43. $4^t \cdot 3^t = 5$

44. $(3.2)^{2y+1}(1 + 3.2) = (3.2)^y$

45. Assume $0 < r < 1$. Without solving equations (I)–(IV) for x, decide which one has

    (a) The largest solution
    (b) The smallest solution
    (c) No solution

    I. $3(1 + r)^x = 7$ II. $3(1 + 2r)^x = 7$
    III. $3(1 + 0.01r)^x = 7$ IV. $3(1 - r)^x = 7$

46. Assume that a, b, and r are positive and that $a < b$. Consider the solution for x to the equation $a(1 + r)^x = b$. Without solving the equation, what is the effect of increasing each of a, r, and b, while keeping each of the other two fixed? Does the solution increase or decrease?

    (a) a **(b)** r **(c)** b

In Problems 47–54, decide for what values of the constant A the equation has

(a) A solution   **(b)** The solution $t = 0$
(c) A positive solution

47. $5^t = A$

48. $3^{-t} = A$

49. $(0.2)^t = A$

50. $A - 2^{-t} = 0$

51. $6.3A - 3 \cdot 7^t = 0$

52. $2 \cdot 3^t + A = 0$

53. $A5^{-t} + 1 = 0$

54. $2(0.7)^t + 0.2A = 0$

Solve the equations in Problems 55–60 for x. Your solutions will involve u.

55. $10^x = 1000^u$

56. $8^x = 2^u$

57. $5^{2x+1} = 125^u$

58. $\left(\dfrac{1}{2}\right)^x = \left(\dfrac{1}{16}\right)^u$

59. $\left(\dfrac{1}{3}\right)^x = 9^u$

60. $5^{-x} = \dfrac{1}{25^u}$

Solve the equations in Problems 61–64 using the following approximations:

$$10^{0.301} = 2, \qquad 10^{0.477} = 3, \qquad 10^{0.699} = 5.$$

Example. Solve $10^x = 6$. *Solution.* We have

$$10^x = 6$$
$$= 2 \cdot 3$$
$$= 10^{0.301} \cdot 10^{0.477}$$
$$= 10^{0.301+0.477}$$
$$= 10^{0.778},$$

so $x = 0.778$.

61. $10^x = 15$

62. $10^x = 9$

63. $10^x = 32$

64. $10^x = 180$

6.5 MODELING WITH EXPONENTIAL FUNCTIONS

Many real-world situations involve growth factors and growth rates, so they lend themselves to being modeled using exponential functions.

Example 1   The quantity, Q, in grams, of a radioactive sample after t days is given by

$$Q = 150 \cdot 0.94^t.$$

Identify the initial value and the growth factor and explain what they mean in terms of the sample.

Solution   The initial value, when $t = 0$, is $Q = 150 \cdot 0.94^0 = 150$. This means we start with 150 g of material. The growth factor is 0.94, which gives a growth rate of -6% per day, since $0.94 - 1 = -0.06$. As we mentioned in the follow-up to Example 3 on page 242, we call a negative rate a *decay rate*, so we say that the substance is decaying by 6% per day. Thus, there is 6% less of the substance remaining after each day passes.

Finding the Formula for an Exponential Function

If we are given the initial value a and the growth factor $b = 1 + r$ of an exponentially growing quantity Q, we can write a formula for Q,

$$Q = ab^t.$$

However, we are not always given a and b directly. Often, we must find them from the available information.

Example 2   A youth soccer league initially has $N = 68$ members. After two years, the league grows to 96 members. If the growth rate continues to be the same, find a formula for $N = f(t)$ where t the number of years after the league was formed.

Solution   Since the number of players is increasing at a constant percent rate, we use the exponential form $N = f(t) = ab^t$. Initially, there are 68 members, so $a = 68$, and we can write $f(t) = 68b^t$. We now use the fact that $f(2) = 96$ to find the value of b:

$$f(t) = 68b^t$$
$$96 = 68b^2$$
$$\frac{96}{68} = b^2$$
$$\pm\sqrt{\frac{96}{68}} = b$$
$$b = \pm 1.188.$$

Since $b > 0$, we have $b = 1.188$. Thus, the function is $N = f(t) = 68(1.188)^t$.

Example 3   Find a formula for the exponential function whose graph is in Figure 6.7.

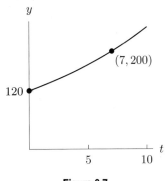

Figure 6.7

Solution   We know that the formula is of the form $y = ab^t$, and we want to solve for a and b. The initial value, a, is the y-intercept, so $a = 120$. Thus

$$y = 120b^t.$$

Because $y = 200$ when $t = 7$, we have

$$200 = 120b^7$$
$$\frac{200}{120} = b^7.$$

Raising both sides to the $1/7$ power,

$$(b^7)^{1/7} = \left(\frac{200}{120}\right)^{1/7}$$
$$b = 1.076.$$

Thus, $y = 120(1.076)^t$ is a formula for the function.

In Section 6.3 we saw how to use exponent laws to rewrite an expression like $4000 \cdot 2^{t/12}$, which represents a quantity that doubles every 12 years, in a form that shows the annual growth rate. The next example shows another method for finding the growth rate from the doubling time.

Example 4   The population of Malaysia is doubling every 38 years.[5]

(a) If the population is G in 2008, what will it be 2046?
(b) When will the population reach $4G$?
(c) What is the annual growth rate?

Solution   (a) Assuming that the population continues to double every 38 years, it will be twice as large in 2046 as it is in 2008, or $2G$. See Figure 6.8.
(b) The population will double yet again, rising from $2G$ to $4G$, after another 38 years, or $2046 + 38 = 2084$. See Figure 6.8.
(c) Let $s(t)$ give the population Malaysia in year t. We have

$$s(t) = ab^t,$$

where a is the initial size in year $t = 0$. The population after 38 years, written $s(38)$, is twice the initial population. This means

$$s(38) = 2a$$
$$ab^{38} = 2a$$
$$b^{38} = 2$$
$$b = 2^{1/38} = 1.018.$$

The population increases by a factor of 1.018 every year, and its annual growth rate is 1.8%.

---

[5]CIA World Factbook at www.cia.gov.

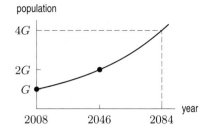

Figure 6.8: The population of Malaysia

Example 5   Hafnium has a half-life of 12.2 hours.

(a) If you begin with 1000 g, when will you have 500 g? 250 g? 125 g?

(b) What is the hourly decay rate?

Solution   (a) The half-life is 12.2 hours, so the level will drop from 1000 g to 500 g in this time. After 12.2 more hours, or 24.4 hours total, the level will drop to half of 500 g, or 250 g. After another 12.2 hours, or 36.6 hours total, the level will drop to half of 250 g, or 125 g. See Figure 6.9.

(b) Let $v(t)$ be the quantity remaining after t hours. Then

$$v(t) = ab^t,$$

where a is the initial amount of hafnium and b is the hourly decay factor. After 12.2 hours, only half the initial amount remains, and so

$$v(12.2) = 0.5a$$
$$ab^{12.2} = 0.5a$$
$$b^{12.2} = 0.5$$
$$\left(b^{12.2}\right)^{1/12.2} = 0.5^{1/12.2}$$
$$b = 0.9448.$$

If r is the hourly growth rate then $r = b - 1 = 0.9448 - 1 = -0.0552 = -5.52\%$. Thus, hafnium decays at a rate of 5.52% per hour.

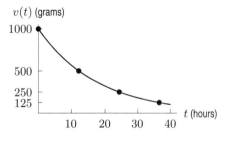

Figure 6.9: Quantity of hafnium remaining

Finding An Exponential Function When We Don't Know The Starting Value

In Examples 2 and 3, we are given the starting value a, so all we need to do is find the growth factor b. However, even if we are not given the starting value, we can sometimes use what we know about exponents to find a formula for an exponential function.

Example 6   An account growing at a constant percent rate contains $5000 in 2004 and $8000 in 2014. What is its annual growth rate? Express the account balance as a function of the number, t, of years since 2000.

Solution   Since the account balance grows at a constant percent rate, it is an exponential function of t. Thus, if $f(t)$ is the account balance t years after 2000, then $f(t) = ab^t$ for positive constants a and b. Since the balance is $5000 in 2004, we know that

$$f(4) = ab^4 = 5000.$$

Since the balance is $8000 in 2014, when $t = 14$, we know that

$$f(14) = ab^{14} = ab^{14} = 8000.$$

Therefore

$$\frac{ab^{14}}{ab^4} = \frac{8000}{5000} = 1.6$$
$$b^{10} = 1.6$$
$$b = 1.6^{1/10} = 1.04812.$$

This tells us the annual growth rate is 4.812%. Solving for a, we use $f(4) = 5000$:

$$ab^4 = 5000$$
$$a = \frac{5000}{1.04812^4} = 4143.$$

This tells us that the initial account balance in 2000 was $4143. Thus, $f(t) = 4143(1.048)^t$.

Example 7   A lab receives a sample of germanium. The sample decays radioactively over time at a constant daily percent rate. After 15 days, 200 grams of germanium remain. After 37 days, 51.253 grams remain. What is the daily percent decay rate of germanium? How much did the lab receive initially?

Solution   If $g(t)$ is the amount of germanium remaining t days after the sample is received, then $g(t) = ab^t$ for positive constants a and b. Since 200 g of germanium remain after 15 days we have

$$g(15) = ab^{15} = 200,$$

and since 51.253 g of germanium remain after 37 days we have

$$g(37) = ab^{37} = 51.253.$$

We can eliminate a by taking ratios:

$$\frac{\text{Amount after 37 days}}{\text{Amount after 15 days}} = \frac{ab^{37}}{ab^{15}} = \frac{51.253}{200}$$
$$\frac{b^{37}}{b^{15}} = 0.2563 \quad a \text{ is eliminated}$$
$$b^{22} = 0.2563.$$

So

$$b = (0.2563)^{1/22} = 0.93999.$$

Since $b = 1 + r$, the decay rate is $r = 1 - 0.93999 = 0.06001$, or about 6% per day. To find a, the amount of germanium initially received by the lab, we substitute for b in either of our two original equations, say, $ab^{15} = 200$:

$$a(0.93999)^{15} = 200$$

$$a = \frac{200}{(0.93999)^{15}}$$

$$a = 506.0.$$

Thus the lab received 506 grams of germanium. You can check that we obtain the same value for a by using the other equation, $ab^{37} = 51.253$. See Figure 6.10.

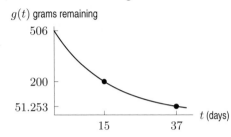

Figure 6.10: Graph showing amount of germanium remaining after t days

Exercises and Problems for Section 6.5

Exercises

Find formulas for the exponential functions described in Exercises 1–4.

1. $f(x)$ if $f(10) = 65$, $f(65) = 20$.

2. An investment initially worth $3000 grows by 30% over a 5-year period.

3. A population of 10,000 declines by 0.42% per year.

4. The graph of g is shown in Figure 6.11.

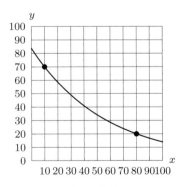

Figure 6.11

In Exercises 5–10, a pair of points on the graph of an exponential function is given. Find a formula for the function.

5. $f(t)$: $(0, 40)$, $(8, 100)$ **6.** $g(t)$: $(0, 80)$, $(25, 10)$

7. $p(x)$: $(5, 20)$, $(35, 60)$ **8.** $q(x)$: $(-12, 72)$, $(6, 8)$

9. $v(x)$: $(0.1, 2)$, $(0.9, 8)$ **10.** $w(x)$: $(5, 20)$, $(20, 5)$

11. The balance in a bank account earning interest at r% per year doubles every 10 years. What is r?

12. Lead-214 has a half-life of 27 minutes. What percent of the initial amount remains after 1 minute?

13. Bismuth-210 has a half-life of 5 days. What is the decay rate per day?

14. An exponentially growing population quadruples in 22 years. How long does it take to double?

15. A radioactive compound decays to 25% of its original quantity in 90 minutes. What is its half-life?

Exercises 16–18 give a population's doubling time in years. Find its growth rate per year.

16. 50   **17.** 143   **18.** 14

19. The doubling time of an investment is 7 years. What is its yearly growth rate?

20. The half-life of nicotine in the body is 2 hours. What is the hourly decay rate?

Find the annual growth rate of the quantities described in Exercises 21–26.

21. A population doubles in size after 8 years.

22. The value of a house triples over a 14-year period.

23. A population goes down by half after 7 years.

24. A stock portfolio drops to one-fifth its former value over a 4-year period.

25. The amount of water used in a community increases by 25% over a 5-year period.

26. The amount of arsenic in a well drops by 24% over 8 years.

Find possible formulas for the exponential functions described in Exercises 27–31.

27. $P = g(t)$ is the size of an animal population that numbers 8000 in year $t = 0$ and 3000 in year $t = 12$.

28. The graph of f contains the points $(24, 120)$ and $(72, 30)$.

29. The value V of an investment in year t if the investment is initially worth $3450 and if it grows by 4.75% annually.

30. A cohort of fruit flies (that is, a group of flies all the same age) initially numbers $P = 800$ and decreases by half every 19 days as the flies age and die.

31. $f(12) = 290$, $f(23) = 175$

32. An investment is worth $V = \$400$ in year $t = 5$ and $V = 1200$ in year $t = 12$. Find a formula for V in terms of t assuming constant percent growth.

33. The value of an investment initially worth $3500 doubles every eight years. Find a formula for $V = u(t)$, the investment's value in year t.

Problems

34. The balance in a bank account grows by a factor of a each year and doubles every 7 years.

   (a) By what factor does the balance change in 14 years? In 21 years?
   (b) Find a.

35. A radioactive substance has a 62 day half-life. Initially there are Q_0 grams of the substance.

   (a) How much remains after 62 days? 124 days?
   (b) When will only 12.5% of the original amount remain?
   (c) How much remains after 1 day?

In Problems 36–38, write a formula for the quantity described.

36. An exponentially growing population which doubles every 9 years.

37. The amount of a radioactive substance whose half-life is 5 days.

38. The balance in an interest-bearing bank account, if the balance triples in 20 years.

39. Between 1994 and 1999, the national health expenditures in the United States were rising at an average of 5.3% per year. The U.S. health expenditures in 1994 were 936.7 billion dollars.

   (a) Express the national health expenditures, P, in billions of dollars, as a function of the year, t, with $t = 0$ corresponding to the year 1994.
   (b) Use this model to estimate the national health expenditures in the year 1999. Compare this number to the actual 1999 expenditures, which were 1210.7 billion dollars.

40. Chicago's population grew from 2.8 million in 1990 to 2.9 million in 2000.

   (a) What was the yearly percent growth rate?
   (b) Assuming the growth rate continues unchanged, what population is predicted for 2010?

41. Let $P(t)$ be the population[6] of Charlotte, NC, in thousands, t years after 1990.

   (a) Interpret $P(0) = 396$ and $P(10) = 541$.
   (b) Find a formula for $P(t)$. Assume the population grows at a constant percentage rate and give your answer to 5 decimal places.
   (c) Find and interpret $P(20)$.

---

[6]Glaeser, J. and Shapiro, N., "City Growth and the 2000 Census: Which Places Grew and Why," www.brookings.edu, accessed on October 24, 2004.

42. Nicotine leaves the body at a constant percent rate. Two hours after smoking a cigarette, 58 mg remains in a person's body; three hours later (5 hours after the cigarette), there were 20 mg. Let $n(t)$ represent the amount of nicotine in the body t hours after smoking the cigarette.

   (a) What is the hourly percent decay rate of nicotine?
   (b) What was the initial quantity of nicotine in the body?
   (c) How much nicotine remains 6 hours after smoking the cigarette?

43. Table 6.10 shows values for an exponential function. Find a formula for the function.

Table 6.10

t	2	3	4	5
$f(t)$	96	76.8	61.44	49.152

The US population is growing by about 1% a year. In 2000, it was 282 million. What is wrong with the statements in Problems 44–46? Correct the equation in the statement.

44. The population will be 300 million t years after 2000, where $282(0.01)^t = 300$.

45. The population, P, in 2020 is the solution to $P - (282 \cdot 1.01)^{20} = 0$.

46. The solution to $282(1.01)^t = 2$ is the number of years it takes for the population to double.

REVIEW EXERCISES AND PROBLEMS FOR CHAPTER SIX

Exercises

The functions in Exercises 1–4 describe different investments. For each investment, what are the starting value and the percent growth rate?

1. $V = 6000(1.052)^t$ **2.** $V = 500(1.232)^t$

3. $V = 14{,}000(1.0088)^t$ **4.** $V = 900(0.989)^t$

5. A young sunflower grows in height by 5% every day; if the sunflower is 6 inches tall on the first day it was measured, write an expression that describes the height h of the sunflower in inches t days later.

6. A radioactive metal weighs 10 grams and loses 1% of its mass every hour. Write an expression that describes its weight w in grams after t hours.

In Exercises 7–14, find the formula for the exponential function whose graph goes through the two points.

7. $(3, 4); (6, 10)$ **8.** $(2, 6); (7, 1)$

9. $(-6, 2); (3, 6)$ **10.** $(-5, 8); (-2, 1)$

11. $(0, 2); (3, 7)$ **12.** $(1, 7); (5, 9)$

13. $(3, 6.2); (6, 5.1)$ **14.** $(2.5, 3.7); (5.1, 9.3)$

Can the expressions in Exercises 15–20 be put in the form ab^t? If so identify the values of a and b.

15. $2 \cdot 5^t$ **16.** 10^{-t} **17.** $1.3t^{4.5}$

18. $10 \cdot 2^{3+2t}$ **19.** $3^{-2t} \cdot 2^t$ **20.** $\dfrac{3}{5^{2+t}}$

In Exercises 21–23, write the expression either as a constant times a power of x, or a constant times an exponential in x. Identify the constant. If a power, identify the exponent, and if an exponential, identify the base.

21. $\dfrac{-21}{2^x}$ **22.** $\dfrac{2^x}{3^x}$ **23.** $\dfrac{x^2}{x^3}$

As x gets larger, are the expressions in Exercises 24–28 increasing or decreasing? By what percent per unit increase of x?

24. $10(1.07)^x$ **25.** $5(0.96)^x$

26. $5(1.13)^{2x}$ **27.** $6\left(\dfrac{1.05}{2}\right)^x$

28. $\dfrac{5(0.97)^x}{3}$

Write expressions representing the quantities described in Exercises 29–34.

29. The balance B doubles n times in a row.

30. The value V increases n times in a row by a factor of 1.04.

31. The investment value begins at V_0 and grows by a factor of k every four years over a twelve-year period.

32. The investment value begins at V_0 and grows by a factor of k every h years over a twenty-year period.

33. The investment value begins at V_0 and grows by a factor of k every h years over an N-year period.

34. The number of people living in a city increases from P_0 by a factor of r for 4 years, and then by a factor of s for 7 years.

In Exercises 35–38 give the growth factor that corresponds to the given growth rate.

35. 60% growth

36. 18% shrinkage

37. 100% growth

38. 99% shrinkage

In Exercises 39–42 give the rate of growth that corresponds to the given growth factor.

39. 1.095   **40.** 0.91   **41.** 2.16   **42.** 0.95

In Exercises 43–47, say whether the quantity is changing in an exponential or linear fashion.

43. An account receives a deposit of $723 per month.

44. A machine depreciates by 17% per year.

45. Every week, 9/10 of a radioactive substance remains from the beginning of the week.

46. One liter of water is added to a trough every day.

47. After 124 minutes, 1/2 of a drug remains in the body.

Assuming that the values in the tables in Exercises 48–50 are values of an exponential function $y = ab^x$, find the function.

48.

x	0	1	2	3
y	200	194	188.18	182.5346

49.

x	0	2	4	6
y	1	2	4	8

50.

x	0	3	12	15
y	10	9	6.561	5.9049

Identify the expressions in Exercises 51–56 as exponential, linear, or quadratic in t.

51. $7 + 2t^2$ **52.** $6m + 7t^2 + 8t$

53. $5q^t$ **54.** $7t + 8 - 4w$

55. $5h^n + t$ **56.** $a(7.08)^t$

Give the initial value a, the growth factor b, and the percentage growth rate r correct to three decimals for the exponential functions in Exercises 57–60.

57. $Q = 1700(1.117)^t$ **58.** $Q = 1250(0.923)^t$

59. $Q = 120 \cdot 3.2^t$ **60.** $Q = 80(0.113)^t$

Find formulas for the exponential functions described in Exercises 61–64.

61. $f(3) = 12, f(20) = 80$

62. An investment is worth $3000 in year $t = 6$ and $7000 in year $t = 14$.

63. The graph of $y = v(x)$ contains the points $(-4, 8)$ and $(20, 40)$.

64. $w(30) = 40, w(80) = 30$

Problems

Write an expression representing the quantities in Problems 65–69.

65. The amount of caffeine at time t hours if there are 90 mg at the start and the quantity decays by 17% per hour.

66. The US population t years after 2000, when it was 281 million. The growth rate is 1% per year.

67. The price of a $30 item in t years if prices increase by 2.2% a year.

68. The difference in value in t years between $1000 invested today at 5% per year and $800 invested at 4% per year.

69. The difference in value in t years between a $1500 investment earning 3% a year and a $1500 investment earning 2% a year.

70. The number of cell phone subscribers in the US has grown exponentially since 1990. There were 16 million subscribers in 1994 and 97 million subscribers in 2000. Find an expression for S, the number of subscribers (in millions) in the US t years after 1990. To the nearest million, how many subscribers are predicted for 2004 by this expression?

71. The number of asthma sufferers in the world was about 84 million in 1990 and 130 million in 2001. Let N represent the number of asthma sufferers (in millions) worldwide t years after 1990. Assuming N grows exponentially, what is the annual growth rate?

In Problems 72–74, correct the mistake in the formula.

72. A population which doubles every 5 years is given by $P = 7 \cdot 2^{5t}$.

73. The quantity of pollutant remaining after t minutes if 10% is removed each minute is given by $Q = Q_0(0.1)^t$.

74. The quantity, Q, which doubles every T years, has a yearly growth factor of $a = 2^T$.

75. For each description (a)–(c), select the expressions (I)–(VI) that could represent it. Assume P_0, r, and t are positive.

I. $P_0(1+r)^t$ 　　　 II. $P_0(1-r)^t$

III. $P_0(1+r)^{-t}$ 　 IV. $P_0(1-r)^{-t}$

V. $P_0 t/(1+r)$ 　　 VI. $P_0 t/(1-r)$

  (a) A population increasing with time.
  (b) A population decreasing with time.
  (c) A population growing linearly with time.

Find possible formulas for the exponential functions in Problems 76–77.

76. $V = f(t)$ is the value of an investment worth $2000 in year $t = 0$ and $5000 in year $t = 5$.

77. The value of the expression $g(8)$ is 80. The solution to the equation $g(t) = 60$ is 11.

78. Without calculating them, put the following quantities in increasing order

  (a) The solution to $2^t = 0.2$
  (b) The solution to $3 \cdot 4^{-x} = 1$
  (c) The solution to $49 = 7 \cdot 5^z$
  (d) The number 0
  (e) The number 1

79. A quantity grows or decays exponentially according to the formula $Q = ab^t$. Match the statements (a)–(b) with the solutions to one or more of the equations (I)–(VI).

I. $b^{2t} = 1$ 　　　　 II. $2b^t = 1$

III. $b^t = 2$ 　　　　 IV. $b^{2t} + 1 = 0$

V. $2b^t + 1 = 0$ 　　 VI. $b^{2t} + 2 = 0$

  (a) The doubling time for an exponentially growing quantity.
  (b) The half-life of an exponentially decaying quantity.

Find the percent rate of change for the quantities described in Problems 80–84.

80. An investment doubles in value every 8 years.

81. A radioactive substance has a half-life of 11 days.

82. After treatment with antibiotic, the blood count of a virus has a half-life of 3 days.

83. The value of an investment triples every seven years.

84. A radioactive substance has a half-life of 25 years.

85. Which grows faster, an investment that doubles every 10 years or one that triples every 15 years?

86. Which disappears faster, a population whose half-life is 12 years or a population that loses $1/3$ of its members every 8 years?

In Problems 87–89, say if 10^a is greater than or less than 10^b.

87. $0 < a < b$

88. $0 < a$ and $b < 0$

89. $a < b < 0$

In Problems 90–92 say if x^a is greater than or less than x^b.

90. $a < b$ and $0 < x < 1$

91. $a < b$ and $x > 1$

92. $x < 0$, a is an even integer, b is an odd integer

In Problems 93–95, say if $\dfrac{1}{10^a}$ is greater than or less than $\dfrac{1}{10^b}$.

93. $0 < a < b$

94. $0 < a$ and $b < 0$

95. $a < b < 0$

Without solving them, indicate the statement that best describes the solution x (if any) to the equations in Problems 96–105. *Example.* The equation $3 \cdot 4^x = 5$ has a positive solution because $4^x > 1$ for $x > 0$, and so the best-fitting statement is (a).

(a) Positive, because $b^x > 1$ for $b > 1$ and $x > 0$.

(b) Negative, because $b^x < 1$ for $b > 1$ and $x < 0$.

(c) Positive, because $b^x < 1$ for $0 < b < 1$ and $x > 0$.

(d) Negative, because $b^x > 1$ for $0 < b < 1$ and $x < 0$.

(e) There is no solution because $b^x > 0$ for $b > 0$.

96. $2 \cdot 5^x = 2070$

97. $3^x = 0.62$

98. $7^x = -3$

99. $6 + 4.6^x = 2$

100. $17 \cdot 1.8^x = 8$

101. $24 \cdot 0.31^x = 85$

102. $0.07 \cdot 0.02^x = 0.13$

103. $240 \cdot 0.55^x = 170$

104. $\dfrac{8}{2^x} = 9$

105. $\dfrac{12}{0.52^x} = 40$

The expressions in Problems 106–109 involve several letters. Think of one letter at a time representing a variable and the rest as non-zero constants. In which cases is the expression linear? In which cases is it exponential? In this case, what is the base?

106. $2^n a^n$

107. $a^n b^n + c^n$

108. $AB^q C^q$

109. Ab^{2t}

Put the expressions in Problems 110–113 in the indicated forms. Give the values of all constants.

110. $\dfrac{540\sqrt{x}}{4(3x)^3}$ in the form kx^p

111. $8^{\frac{2x}{3}+2}$ in the form ab^x

112. $\sqrt[3]{\sqrt[3]{x^2}\sqrt{x^3}}$ in the form kx^p.

113. $500 - 500 \cdot 2^{-t/3}$ in the form $a\left(1 - 2^{-kt}\right)$.

114. A container of ice cream is taken from the freezer and sits in a room for t minutes. Its temperature in degrees Fahrenheit is $a - b \cdot 2^{-t} + b$, where a and b are positive constants.

(a) Write this expression in a form that shows that the temperature is always

   (i) Less than $a + b$ **(ii)** Greater than a

(b) What are reasonable values for a and b?

A spider egg sac opens releasing hundreds of baby spiders, most of which are eaten after a few days. The expression

$$600\left(\frac{1}{2}\right)^t$$

describes the number of surviving spiders t days after the egg sac opens. Thus there are initially 600 baby spiders, and each day the number of survivors falls by one-half. The expressions in Problems 115–118 also describe the number of survivors, but use different variables ($T, n, s,$ and r). Rewrite the expressions in a way that shows the relationship between these variables and t, and describe what the expressions tell you in terms of these variables.

115. $600 \cdot 2^{-7T}$

116. $600 \cdot 2^{-n/24}$

117. $300\left(\dfrac{1}{2}\right)^s$

118. $600\left(\dfrac{1}{32}\right)^r$

Chapter Seven

Exponential Equations and Logarithms

Contents

7.1   INTRODUCTION TO LOGARITHMS

Suppose that a population of bacteria is given by $P = 10^t$ after t hours. When will the population reach 4000? To find out, we must solve the following equation for the exponent t:

$$10^t = 4000.$$

Since 4000 falls between $10^3 = 1000$ and $10^4 = 10,000$, the exponent must fall between 3 and 4. We can narrow down the approximate value of the exponent by calculating 10^t for values of t between 3 and 4.

Example 1   Find the solution to $10^t = 4000$ to within one decimal place.

Solution   Table 7.1 shows values of 10^t for t between 3.2 and 3.8.

Table 7.1   *Solve* $10^t = 4000$

x	3.2	3.3	3.4	3.5	3.6	3.7	3.8
10^x	1584.89	1995.26	2511.89	3162.28	3981.07	5011.87	6309.57

Since 4000 falls between $10^{3.6}$ and $10^{3.7}$, the solution to $10^t = 4000$ must be between 3.6 and 3.7. We conclude that it takes about 3.6 hours for the population to reach 4000.

What is a Logarithm?

We often need to solve equations involving exponents. We define the *common logarithm*, or simply the *log*, written $\log_{10} x$, or $\log x$, as follows.

If x is a positive number,

$$\log x \text{ is the exponent of 10 that gives } x.$$

In other words,

$$\text{if} \quad y = \log x \quad \text{then} \quad 10^y = x.$$

In Example 7 on page 272, we return to the equation $10^t = 4000$ from Example 1, and solve it using logarithms. But first, let's look at some straightforward examples to help us understand this new operation.

Example 2   Find $\log 1000$, $\log 100$, $\log 10$.

Solution   We have

$$\log 1000 = 3 \quad \text{because} \quad 10^3 = 1000$$
$$\log 100 = 2 \quad \text{because} \quad 10^2 = 100$$
$$\log 10 = 1 \quad \text{because} \quad 10^1 = 10.$$

Example 3   Rewrite the following statements using exponents instead of logs.

(a) $\log 100 = 2$ (b) $\log 0.1 = -1$ (c) $\log 40 = 1.602$

Solution   For each statement, we use the fact that if $y = \log x$ then $10^y = x$.

(a) $\log 100 = 2$ means that $10^2 = 100$.
(b) $\log 0.1 = -1$ means that $10^{-1} = 0.1$.
(c) $\log 40 = 1.602$ means that $10^{1.602} = 40$.

Example 4   Rewrite the following statements using logs instead of exponents.

(a) $10^4 = 10{,}000$ (b) $10^{-3} = 0.001$ (c) $10^{0.6} = 3.981$

Solution   For each statement, we use the fact that if $10^y = x$, then $y = \log x$.

(a) $10^4 = 10{,}000$ means that $\log 10{,}000 = 4$.
(b) $10^{-3} = 0.001$ means that $\log 0.001 = -3$.
(c) $10^{0.6} = 3.981$ means that $\log 3.981 = 0.6$.

Logarithms Are Exponents

It is useful to keep in mind that a logarithm is nothing more than the exponent in a power 10. Thinking in terms of exponents is often a good way to answer a logarithm problem.

Example 5   Without a calculator, evaluate the following, if possible:

(a) $\log 1$ (b) $\log \sqrt{10}$ (c) $\log \dfrac{1}{100{,}000}$

(d) $\log 0.01$ (e) $\log \dfrac{1}{\sqrt{1000}}$ (f) $\log(-10)$

Solution
(a) We have $\log 1 = 0$ because $10^0 = 1$.
(b) We have $\log \sqrt{10} = 0.5$ because $10^{0.5} = \sqrt{10}$.
(c) We have $\log \dfrac{1}{100{,}000} = -5$ because $10^{-5} = \dfrac{1}{100{,}000}$.
(d) We have $\log 0.01 = -2$ because $10^{-2} = 0.01$.
(e) We have

$$\log \frac{1}{\sqrt{1000}} = -\frac{3}{2} \quad \text{because} \quad 10^{-3/2} = \frac{1}{(10^3)^{1/2}} = \frac{1}{\sqrt{1000}}.$$

(f) Since 10 to any real-valued power is positive, -10 cannot be written as a power of 10. Thus, $\log(-10)$ is undefined.

The Log Operation Undoes Raising 10 to a Power

We know that the operation of subtraction undoes the operation of addition and that the square root operation undoes the operation of squaring. Likewise, the operation of taking a logarithm undoes the operation of raising 10 to a power. For example,

$$\log 10^5 = \log 100{,}000 = 5 \quad \longrightarrow \quad \text{the logarithm undoes raising 10 to a power, leaving 5}$$

$$10^{\log 5} = 10^{0.69897} = 5 \quad \longrightarrow \quad \text{raising 10 to the power of a logarithm undoes the logarithm, leaving 5.}$$

In general:

For any N,

$$\log\left(10^N\right) = N$$

because the log operation undoes the operation of raising 10 to a power. Also, for $N > 0$,

$$10^{\log N} = N$$

because the operation of raising 10 to a power undoes the log operation.

Example 6 Evaluate without using a calculator:

(a) $\log\left(10^{6.7}\right)$ 　　　　　(b) $10^{\log 2.5}$ 　　　　　(c) $10^{\log(x+1)}, \quad x > -1$

Solution We have:

(a) $\log\left(10^{6.7}\right) = 6.7$ 　　(b) $10^{\log 2.5} = 2.5$ 　　(c) $10^{\log(x+1)} = x + 1.$

We can use a calculator to check the first two results of Example 6. For instance, we see that $10^{6.7} = 5{,}011{,}872.34$ and that $\log 5{,}011{,}872.34 = 6.7$, confirming that $\log(10^{6.7}) = 6.7$. In general, we will use a calculator to evaluate logarithms.

Solving Equations Using Logarithms

Knowing how to evaluate logarithms can help us to solve certain equations involving a variable in the exponent, such as the equation from Example 1.

Example 7 Solve the equation $10^t = 4000$ using logarithms.

Solution To solve the equation $10^t = 4000$, we must find the power of 10 that gives 4000. We know that the solution to $10^t = 4000$ is $t = \log 4000 = 3.602$.

When using logs to solve equations, we must sometimes simplify the equation first, as shown in Example 8.

Example 8 Solve the equation $3 \cdot 10^t - 8 = 13$.

Solution

$$3 \cdot 10^t - 8 = 13$$
$$3 \cdot 10^t = 21 \qquad \text{add 8 to both sides}$$
$$10^t = 7 \qquad \text{divide both sides by 3}$$
$$t = \log 7 = 0.845 \quad \text{use the definition.}$$

In Example 7, we solved an equation of the form $10^t = N$. In the next example, we see how to solve an equation of the form $\log N = t$.

Example 9 Solve the equation $\log N = 1.9294$.

Solution Using the definition of the logarithm gives

$$N = 10^{1.9294} = 84.996.$$

In Example 9, we used the definition of the logarithm directly. We take the same approach in Example 10 after an initial algebraic operation.

Example 10   Solve $4 \log(2x - 6) = 8$.

Solution   We have

$$4 \log(2x - 6) = 8$$
$$\log(2x - 6) = 2 \qquad \text{divide both sides by 4}$$
$$2x - 6 = 10^2 \quad \text{by the definition}$$
$$2x - 6 = 100$$
$$2x = 106 \quad \text{add 6 to both sides}$$
$$x = 53.$$

The Logarithmic Function

Just as we can use the square root operation to define the function $y = \sqrt{x}$, we can use logarithms to define the so-called *logarithmic function*, $y = \log x$.

The Domain of the Logarithmic Function

Just as the square root operation is undefined for certain numbers, so is the logarithm. We see in part (f) of Example 5, that the logarithm of x is undefined if x is negative. Since the logarithm is also undefined if $x = 0$, we see that the domain of the function $y = \log x$ is $x > 0$.

Graphs of Logarithmic Functions

Data for the values of $y = \log x$ that we have previously obtained are given in Table 7.2, and the graph of the logarithm function is shown in Figure 7.1.

Table 7.2   *Values of* $\log x$

x	0.001	0.01	0.1	1	10	100	1000
$\log x$	-3	-2	-1	0	1	2	3

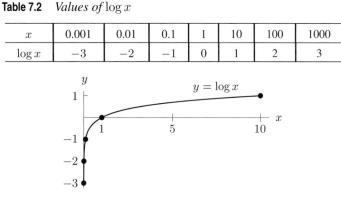

Figure 7.1: Graph $\log x$

Looking at the graph in Figure 7.1, we determine the range of $y = \log x$. Note that the graph is increasing and has an x-intercept at $x = 1$. The values of $\log x$ are positive for x greater than one and are negative for x less than one. The graph climbs to $y = 1$ at $x = 10$ and to $y = 2$ at $x = 100$.

To reach the modest height of $y = 10$ requires x to equal 10^{10}, or 10 billion! This suggests that the logarithmic function increases very slowly. Nonetheless, the graph of $y = \log x$ eventually climbs to any value we choose. Although x cannot equal zero in the logarithmic function, we can choose $x > 0$ to be as small as we like. As x decreases toward zero, the values of $\log x$ become more and more negative.

How Are Logarithms Different From Roots?

Logarithms and roots both involve bases and exponents, but they are entirely different operations. Consider the square root operation:

$$\sqrt{4} = 2 \quad \text{because} \quad 2^2 = 4 \quad \longrightarrow \quad \text{base is 2, exponent is 2}$$
$$\sqrt{9} = 3 \quad \text{because} \quad 3^2 = 9 \quad \longrightarrow \quad \text{base is 3, exponent is 2}$$
$$\sqrt{16} = 4 \quad \text{because} \quad 4^2 = 16 \quad \longrightarrow \quad \text{base is 4, exponent is 2.}$$

We see that with the square root operation, the base is different in each case, but the exponent always equals 2. The situation is similar with the cube root operation:

$$\sqrt[3]{8} = 2 \quad \text{because} \quad 2^3 = 8 \quad \longrightarrow \quad \text{base is 2, exponent is 3}$$
$$\sqrt[3]{27} = 3 \quad \text{because} \quad 3^3 = 27 \quad \longrightarrow \quad \text{base is 3, exponent is 3}$$
$$\sqrt[3]{64} = 4 \quad \text{because} \quad 4^3 = 64 \quad \longrightarrow \quad \text{base is 4, exponent is 3.}$$

Again, the base is different in each case, but the exponent always equals 3. Logarithms are different. Unlike roots, the base is the same in each case. It is the exponent that changes:

$$\log 0.1 = -1 \quad \text{because} \quad 10^{-1} = 0.1 \quad \longrightarrow \quad \text{base is 10, exponent is } -1$$
$$\log 10 = 1 \quad \text{because} \quad 10^1 = 10 \quad \longrightarrow \quad \text{base is 10, exponent is 1}$$
$$\log 100 = 2 \quad \text{because} \quad 10^2 = 100 \quad \longrightarrow \quad \text{base is 10, exponent is 2.}$$

In general, when finding the nth root of x, we are looking for the *base* that when raised to n gives us x. But when finding the logarithm of x, we are looking for the *exponent* in the power of 10 that gives us x.

Exercises and Problems for Section 7.1

Exercises

In Exercises 1–8, rewrite the equation using exponents instead of logarithms.

1. $\log 0.01 = -2$

2. $\log 1000 = 3$

3. $\log 20 = 1.301$

4. $\log A^2 = B$

5. $\log 5000 = 3.699$

6. $\log \dfrac{1}{10^3} = -3$

7. $\log(\alpha\beta) = 3x^2 + 2y^2$

8. $\log\left(\dfrac{a}{b}\right) = 9$

In Exercises 9–16, rewrite the equation using logarithms instead of exponents.

9. $10^5 = 100{,}000$

10. $10^{-4} = 0.0001$

11. $10^{2.301} = 200$

12. $10^m = n$

13. $100^{2.301} = 39{,}994$

14. $10^{-0.08} = 0.832$

15. $10^{a^2 b} = 97$

16. $10^{qp} = nR$

Solve the equations in Exercises 17–24.

17. $10^x = 10{,}000$ **18.** $10^x = 421$

19. $10^x = 0.7162$ **20.** $10^x = \dfrac{23}{37}$

21. $10^x = 17.717$ **22.** $10^x = \dfrac{3}{\sqrt{17}}$

23. $10^x = 57$ **24.** $15.5 = 10^y$

In Exercises 25–34, evaluate without a calculator, or say if the expression is undefined.

25. $\log 0.001$ **26.** $\log \dfrac{1}{10}$

27. $\log \sqrt{1000}$ **28.** $\log \dfrac{1}{\sqrt{100{,}000}}$

29. $\log(-100)$ **30.** $10^{\log(-1)}$

31. $10^{\log 1}$ **32.** $\log 10^{-5.4}$

33. $\log 10^7$ **34.** $10^{\log 100}$

Solve the equations in Exercises 35–42.

35. $\log x = \dfrac{1}{2}$ **36.** $\log x = -3$

37. $\log x = 5$ **38.** $\log x = 1.172$

39. $3\log x = 6$ **40.** $7 - 2\log x = 10$

41. $2(\log x - 1) = 9$ **42.** $\log(x - 3) = 1$

43. A trillion is one million million. What is the logarithm of a trillion?

Problems

In Problems 44–49, without using a calculator, find two consecutive integers lying above and below the logarithm of the number. For example, if the number is 63, we use the fact that $10 < 63 < 100$. Since $10^1 = 10$ and $10^2 = 100$, we can say that $1 < \log 63 < 2$.

44. 205 **45.** 8991

46. $1.22 \cdot 10^4$ **47.** $0.99 \cdot 10^5$

48. 0.6 **49.** 0.012

50. Complete the following table. Based on your answers, which table entry is closest to $\log 8$?

x	0.901	0.902	0.903	0.904	0.905
10^x					

51. What does the table tell you about $\log 27$?

x	1.40	1.42	1.44	1.46
10^x	25.1	26.3	27.5	28.8

Solve the equations in Problems 52–54, first approximately, as in Example 1, by filling in the given table, and then to four decimal places by using logarithms.

52.

Table 7.3 *Solve* $10^x = 500$

x	2.6	2.7	2.8	2.9
10^x				

53.

Table 7.4 *Solve* $10^x = 3200$

x	3.4	3.5	3.6	3.7
10^x				

54.

Table 7.5 *Solve* $10^x = 0.03$

x	-1.6	-1.5	-1.4	-1.3
10^x				

55. Complete the following table. Use your answer to estimate $\log 0.5$.

x	-0.296	-0.298	-0.300	-0.302
10^x				

7.2 PROPERTIES OF LOGARITHMS

In the last section, we used logs to solve the equation $10^t = 4000$. But what about an equation like $3^t = 4000$?

Example 1   Estimate the solution to $3^t = 4000$.

Solution   We have:

$$3^7 = 2187 \qquad \longrightarrow \qquad 7 \text{ is too small}$$
$$3^8 = 6561 \qquad \longrightarrow \qquad 8 \text{ is too big}$$
$$3^{7.5} = 3787.995 \qquad \longrightarrow \qquad 7.5 \text{ is too small}$$
$$3^{7.6} = 4227.869 \qquad \longrightarrow \qquad 7.6 \text{ is too big}$$

Thus, the exponent of 3 that gives 4000 is between 7.5 and 7.6.

It turns out that we can use logs to solve the equation in Example 1, even though the base is 3. This is because logarithms satisfy a number of useful properties having to do with *all* bases, not just base 10.

Example 2   Using logs, find the exact solution to $3^t = 4000$.

Solution   We can rewrite this equation using base 10 instead of base 3, using the fact that $3 = 10^{\log 3}$:

$$3^t = 4000 \quad \text{original equation}$$
$$(10^{\log 3})^t = 4000 \quad \text{because } 3 = 10^{\log 3}.$$

Now rewrite $(10^{\log 3})^t$ as $10^{t \log 3}$ using the exponent laws. This gives

$$10^{t \log 3} = 4000.$$

We now use the definition to write:

$$t \log 3 = \log 4000 \qquad\qquad \text{the variable } t \text{ is no longer in the exponent}$$
$$t = \frac{\log 4000}{\log 3} = 7.5496 \quad \text{dividing through by } \log 3.$$

The solution, $t = \log(4000)/\log 3 = 7.5496$, agrees with our earlier result.

Properties Of Logarithms

We often turn to logarithms when solving equations involving exponents. In Example 2, we used the definition of the logarithm to change the form of an equation. The results foreshadow some properties of logarithms which are useful in solving equations. We summarize these properties here, and justify them at the end of this section. (See page 279.)

Properties of the Common Logarithm
- The operation of raising 10 to a power and the log operation undo each other:

$$\log(10^N) = N \qquad \text{for all } N$$
$$10^{\log N} = N \qquad \text{for } N > 0$$

- For a and b both positive and any value of t,

$$\log(ab) = \log a + \log b$$
$$\log\left(\frac{a}{b}\right) = \log a - \log b$$
$$\log(b^t) = t \cdot \log b$$

- Since $10^0 = 1$,

$$\log 1 = 0.$$

Let's see how to use these properties to solve the equation in Example 2 more efficiently.

Example 3   Solve $3^t = 4000$ using the log property $\log(b^t) = t \cdot \log b$.

Solution   We have

$$3^t = 4000$$
$$\log(3^t) = \log 4000 \qquad \text{taking logs of both sides}$$
$$t \cdot \log 3 = \log 4000 \qquad \text{using } \log(b^t) = t \cdot \log b$$
$$t = \frac{\log 4000}{\log 3} = 7.5496 \quad \text{dividing by } \log 3.$$

This is the same answer that we got in Example 2.

Example 4   In Example 3 on page 247, the value of an investment t years after 2004 is given by $V = 4000 \cdot 2^{t/12}$. Solve the equation
$$4000 \cdot 2^{t/12} = 7000,$$
and say what it tells you about the investment.

Solution   We have

$$4000 \cdot 2^{t/12} = 7000$$
$$2^{t/12} = \frac{7000}{4000} = 1.75$$
$$\log\left(2^{t/12}\right) = \log 1.75$$
$$\frac{t}{12} \cdot \log 2 = \log 1.75 \qquad \text{Using log property } \log(b^t) = t \cdot \log b$$
$$t = \frac{12 \log 1.75}{\log 2}$$
$$= 9.688.$$

This tells us that the investment's value will reach $7000 after about 9.7 years, or midway into 2013.

Using Log Properties to Rewrite Expressions

We can also use the properties of logarithms to rewrite and simplify expressions.

Example 5   Show that the expressions $\log(1/b)$ and $-\log b$ are equivalent.

Solution   There are at least two ways to see this using the log rules we know. First, we know that $\log(a/b) = \log a - \log b$, so $\log(1/b) = \log 1 - \log b$. Since $\log 1 = 0$, we have

$$\log \frac{1}{b} = \log 1 - \log b = 0 - \log b = -\log b.$$

Alternatively, we know that

$$\log \frac{1}{b} = \log \left(b^{-1}\right)$$
$$= -1 \cdot \log b \quad \text{Using } \log \left(b^t\right) = t \cdot \log b.$$

Example 6   Simplify the expression $\log 10^{\sqrt[3]{8-x^3}}$.

Solution   We know that $\log 10^N = N$, so letting $N = \sqrt[3]{8 - x^3}$, we have

$$\log 10^{\sqrt[3]{8-x^3}} = \sqrt[3]{8 - x^3}.$$

Note that this cannot be simplified further.

Example 7   Let $u = \log x$ and $v = \log y$. Rewrite the following expressions in terms of u and v, or state that this is not possible.

(a) $\log(4x/y)$ (b) $\log(4xy^3)$ (c) $\log(4x - y)$

Solution   (a) We have

$$\log(4x/y) = \log(4x) - \log y \qquad \text{using } \log \left(\frac{a}{b}\right) = \log a - \log b$$
$$= \log 4 + \log x - \log y \quad \text{using } \log(ab) = \log a + \log b$$
$$= \log 4 + u - v.$$

(b) We have

$$\log(4xy^3) = \log(4x) + \log(y^3) \qquad \text{using } \log(ab) = \log a + \log b$$
$$= \log 4 + \log x + \log(y^3) \quad \text{again using } \log(ab) = \log a + \log b$$
$$= \log 4 + \log x + 3 \log y \quad \text{using } \log(b^t) = t \cdot \log b$$
$$= \log 4 + u + 3v.$$

Note that it would be incorrect to write

$$\log(4xy^3) = 3 \log(4xy).$$

This is because the exponent of 3 applies only to the y and not to the expression $4xy$.

(c) It is not possible to rewrite $\log(4x - y)$ in terms of u and v.

Misconceptions and Calculator Errors Involving Logs

It is important to know how to use the properties of logarithms. It is equally important to recognize statements that are *not* true. Beware of the following:

- $\log(a + b)$ is not the same as $\log a + \log b$
- $\log(a - b)$ is not the same as $\log a - \log b$
- $\log(ab)$ is not the same as $(\log a)(\log b)$
- $\log\left(\dfrac{a}{b}\right)$ is not the same as $\dfrac{\log a}{\log b}$
- $\log\left(\dfrac{1}{b}\right)$ is not the same as $\dfrac{1}{\log b}$.

In fact, there are no formulas that simplify the expressions $\log(a + b)$, $\log(a - b)$, $\dfrac{\log a}{\log b}$, and $\dfrac{1}{\log b}$. As for $\log(1/b)$, we saw in Example 5 that it equals $-\log b$.

Example 8

Evaluate $\log\left(\dfrac{16}{5}\right)$ using a calculator.

Solution

Using a calculator to evaluate this sort of expression requires care. We have

$$\log\left(\frac{16}{5}\right) = \log 3.2 = 0.505.$$

Note that on some calculators, entering $\log 16/5$ gives 0.241, which is incorrect. This is because the calculator assumes that you mean $(\log 16)/5$, which is not the same as $\log(16/5)$. Notice also that

$$\frac{\log 16}{\log 5} = \frac{1.2041}{0.6990} = 1.723,$$

which is not the same as either $(\log 16)/5$ or $\log(16/5)$. Thus, the following expressions, though similar in appearance, are all different:

$$\log\frac{16}{5} = 0.505, \qquad \frac{\log 16}{5} = 0.241, \qquad \text{and} \qquad \frac{\log 16}{\log 5} = 1.723.$$

We now show that our suggestion that $\log(a \cdot b) = \log a + \log b$ is correct.

Justification of $\log(a \cdot b) = \log a + \log b$ and $\log(a/b) = \log a - \log b$

If a and b are both positive, we can write $a = 10^m$ and $b = 10^n$, so $\log a = m$ and $\log b = n$. Then, the product $a \cdot b$ can be written

$$a \cdot b = 10^m \cdot 10^n = 10^{m+n}.$$

Therefore $m + n$ is the power of 10 needed to give $a \cdot b$, so

$$\log(a \cdot b) = m + n,$$

which gives

$$\log(a \cdot b) = \log a + \log b.$$

Similarly, the quotient a/b can be written as

$$\frac{a}{b} = \frac{10^m}{10^n} = 10^{m-n}.$$

Therefore $m - n$ is the power of 10 needed to give a/b, so

$$\log\left(\frac{a}{b}\right) = m - n,$$

and thus

$$\boxed{\log\left(\frac{a}{b}\right) = \log a - \log b.}$$

Justification of $\log(b^t) = t \cdot \log b$

Suppose that b is positive, so we can write $b = 10^k$ for some value of k. Then

$$b^t = (10^k)^t.$$

We have rewritten the expression b^t so that the base is a power of 10. Using a property of exponents, we can write $(10^k)^t$ as 10^{kt}, so

$$b^t = (10^k)^t = 10^{kt}.$$

Therefore kt is the power of 10 which gives b^t, so

$$\log(b^t) = kt.$$

But since $b = 10^k$, we know $k = \log b$. This means

$$\log(b^t) = (\log b)t = t \cdot \log b.$$

Thus, for $b > 0$ we have

$$\boxed{\log\left(b^t\right) = t \cdot \log b.}$$

Exercises and Problems for Section 7.2

Exercises

Solve the equations in Exercises 1–3, first approximately by filling in the given table, then to four decimal places by using logarithms.

1.

Table 7.6   *Solve* $2^x = 20$

x	4.1	4.2	4.3	4.4
2^x				

2.

Table 7.7   *Solve* $5^x = 130$

x	2.9	3	3.1	3.2
5^x				

3.

Table 7.8   *Solve* $0.5^x = 0.1$

x	3.1	3.2	3.3	3.4
0.5^x				

Solve the exponential equations in Exercises 4–8 without using logarithms, then use logarithms to confirm your answer.

4. $2^4 = 4^x$

5. $\dfrac{1}{81} = 3^x$

6. $1024 = 2^x$

7. $5^3 = 25^x$

8. $64^3 = 4^x$

Solve the equations in Exercises 9–18.

9. $2^t = 19$

10. $1.071^x = 3.25$

11. $17^z = 12$

12. $\left(\dfrac{2}{3}\right)^y = \dfrac{5}{7}$

13. $80^w = 100$

14. $0.088^a = 0.54$

15. $(1.041)^t = 520$

16. $2^p = 90$

17. $(1.033)^q = 600$

18. $(0.988)^r = 55$

19. As of December 13, 2005, the US national debt[1] was $8,136,066,508,140.10, or about $8.136 trillion. What is the logarithm of this figure, correct to three decimals? **Hint**: $\log 8.136 = 0.9104$.

Problems

Are the expressions in Problems 20–22 equivalent for positive a and b? If so, explain why. If not, give values for a and b that lead to different values for the two expressions.

20. $\log(10b)$ and $1 + \log b$

21. $\log(10^{a+b})$ and $\log(a + b)$

22. $10^{\log(a+b)}$ and $a + b$

In Problems 23–30, rewrite the expression in terms of $\log A$ and $\log B$, or state that this is not possible.

23. $\log(AB^2)$

24. $\log(2A/B)$

25. $\log(A + B/A)$

26. $\log(1/(AB))$

27. $\log(A + B)$

28. $\log(AB^2 + B)$

29. $\log\left(A\sqrt{B}\right) + \log\left(A^2\right)$

30. $\log(A(A + B)) - \log(A + B)$

If possible, use log properties to rewrite the expressions in Problems 31–36 in terms of u, v, w given that

$$u = \log x, v = \log y, w = \log z.$$

Your answers should not involve logs.

31. $\log xy$

32. $\log \dfrac{x}{z}$

33. $\log x^2 y^3 \sqrt{z}$

34. $\dfrac{\log \sqrt{x^3}}{\log \frac{y}{z^2}}$

35. $\left(\log \dfrac{1}{y^3}\right)^2$

36. $\log\left(x^2 + y^2\right)$

Simplify the expressions in Problems 37–40.

37. $10^{\log(2x+1)}$

38. $10^{\log \sqrt{x^2+1}}$

39. $100^{\log(x+1)}$

40. $10^{\log x + \log(1/x)}$

41. (a) Calculate $\log 2$, $\log 20$, $\log 200$ and $\log 2000$ and describe the pattern.
   (b) Using the pattern in part (a) make a guess about the values of $\log 20{,}000$ and $\log 0.2$.
   (c) Justify the guess you made in part (b) using the properties of logarithms.

42. What properties are required of a and b if $\log(a/b) = r$ has a solution for
   (a) $r > 0$
   (b) $r < 0$
   (c) $r = 0$

43. Write the expression $28 \cdot 1.121^t$ in the form $a \cdot 10^{kt}$ and give the values of a and k.

44. Write the expression $30 \cdot 1.21^t$ in the form $a \cdot 10^{kt}$, and give the values of a and k.

45. Write the expression $210 \cdot 2^{1-t/5}$ in the form $a \cdot 10^{kt}$, and give the values of a and k.

46. Suppose

$$\log A = 2 + \log B.$$

How many times as large as B is A?

---

[1]The US Bureau of the Public Debt posts the value of the debt to the nearest penny at its website, www.publicdebt.treas.gov.

7.3   SOLVING EQUATIONS USING LOGARITHMS

In this section we use the properties of logarithms to solve equations involving exponents. Often it is advantageous to perform some algebraic manipulation before using logarithms. This is evident in the following examples.

Example 1   Solve $5 \cdot 7^x = 20$.

Solution   We have

$$
\begin{aligned}
5 \cdot 7^x &= 20 \\
7^x &= 4 && \text{dividing through by 5} \\
\log(7^x) &= \log 4 && \text{taking logs of both sides} \\
x \cdot \log 7 &= \log 4 && \text{using } \log(b^t) = t \cdot \log b \\
x &= \frac{\log 4}{\log 7} = 0.712 && \text{dividing by } \log 7.
\end{aligned}
$$

In the solution to the last example, we first divide through by 5 and then take logs of both sides. However, we can also solve the equation by taking logs right away:

$$
\begin{aligned}
5 \cdot 7^x &= 20 \\
\log(5 \cdot 7^x) &= \log 20 && \text{taking logs of both sides} \\
\log 5 + \log(7^x) &= \log 20 && \text{using } \log(ab) = \log a + \log b \\
\log(7^x) &= \log 20 - \log 5 && \text{subtracting } \log 5 \text{ from both sides} \\
x \cdot \log 7 &= \log 20 - \log 5 && \text{using } \log(b^t) = t \cdot \log b \\
x &= \frac{\log 20 - \log 5}{\log 7} = 0.712 && \text{dividing by } \log 7.
\end{aligned}
$$

Although our answer looks different, it works out to the same value, 0.712. To see that the answers are the same, notice that

$$
\log 20 - \log 5 = \log\left(\frac{20}{5}\right) = \log 4 \quad \text{using } \log a - \log b = \log\left(\frac{a}{b}\right).
$$

Example 2   Solve the equation $300 \cdot 7^t = 45$.

Solution   We have:

$$
\begin{aligned}
300 \cdot 7^t &= 45 \\
7^t &= \frac{45}{300} = 0.15 \\
\log(7^t) &= \log 0.15 && \text{take logs of both sides} \\
t \log 7 &= \log 0.15 && \text{using } \log b^t = t \cdot \log b \\
t &= \frac{\log 0.15}{\log 7} = -0.975.
\end{aligned}
$$

Notice that the solution to Example 1 is positive, whereas the solution to Example 2 is negative. In Example 1, we have

$$(\text{Smaller number}) \cdot 7^x = \text{Larger number}.$$

This means that 7^x must be greater than 1, so $x > 0$. In contrast, in Example 2 we have

$$(\text{Larger number}) \cdot 7^x = \text{Smaller number}.$$

This means that 7^x must be less than 1, so $x < 0$. See Figures 7.2 and 7.3 for a comparison.

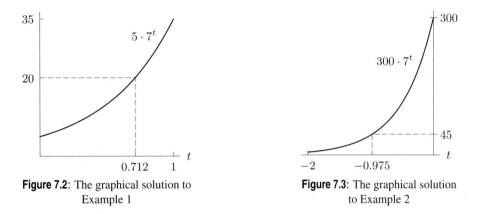

Figure 7.2: The graphical solution to Example 1

Figure 7.3: The graphical solution to Example 2

In Examples 1 and 2, we use logs to solve an equation whose left-hand side involves a variable in the exponent. In Example 3, we solve an equation in which both sides involve the variable in the exponent.

Example 3   Solve $1200(1.051)^x = 2500(1.044)^x$.

Solution   Here is one approach:

$$1200(1.051)^x = 2500(1.044)^x$$

$$\frac{(1.051)^x}{(1.044)^x} = \frac{2500}{1200} \qquad\qquad \text{dividing by 1200 and by } (1.044)^x$$

$$\left(\frac{1.051}{1.044}\right)^x = \frac{25}{12} \qquad\qquad \text{rewrite left-hand side using exponent rule}$$

$$\log\left(\frac{1.051}{1.044}\right)^x = \log\frac{25}{12} \qquad\qquad \text{take logs of both sides}$$

$$x \cdot \log\frac{1.051}{1.044} = \log\frac{25}{12} \qquad\qquad \text{using } \log(b^t) = t \cdot \log b$$

$$x = \frac{\log(25/12)}{\log(1.051/1.044)} = 109.833. \quad \text{dividing by } \log(1.051/1.044)$$

Example 4   Solve $\log(x + 198) - \log x = 2$ for x.

Solution   Using the properties of logarithms, we have

$$\log(x + 198) - \log x = 2$$
$$\log \frac{(x + 198)}{x} = 2 \qquad \text{using } \log\left(\frac{a}{b}\right) = \log a - \log b$$
$$\frac{(x + 198)}{x} = 100 \qquad \text{using the definition}$$
$$x + 198 = 100x \qquad \text{multiplying both sides by } x$$
$$198 = 99x \qquad \text{gathering terms}$$
$$x = 2.$$

Substituting $x = 2$ into the original equation gives $\log(2 + 198) - \log(2) = 2$, which is valid.

Example 5   Solve $\log x + \log(x - 3) = 1$ for x.

Solution   Using the properties of logarithms, we have

$$\log x + \log(x - 3) = 1$$
$$\log(x(x - 3)) = 1 \qquad \text{using } \log(ab) = \log a + \log b$$
$$x(x - 3) = 10 \qquad \text{using the definition}$$
$$x^2 - 3x - 10 = 0 \qquad \text{subtracting 10 from both sides}$$
$$(x - 5)(x + 2) = 0 \qquad \text{factoring.}$$

So, solving for x gives $x = 5$ or $x = -2$.

Although $x = 5$ and $x = -2$ satisfy the quadratic equation $x^2 - 3x - 10 = 0$, we must check both values of x to see if they satisfy our original equation.

For $x = 5$, we have $\log(5) + \log(2) = 1$, which is a valid statement. For $x = -2$, we have $\log(-2) + \log((-5) = 1$. Note that this is not valid, since $\log$ is not defined for negative input values. Thus, $x = 5$ is the only solution. The extraneous solution, $x = -2$, is introduced when we combine the two logarithm expressions into one.

Thus far, all our work has been done with logarithms with a base of 10. That is, if $y = \log x$ then $10^y = x$. Other bases may also be used. For any base b such that $b > 0, b \neq 1$, we define the following.

If x is a positive number,

$$\log_b x \text{ is the exponent of } b \text{ that gives } x.$$

In other words,

$$\text{if} \quad y = \log_b x \quad \text{then} \quad b^y = x.$$

Example 6   Use the definition to rewrite $3^t = 4000$ as a logarithm.

Solution   Using the definition,

$$3^t = 4000$$

is equivalent to

$$t = \log_3(4000).$$

In Example 3 on page 277, we solved $3^t = 4000$ using log base 10 to find

$$t = \frac{\log 4000}{\log 3} = 7.5496.$$

Example 7   Show that $\log_3(4000)$ is equivalent to $\dfrac{\log 4000}{\log 3}$.

Solution   Starting with

$$\log_3(4000) = t,$$

we have

$$3^t = 4000 \qquad \text{using the definition}$$
$$\log 3^t = \log 4000 \qquad \text{taking log of both sides}$$
$$t \log 3 = \log 4000 \qquad \text{using log property}$$

$$t = \frac{\log 4000}{\log 3} \qquad \text{dividing both sides by } \log 3.$$

Example 7 suggests the following result and enables us to evaluate $\log_b$ for any positive number in terms of $\log_{10}$ and thus with a calculator.

If a and b are positive numbers,

$$\log_b a = \frac{\log_{10} a}{\log_{10} b} = \frac{\log a}{\log b}$$

The properties shown for the common logarithm in Section 7.2 hold for log to any base b ($b > 0$).

Exercises and Problems for Section 7.3

Exercises

Solve the equations in Exercises 1–24.

1. $3 \cdot 10^t = 99$

2. $5 \cdot 10^x - 35 = 0$

3. $2.6 = 50 \cdot 10^t$

4. $10^t + 5 = 0$

5. $10^t - 5.4 = 7.2 - 10^t$

6. $2(10^y + 3) = 4 - 5(1 - 10^y)$

7. $100(1.041)^t = 520$

8. $5 \cdot 2^p = 90$

9. $40(1.033)^q = 600$

10. $100(0.988)^r = 55$

11. $8 \cdot \left(\frac{2}{3}\right)^t = 20$

12. $500 \cdot 1.31^b = 3200$

13. $700 \cdot 0.882^t = 80$

14. $50 \cdot 2^{x/12} = 320$

15. $120 \left(\frac{2}{3}\right)^{z/17} = 15$

16. $130 \cdot 1.031^w = 220 \cdot 1.022^w$

17. $700 \cdot 0.923^p = 300 \cdot 0.891^p$

18. $820(1.031)^t = 1140(1.029)^t$

19. $84.2(0.982)^y = 97(0.891)^y$

20. $1320(1.045)^q = 1700(1.067)^q$

21. $0.315(0.782)^x = 0.877(0.916)^x$

22. $500(1.032)^t = 750$

23. $2300(1.0417)^t = 8400$

24. $2215(0.944)^t = 800$

Problems

Solve the equations in Problems 25–31.

25. $50 \cdot 3^{t/4} = 175.$

26. $0.35 \cdot 2^{-12t} = 0.05$

27. $3 \log(x/2) = 6$

28. $10^{x+5} = 17$

29. $25 = 2 \cdot 10^{2x+1}$

30. $3 - 10^{x-0.5} = 10^{x-0.5}$

31. $2(10^{5x+1} + 2) = 3(1 + 10^{5x+1})$

Solve the equations in Problems 32–37.

32. $\log(2x) - 1 = 3$

33. $2 \log x - 1 = 5$

34. $2 \log(x - 1) = 6$

35. $\frac{1}{2} \log(2x - 1) + 2 = 5$

36. $2(\log(x - 1) - 1) = 4$

37. $\log(\log x) = 1$

In Problems 38–45, assume a and b are positive constants. Imagine solving for x (but don't actually do so). Will your answer involve logarithms? Explain how you can tell.

38. $10^x = a$

39. $10^2 x = 10^3 + 10^2$

40. $x^{10} - 10a = 0$

41. $2^{x+1} - 3 = 0$

42. $Q = b^x$

43. $a = \log x$

44. $3(\log x) + a = a^2 + \log x$

45. $Pa^{-kx} = Q$

7.4   APPLICATIONS OF LOGARITHMS TO MODELING

Logarithms provide a useful tool for analyzing exponential models. For instance, if a quantity such as a population or investment balance is modeled using exponential functions, we can use logs to find when the quantity reaches a certain value.

Example 1   The balance of a bank account begins at \$32,000 and earns 3.5% interest per year. When will it reach \$70,000?

Solution   The balance in year t is given by $32{,}000(1.035)^t$ dollars. We have

$$32{,}000(1.035)^t = 70{,}000$$
$$1.035^t = \frac{70}{32}$$
$$\log 1.035^t = \log\left(\frac{70}{32}\right) \qquad \text{taking logs of both sides}$$
$$t \cdot \log 1.035 = \log\left(\frac{70}{32}\right) \qquad \text{using } \log(b^t) = t \cdot \log b$$
$$t = \frac{\log(70/32)}{\log 1.035} = 22.754.$$

The balance will reach \$70,000 in 22.754 years.

Example 2   Between the years 1991 and 2001, the population of Naples, Italy decreased, on average, by 0.605% per year.[2] In 2001, the population was 1,004,500. If the population continues to shrink by 0.605% per year:

(a) What will the population be in 2010?
(b) When will the population reach half its 2001 level?

Solution   The population t years after 2001 is given by $1{,}004{,}500(0.99395)^t$.

(a) The year 2010 corresponds to $t = 9$. We have

$$\text{Population in year } 9 = 1{,}004{,}500(0.99395)^9 = 951{,}110.$$

(b) Half the 2001 level is $1{,}004{,}500/2 = 502{,}250$. Solving, we have:

$$1{,}004{,}500(0.99395)^t = 502{,}250$$
$$0.99395^t = \frac{502{,}250}{1{,}004{,}500} = \frac{1}{2}$$
$$\log 0.99395^t = \log\left(\frac{1}{2}\right) \qquad \text{taking logs of both sides}$$
$$t \log 0.99395 = \log\left(\frac{1}{2}\right) \qquad \text{using } \log b^t = t \log b$$
$$t = \frac{\log(1/2)}{\log 0.99395} = 114.223,$$

so Naples' population will halve in about 114 years.

In Examples 1 and 2, we used logs to solve equations of the form

$$f(x) = N, \qquad \text{where } f \text{ is an exponential function.}$$

In Example 3, we use logs to solve an equation of the form

$$f(x) = g(x), \qquad \text{where } f \text{ and } g \text{ are exponential functions.}$$

[2]http://www.citypopulation.de/Italien.html, accessed May 30, 2003.

Example 3 The population of City A begins with 25,000 people and grows at 2.5% per year. The population of City B begins with a larger population of 200,000 people but grows at the slower rate of 1.2% per year. Assuming that these growth rates hold constant, will the population of City A ever catch up to the population of City B? If so, when?

Solution In year t, the population (in 1000s) of City A is $25(1.025)^t$ and the population of City B is $200(1.012)^t$. City A starts out smaller, but it grows at a faster percent rate than City B. To find out when the population of City A catches up to City B, we solve the equation

$$\underbrace{25(1.025)^t}_{\text{population of } A} = \underbrace{200(1.012)^t}_{\text{population of } B}.$$

We first collect the terms in t before applying logs:

$$25(1.025)^t = 200(1.012)^t$$
$$\frac{(1.025)^t}{(1.012)^t} = \frac{200}{25} = 8$$
$$\left(\frac{1.025}{1.012}\right)^t = 8 \qquad\qquad \text{using } \frac{a^t}{b^t} = \left(\frac{a}{b}\right)^t$$
$$\log\left(\frac{1.025}{1.012}\right)^t = \log 8 \qquad\qquad \text{taking logs of both sides}$$
$$t\log\left(\frac{1.025}{1.012}\right) = \log 8 \qquad\qquad \text{using } \log b^t = t\log b$$
$$t = \frac{\log 8}{\log\left(\frac{1.025}{1.012}\right)} = 162.914.$$

The cities' populations will be the same in about 163 years. We can verify our solution by finding the populations of Cities A and B in year $t = 162.914$:

$$\text{Population of City } A = 25(1.025)^{162.914} = 1{,}396.393 \text{ thousand}$$
$$\text{Population of City } B = 200(1.012)^{162.914} = 1{,}396.394 \text{ thousand}.$$

The answers are not exactly equal because we rounded the value of t.

Doubling Times and Half-Lives

We can use logs to calculate doubling times and half-lives.

Example 4 The size of a population of frogs is given by $185(1.135)^t$ where t is in years.

(a) What is the initial size of the population?
(b) How long does it take the population to double in size?
(c) How long does this population take to quadruple in size? To increase by a factor of 8?

Solution

(a) The initial size is $185(1.135)^0 = 185$ frogs.

(b) In order to double, the population must reach 370 frogs. We need to solve the following equation for t:

$$185(1.135)^t = 370$$
$$1.135^t = 2$$
$$\log\left(1.135^t\right) = \log 2$$
$$t \cdot \log 1.135 = \log 2$$
$$t = \frac{\log 2}{\log 1.135} = 5.474 \text{ years.}$$

(c) We have seen that the population increases by 100% during the first 5.474 years. But since the population is growing at a constant percent rate, it will increase by 100% every 5.474 years, not just the first 5.474 years. Thus it doubles its initial size in the first 5.474 years, quadruples its initial size in two 5.474 year periods, or 10.948 years, and increases by a factor of 8 in three 5.474 year periods, or 16.422 years. We could also have used logs to solve directly for these time periods.

We can verify the results of the last example as follows:

$$185(1.135)^{5.474} = 370 \quad \text{twice the initial size}$$
$$185(1.135)^{10.948} = 740 \quad \text{4 times the initial size}$$
$$185(1.135)^{16.422} = 1480 \quad \text{8 times the initial size.}$$

Example 5   Polonium-218 decays at a rate of 16.578% per day. What is the half-life if you start with:

(a)   100 grams   (b) a grams

Solution   Decaying at a rate of 16.578% per day means that there is 83.422% remaining per day.

In part (a), we start with 100 g of Polonium-218 and need to find how long it takes for there to be 50 g, which means we must solve the equation

$$100(0.83422)^t = 50.$$

In part (b), we start with a grams and need to find how long it takes for there to be $a/2$ grams, which means we must solve the equation

$$a(0.83422)^t = \frac{a}{2}.$$

We solve these equations side-by-side to highlight the equivalent steps:

$$100(0.83422)^t = 50 \qquad\qquad\qquad a(0.83422)^t = \frac{a}{2}$$

$$0.83422^t = \frac{50}{100} \qquad\qquad\qquad 0.83422^t = \frac{\left(\frac{a}{2}\right)}{a}$$

$$0.83422^t = \frac{1}{2} \qquad\qquad\qquad 0.83422^t = \frac{1}{2}$$

$$t \log 0.83422 = \log\frac{1}{2} \qquad\qquad\qquad t \log 0.83422 = \log\frac{1}{2}$$

$$t = \frac{\log\frac{1}{2}}{\log 0.83422} = 3.824 \qquad\qquad\qquad t = \frac{\log\frac{1}{2}}{\log 0.83422} = 3.824,$$

so Polonium-218 has a half-life of about 3.824 days. Notice that from the third step on, the calculations are exactly the same, so we arrive at the same answer. The point of using a letter like a instead of a number like 100 is to show that the half-life does not depend on how much Polonium-218 we start with.

For another way to think about Example 5, notice that since any exponential function is of the form

$$\text{(Initial value)} \cdot \text{(Growth factor)}^t,$$

when we reach half the initial value, we know that the (Growth factor)^t component must equal $1/2$:

$$\text{(Initial value)} \cdot \underbrace{\text{(Growth factor)}^t}_{\frac{1}{2}} = \frac{1}{2} \cdot \text{Initial value}.$$

This means that in order to find the half-life, we must solve an equation of the form

$$\text{(Growth factor)}^t = \frac{1}{2}.$$

This is why the third step of both equations in Example 5 is

$$0.83422^t = \frac{1}{2}.$$

Logarithms and Measurement

We can compare sizes of quantities by computing their ratios. For example, if A is twice as big as B, then

$$\frac{\text{Size of A}}{\text{Size of B}} = 2.$$

Scientists often use this method of description to compare quantities that vary over a very wide range.

Orders of Magnitude

One way to describe the ratio between two numbers is as an *order of magnitude*. If one object is 10 times heavier than another, we say it is an order of magnitude heavier. If it is two factors of 10 heavier, that is $10 \cdot 10 = 100$ times heavier, we say it is two orders of magnitude heavier, and so on. For example,

$$\frac{\text{Distance from Sun to nearest star}}{\text{Distance from Earth to Sun}} = \frac{10^{16} \text{ meters}}{10^{11} \text{ meters}} = 10^{16-11} = 10^5 = \text{Five factors of 10}.$$

Thus, the distance from the Sun to the nearest star is five orders of magnitude greater than the distance from Earth to the Sun. Note that the order of magnitude is the logarithm of the ratio of the distances.

Example 6   The sound intensity of a rock concert is 10^{-4} watts/cm^2. A typical conversation has sound intensity 10^{-10} watts/cm^2. How many orders of magnitude more intense is the sound of a rock concert?

Solution   To compare the two intensities, we compute the ratio:

$$\frac{\text{Sound intensity of rock concert}}{\text{Sound intensity of normal conversation}} = \frac{10^{-4}}{10^{-10}} = 10^{-4-(-10)} = 10^{6}.$$

Thus, the sound intensity of the rock concert is 1,000,000 times greater than the sound intensity of a normal conversation. The log of this ratio is 6, so we say the sound intensity of the concert is 6 orders of magnitude greater than the sound intensity of a normal conversation.

Decibels

The intensity of audible sound varies over an enormous range. The scale used to describe sound is the *decibel* scale (abbreviated dB). The range of sound is so enormous that we consider the logarithm of the sound intensity. To measure a sound in decibels, the sound's intensity, I, is compared to the intensity of a standard benchmark sound, I_0. The intensity of I_0 is defined to be 10^{-16} watts/cm^2, roughly the lowest intensity audible to humans. The comparison between a sound intensity I and the benchmark sound intensity I_0 is made as follows:

$$\text{Noise level in decibels} = 10 \cdot \log\left(\frac{I}{I_0}\right).$$

Note that the decibel rating is ten times the number of orders of magnitude found by taking the ratio of the sound intensity of the object and the benchmark sound intensity.

Example 7   Using the sound intensities given in Example 6, find the decibel rating of
(a) A normal conversation,   (b) A rock concert,

Solution   (a) The decimal rating of a normal conversation is found by taking

$$10 \cdot \log\left(\frac{\text{Sound intensity of normal conversation}}{\text{Benchmark sound intensity}}\right) = 10 \cdot \log\left(\frac{10^{-10}}{10^{-16}}\right)$$
$$= 10 \cdot \log 10^{-10-(-16)}$$
$$= 10 \cdot \log 10^{6} = 60 \text{ dB}.$$

(b) The decimal rating of a rock concert is given by

$$10 \cdot \log\left(\frac{\text{Sound intensity of rock concert}}{\text{Benchmark sound intensity}}\right) = 10 \cdot \log\left(\frac{10^{-4}}{10^{-16}}\right)$$
$$= 10 \cdot \log 10^{-4-(-16)}$$
$$= 10 \cdot \log 10^{12} = 120 \text{ dB}.$$

The decibel rating is a type of *logarithmic scale.* Other types of logarithmic scales are often used to compare numbers over a very large range, for example, the strength of earthquakes, acidity (pH) levels, or the magnitude of stellar objects.

Exercises and Problems for Section 7.4

Exercises

1. A travel mug of $90°C$ coffee is left on the roof of a parked car on a $0°C$ winter day. The temperature of the coffee after t minutes is given by $H = 90(0.5)^{t/10}$. When will the coffee be only lukewarm ($30°C$)?

2. An investment initially worth $1200 grows by 11.2% per year. When will the investment be worth $5000?

3. A 10 kg radioactive metal loses 5% of its mass every 14 days. How many days will it take until the metal weighs 1 kg?

4. An investment begins with $500 and earns 10.1% per year. When will it be worth $1200?

5. In 2005, the population of Turkey was about 70 million and growing at 1.09% per year, and the population of the European Union (EU) was about 457 million and growing at 0.15% per year.[3] If current growth rates continue, when will Turkey's population equal that of the EU?

6. An abandoned building's value is $250(0.95)^t$ thousand dollars. When will Etienne be able to buy the building with his savings account which has $180,000 at year $t = 0$ and is growing by 12% per year?

7. City A, with population 60,000, is growing at a rate of 3% per year, while city B, with population 100,000, is losing its population at a rate of 4% per year. After how many years are their populations equal?

8. In 2003 the number of Sport Utility Vehicles sold in China was predicted to grow by 30% a year for the next 5 years.[4] What is the doubling time?

9. A substance decays at 4% per hour. Suppose we start with 100 grams of the substance. What is the half-life?

10. A population grows at a rate of 11% per day. Find its doubling time. How long does it take for the population to quadruple?

11. Cesium-137, which is found in spent nuclear fuel, decays at 2.27% per year. What is its half-life?

In Exercises 12–15, find the doubling time (for increasing quantities) or half-life (for decreasing quantities).

12. $Q = 200 \cdot 1.21^t$

13. $Q = 450 \cdot 0.81^t$

14. $Q = 55 \cdot 5^t$

15. $Q = 80 \cdot 0.22^t$

16. An investment initially worth $1900 grows at an annual rate of 6.1%. When will the investment be worth $5000?

Problems

17. By how many orders of magnitude is the radius of the universe (10^{26} meters) greater than the radius of the solar system (10^{12} meters)?

18. The population of a city t years after 1990 is given by $f(t) = 1800(1.055)^t$. Solve the equation $f(t) = 2500$. What does your answer tell you about the city?

19. The number of working lightbulbs in a large office building after t months is given by $g(t) = 4000(0.8)^t$. Solve the equation $g(t) = 1000$. What does your answer tell you about the lightbulbs?

20. The number of acres of wetland infested by an invasive plant species after t years is given by $u(t) = 12 \cdot 2^{t/3}$. Solve the equation $u(t) = 50$. What does your answer tell you about the wetland?

21. The amount of natural gas in thousands of cubic feet per day delivered by a well after t months is given by $v(t) = 80 \cdot 2^{-t/5}$. Solve the equation $v(t) = 5$. What does your answer tell you about the well?

22. A population of prairie dogs grows exponentially. The colony begins with 35 prairie dogs; three years later there are 200 prairie dogs.

   (a) Give a formula for the population as a function of time.

   (b) Use logarithms to find, to the nearest year, when the population reaches 1000 prairie dogs.

23. The dollar value of two investments after t years is given by $f(t) = 5000(1.062)^t$ and $g(t) = 9500(1.041)^t$. Solve the equation $f(t) = g(t)$. What does your solution tell you about the investments?

24. The amount of contamination remaining in two different wells t days after a chemical spill, in parts per billion (ppb), is given by $r(t) = 320(0.94)^t$ and $s(t) = 540(0.91)^t$. Solve the equation $r(t) = s(t)$. What does the solution tell you about the wells?

25. The size of two towns t years after 2000 is given by $u(t) = 1200(1.019)^t$ and $v(t) = 1550(1.038)^t$. Solve

---

[3]http://www.cia.gov/cia/publications/factbook, accessed October 14, 2005.
[4]"They're Huge, Heavy, and Loveable", *Business Week*, p. 58, (December 29, 2003).

the equation $u(t) = v(t)$. What does the solution tell you about the towns?

You deposit $10,000 into a bank account. In Problems 26–29, use the annual interest rate given to find the number of years before the account holds

(a) $15,000   (b) $20,000

26. 2%   **27.** 3.5%   **28.** 7%   **29.** 14%

30. A balloon is inflated in such a way that each minute its volume doubles in size. After 10 minutes, the volume of the balloon is 512 cm³.

(a) What is the volume of the balloon initially?
(b) When is the volume of the balloon 256 cm³, half of what it is at 10 minutes?
(c) When is the volume of the balloon 100 cm³?

31. A radioactive substance decays at a constant percentage rate per year.

(a) Find the half-life if it decays at a rate of
   (i)   10% per year.   (ii)   19% per year.
(b) Compare your answers in parts (i) and (ii). Why is one exactly half the other?

32. Iodine-131, used in medicine, has a half-life of 8 days.

(a) If 5 mg are stored for a week, how much is left?
(b) How many days does it take before only 1 mg remains?

33. The population of Nevada, the fastest growing state, increased by 12.2% between April 2000 and July 2003.[5]

(a) What was the yearly percent growth rate?
(b) What is the doubling time?

34. The population of Florida is 17 million and growing at 1.96% per year.[6]

(a) What is the doubling time?
(b) If the population continues to grow at the same rate, in how many years will the population reach 100 million?

35. When the minimum wage was enacted in 1938, it was $0.25 per hour.[7]

(a) If the minimum wage increased at the average rate of inflation, 4.3% per year, how much would it be in 2004?

(b) The minimum wage in 2004 was $5.15. Has the minimum wage been increasing faster or slower than inflation?
(c) If the minimum wage grows at 4.3% per year from 1938, when does it reach $10 per hour?

36. Three cities have the populations and annual growth rates in the table.

(a) Without any calculation, how can you tell which pairs of cities will have the same population at some time in the future?
(b) Find the time(s) at which each pair of cities has the same population.

City	A	B	C
Population (millions)	1.2	3.1	1.5
Growth rate (%/yr)	2.1	1.1	2.9

37. Water is passed through a pipe containing porous material to filter out pollutants. Each inch of pipe removes 50% of the pollutant entering.

(a) True or false: The first inch of the pipe removes 50% of the pollutants and the second inch removes the remaining 50%.
(b) What length of pipe is needed to remove all but 10% of the pollutants?

38. The population in millions of a bacteria culture after t hours is given by $y = 20 \cdot 3^t$.

(a) What is the initial population?
(b) What is the population after 2 hours?
(c) How long does it take for the population to reach 1000 million bacteria?
(d) What is the doubling time of the population?

39. The population in millions of a bacteria culture after t hours is given by $y = 30 \cdot 4^t$.

(a) What is the initial population?
(b) What is the population after 2 hours?
(c) How long does it take for the population to reach 1000 million bacteria?
(d) What is the doubling time of the population?

40. What is the doubling time of the population of a bacteria culture whose growth is modeled by $y = ab^t$, with t in hours?

---

[5]http://quickfacts.census.gov/qfd/states, accessed October 18, 2004.
[6]http://quickfacts.census.gov/qfd/states, accessed October 18, 2004.
[7]www.dol.gov, accessed October 18, 2004.

REVIEW EXERCISES AND PROBLEMS FOR CHAPTER SEVEN

Exercises

In Exercises 1–6, rewrite the equation using exponents instead of logarithms.

1. $\log 10 = 1$

2. $\log 0.001 = -3$

3. $\log 54.1 = 1.733$

4. $\log 0.328 = -0.484$

5. $\log w = r$

6. $\log 384 = n$

In Exercises 7–12, rewrite the equation using logarithms instead of exponents.

7. $10^6 = 1,000,000$

8. $10^0 = 1$

9. $10^{-1} = 0.1$

10. $10^{1.952} = 89.536$

11. $10^{-0.253} = 0.558$

12. $10^a = b$

13. (a) Rewrite the statement $10^2 = 100$ using logs.
   (b) Rewrite the statement $10^2 = 100$ using the square root operation.

14. (a) Rewrite the statement $10^3 = 1000$ using logs.
   (b) Rewrite the statement $10^3 = 1000$ using the cube root operation.

In Exercises 15–30, evaluate without a calculator, or say if the expression is undefined.

15. $\log 100$

16. $\log(1/100)$

17. $\log 1$

18. $\log 0$

19. $\log(-1)$

20. $\log \sqrt{10}$

21. $\log \sqrt[3]{10}$

22. $\log \dfrac{1}{\sqrt[4]{10}}$

23. $\log 10^{3.68}$

24. $\log 10^{-0.584}$

25. $\log 10^{2n+1}$

26. $10^{\log 10}$

27. $10^{\log(-10)}$

28. $10^{\log(1/100)}$

29. $10^{\log(5.9)}$

30. $10^{\log(3a-b)}$

Solve the equations in Exercises 31–68.

31. $5^t = 20$

32. $\pi^x = 100$

33. $(\sqrt{2})^x = 30$

34. $8(1.5)^t = 60.75$

35. $127(16)^t = 26$

36. $40(15)^t = 6000$

37. $3(64)^t = 12$

38. $16\left(\dfrac{1}{2}\right)^t = 4$

39. $100 \cdot 4^t = 254$

40. $(0.03)^{t/3} = 0.0027$

41. $155/3^t = 15$

42. $5 \cdot 6^x = 2$

43. $(0.9)^{2t} = 3/2$

44. $15^{t-2} = 75$

45. $2^{2t-5} = 1024$

46. $4(3^{t^2-7t+19}) = 4 \cdot 3^9$

47. $2(17)^t - 5 = 25$

48. $\dfrac{1}{4}(5)^t + 86.75 = 3^5$

49. $80 - (1.02)^t = 20$

50. $4(0.6)^t = 10(0.6)^{3t}$

51. $250(0.8)^t = 800(1.1)^t$

52. $AB^t = C, A \neq 0$

53. $10^x = 1,000,000,000$

54. $10^x = 0.00000001$

55. $10^x = \dfrac{1}{\sqrt[3]{100}}$

56. $10^x = 100^3$

57. $8 \cdot 10^x - 5 = 4$

58. $10 + 2(1 - 10^x) = 8$

59. $10^x = 3 \cdot 10^x - 3$

60. $10^x \sqrt{5} = 2$

61. $12 - 4(1.117)^t = 10$

62. $50 \cdot 2^t = 1500$

63. $30\left(1 - 0.5^t\right) = 25$

64. $800\left(\dfrac{9}{8}\right)^t = 1000$

65. $r + b^t = z$

66. $v + wb^t = L$

67. $40(1.118)^t = 90(1.007)^t$

68. $0.0315(0.988)^t = 0.0422(0.976)^t$

Simplify the expressions in Exercises 69–75.

69. $\log 10^{5x-3}$

70. $\log \sqrt{10^x}$

71. $\log\left(1000^{2x^3}\right)$

72. $(0.1)^{-\log(1+x^3)}$

73. $\log \dfrac{1}{\sqrt[3]{10^x}}$

74. $10^{2\log(3x+1)}$

75. $\log \sqrt{1000^x}$

Solve the equations in Exercises 76–85.

76. $\log x = 2\log\sqrt{2}$

77. $2\log x = \log 9$

78. $\log(x-1) = 3$

79. $\log(x^2) = 6$

80. $\log x = -5$

81. $\log x = \dfrac{1}{3}$

82. $\log x = 4.41$

83. $2\log x = 5$

84. $15 - 3\log x = 10$

85. $2\log x + 1 = 5 - \log x$

A population grows at the constant percent growth rate given in Exercises 86–88. Find the doubling time.

86. 5%   **87.** 10%   **88.** 25%

Problems

89. Percy Weasley's first job with the Ministry of International Magical Cooperation was to write a report trying to standardize the thickness of cauldrons, because "leakages have been increasing at a rate of almost three percent a year."[8] If this continues, when will there be twice as many leakages as there are now?

90. The population of a town decreases from an initial level of 800 by 6.2% each year. How long until the town is half its original size? 10% of its original size?

91. The number of regular radio listeners who haven't heard a new song begins at 600 and drops by half every two weeks. How long will it be until there are only 50 people who haven't heard the new song?

92. A predatory species of fish is introduced into a lake. The number of these fish after t years is given by $N = 20 \cdot 2^{t/3}$.

  (a) What is the initial size of the fish population?
  (b) How long does it take the population to double in size? To triple in size?
  (c) The population will stop growing when there are 200 fish. When will this occur?

93. One investment begins with $5000 and earns 6.2% per year. A second investment begins with less money, $2000, but it earns 9.2% per year. Will the second investment ever be worth as much as the first? If so, when?

94. Complete the following table. Based on your answers, which table entry is closest to $\log 15$?

x	1.175	1.176	1.177	1.178	1.179
10^x					

95. What does the table tell you about $\log 45$?

x	1.64	1.66	1.68	1.70
10^x	43.7	45.7	47.9	50.1

Without calculating them, explain why the two numbers in Problems 96–99 are equal.

96. $\log(1/2)$ and $-\log 2$ **97.** $\log(0.2)$ and $-\log 5$

98. $10^{\frac{1}{2}\log 5}$ and $\sqrt{5}$ **99.** $\log 303$ and $2+\log 3.03$

In Problems 100–105, assume x and b are positive. Which of statements (I)–(VI) is equivalent to the given statement?

I. $x = b$ II. $\log b = x$
III. $\log x = b$ IV. $x = \frac{1}{10}\log b$
V. $x = b/10$ VI. $x = b^{1/10}$

100. $10^b = x$ **101.** $x^{10} = b$

102. $x^3 - b^3 = 0$ **103.** $b^3 = 10^{3x}$

104. $10^x = b^{1/10}$ **105.** $10^{b/10} = 10^x$

If possible, use log properties to rewrite the expressions in Problems 106–111 in terms of u, v, w given that

$$u = \log x, v = \log y, w = \log z.$$

Your answers may involve constants such as $\log 2$ but non-constant logs and exponents are not allowed.

106. $\log xy$ **107.** $\log \dfrac{y}{x}$

108. $\log\left(2xz^3\sqrt{y}\right)$ **109.** $\dfrac{\log\left(x^2\sqrt{y}\right)}{\log\left(z^3\sqrt[3]{x}\right)}$

110. $\log\left(\log\left(3^x\right)\right)$

111. $\left(\log\dfrac{1}{x}\right)\left(\log\left(2y\right)\right)\left(\log\left(z^5\right)\right)$

---

[8]J.K. Rowling, *Harry Potter and the Goblet of Fire* (Scholastic Press 2000).

Chapter Eight

Polynomials

Contents

8.1   POLYNOMIALS

What is a Polynomial?

Have you ever wondered how your calculator works? The electronic circuits inside a calculator can perform only the basic operations of arithmetic. But what about the square-root button, or the x^y button? How can a calculator evaluate expressions like $\sqrt[9]{7}$ or $(3/2)^{2/3}$ using only addition and multiplication?

The answer is that some calculators use special built-in formulas that return excellent approximations for $\sqrt{x}$ or x^y. These formulas require only simple arithmetic, so they can be written using only the low-level instructions understood by computer chips.

Example 1   The expression

$$0.05x^3 - 0.26x^2 + 1.04x + 0.17$$

can be used to approximate the values of $x^{2/3}$ near $x = 1$.

(a) Show that this expression gives a good estimate for the value of $x^{2/3}$ for $x = 0.9$ and $x = 1.5$.
(b) What operations are used in evaluating these two expressions?

Solution   (a) We have

$$0.05(0.9)^3 - 0.26(0.9)^2 + 1.04(0.9) + 0.17 = 0.9319$$
$$(0.9)^{2/3} = 0.9322,$$

so the expression $0.05x^3 - 0.26x^2 + 1.04x + 0.17$ gives an excellent approximation of $x^{2/3}$ at $x = 0.9$. Likewise,

$$0.05(1.5)^3 - 0.26(1.5)^2 + 1.04(1.5) + 0.17 = 1.3138$$
$$(1.5)^{2/3} = 1.3104.$$

Once again, the approximation is quite good.
(b) Although the expression $0.05x^3 - 0.26x^2 + 1.04x + 0.17$ looks more complicated than $x^{2/3}$, it is simpler in one important way. You can calculate the values of the expression using only multiplication, addition, and subtraction. For instance, at $x = 0.9$, the first two terms are

$$0.05x^3 = 0.05(0.9)^3 = (0.05)(0.9)(0.9)(0.9) = 0.03645$$
$$-0.26x^2 = -0.26(0.9)^2 = (-1)(0.26)(0.9)(0.9) = -0.2106.$$

The other terms are calculated in a similar way, then added together. In contrast, if we want to evaluate $(0.9)^{2/3}$ directly, we have to take cube roots.[1]

The formulas used by a calculator to approximate the value of expressions like $x^{2/3}$ are more complicated than the one we have shown here. However, the basic principle is the same: the calculator uses only addition and multiplication.

An expression like the one in Example 1, that can be formed by adding together non-negative integer powers of x multiplied by constants, is an example of a *polynomial*. In general, a polynomial in x is any expression that can be formed from x and from constants using only the operations of addition and multiplication.

---

[1]In fact, there are methods for extracting cube roots similar to (though far more complicated than) the method of long division. See http://en.wikipedia.org/wiki/Shifting_nth-root_algorithm for details.

Example 2 Which of the following are polynomials in x?

(a) $x^2 + 3x^4 - \dfrac{x}{3}$ (b) $5\sqrt{x} - 25x^3$ (c) $x\sqrt{5} - (3x)^3$

(d) $9 \cdot 2^x + 4 \cdot 3^x + 15x$ (e) $(x-2)(x-4)$ (f) $\dfrac{3}{x} - \dfrac{x}{7}$

Solution (a) This is a polynomial. Writing it as

$$x^2 + 3x^4 - \frac{x}{3} = x^2 + 3x^4 - \frac{1}{3}x,$$

we see that the operations include raising x to positive integer powers (x^2 and x^4) and multiplying by constants (including the constant $-1/3$).

(b) This is not a polynomial because it involves taking the square root of x. We have $\sqrt{x} = x^{1/2}$, and $1/2$ is not an integer.

(c) This is a polynomial. Writing it as

$$x \cdot \sqrt{5} - 27x^3,$$

we see that the operations include multiplying x by the constant $\sqrt{5}$ and raising x to the power of 3.

(d) While a polynomial can have integer powers of x (like x^2 or x^3), it cannot have expressions like 2^x, where the variable appears in the exponent.

(e) This is a polynomial because it is formed by multiplying $x - 2$ and $x - 4$, and each of these is formed by adding x to constants. Multiplying out, we can put it in a form like that in Example 1:

$$(x-2)(x-4) = x^2 - 4x - 2x + 8 = x^2 - 6x + 8.$$

(f) This is not a polynomial because x appears in the denominator. If we write this as

$$3x^{-1} - \frac{x}{7},$$

we see that one of the operations is raising x to a negative power.

Polynomial Functions

Just as a linear function is one defined by a linear expression, and a quadratic function is one defined by a quadratic expression, a polynomial function is a function of the form

$$y = p(x) = \text{Polynomial in } x.$$

Example 3 A box is 4 ft longer than it is wide and twice as high as it is wide. Express the volume of the box as a polynomial function of the width.

Solution Let w be the width of the box in feet, and let V be its volume in cubic feet. Then its length is $w + 4$, and its height is $2w$, so

$$V = w(w+4)(2w) = 2w^2(w+4) = 2w^3 + 8w^2.$$

Example 4 A deposit of $800 is made into a bank account with an annual interest rate of r. Express B, the balance after 4 years, as a polynomial function of r.

Solution Each year the balance is multiplied by $1 + r$, so at the end of four years we have

$$B = 800(1+r)^4.$$

This is probably the most convenient form for the expression describing this function. However, by expanding and collecting like terms, we could express it as a sum of terms involving powers up to r^4. We do this in Problem 13 on page 300.

Example 5   Which of the following describe polynomial functions?

(a) The surface area of a cylinder of radius R and height h is $f(R) = 2\pi Rh + 2\pi R^2$.

(b) The balance after t years in a bank account with an annual interest rate of r is $B(t) = P(1+r)^t$.

(c) The effective potential energy of a sun-planet system is a function of the distance r between the planet and the sun given by $g(r) = -3r^{-1} + r^{-2}$.

Solution   (a) This is a polynomial function because it is expressed as a sum of a constant times R plus another constant times R^2.

(b) This is not a polynomial function because the independent variable occurs in the exponent.

(c) This is not a polynomial because r appears to a negative power.

Exercises and Problems for Section 8.1

Exercises

Based on the operations involved, which of the expressions in Exercises 1–6 are polynomials in x? If an expression is not a polynomial in x, what operation(s) rule it out?

1. $\dfrac{2x^3}{5} - 2x^7$

2. $x(x-1) - x^2(1-x^3)$

3. $\sqrt{2}x - x^2 + x^4$

4. $\sqrt{2x} - x^3 + x^5$

5. $\dfrac{1}{x} + \dfrac{x^2}{3} + 4x^3$

6. $14x^4 + 7x^3 - 3 \cdot 2^x + 2x + 15$

In Exercises 7–12, is the expression a polynomial in the given variable?

7. $(4 - 2p^2)p + 3p - (p+2)^2$, in p

8. $(x-1)(x-2)(x-3)(x-4) + 29$, in x

9. $\left(a + \dfrac{1}{x}\right)^2 - \left(a - \dfrac{1}{x}\right)^2$, in x

10. $\left(a + \dfrac{1}{x}\right)^2 - \left(a - \dfrac{1}{x}\right)^2$, in a

11. $\dfrac{n(n+1)(n+2)}{6}$, in n

12. $P\left(1 + \dfrac{r}{12}\right)^{10}$, in r

Problems

13. Expand the polynomial in Example 4.

Problems 14–17 refer to the functions $f(x)$ and $g(x)$: For values of x near 0, approximate values of the function

$$f(x) = \frac{1}{\sqrt{1-x}}$$

can be found using the function

$$g(x) = 1 + \frac{1}{2}x + \frac{3}{8}x^2 + \frac{5}{16}x^3.$$

14. Evaluate f and g at $x = 0$. What does this tell you about the graphs of these two functions?

15. Evaluate $f(x)$ and $g(x)$ at $x = 0.1, 0.2, 0.3$ and record your answers to three decimal places in a table. Does your table support the claim that $g(x)$ gives a good approximation for $f(x)$ for values of x near 0?

16. Show that $f(x)$ is undefined at $x = 1$ and $x = 2$, but that $g(x)$ is defined at these values. Explain why the algebraic operations used to define f may lead to undefined values, whereas the operations used to define g will not.

17. Given that

$$f(1/2) = \frac{1}{\sqrt{1-\frac{1}{2}}} = \frac{1}{\sqrt{\frac{1}{2}}} = \frac{1}{\frac{1}{\sqrt{2}}} = \sqrt{2},$$

use $g(x)$ to find a rational number (a fraction) that approximately equals $\sqrt{2}$.

8.2   THE FORM OF A POLYNOMIAL

Linear and Quadratic Polynomials

Any linear expression is a polynomial, because $b + mx$ involves adding the constant b to a constant multiple of the first power of x. Likewise, any quadratic expression is a polynomial, because $ax^2 + bx + c$ involves adding a constant c to constant multiples of the first and second powers of x.

The Degree of a Polynomial

Polynomials can be classified into a progression of expressions that include ever-higher powers of x. For example, *cubic* polynomials come next in line after quadratics, and can be written $ax^3 + bx^2 + cx + d$. *Quartic* polynomials come after cubics and include a term in x^4, and *quintic* polynomials come after quartics and include a term in x^5.

    The highest power of the variable occurring in the polynomial is called the *degree* of the polynomial. Thus, a linear expression has degree 1 and is a first-degree polynomial, a quadratic expression has degree 2 and is a second-degree polynomial, and so on.[2]

Example 1   Give the degree of

(a) $4x^5 - x^3 + 3x^2 + x + 1$ (b) $9 - x^3 - x^2 + 3(x^3 - 1)$

(c) $(1 - x)(1 + x^2)^2$

Solution   (a) The polynomial $4x^5 - x^3 + 3x^2 + x + 1$ is of degree 5 because the term with the highest power is $4x^5$.

(b) Expanding and collecting like terms, we have

$$9 - x^3 - x^2 + 3(x^3 - 1) = 9 - x^3 - x^2 + 3x^3 - 3 = 2x^3 - x^2 + 6,$$

so this is a polynomial of degree 3.

(c) Expanding and collecting like terms, we have

$$(1 - x)(1 + x^2)^2 = (1 - x)(1 + 2x^2 + x^4)$$
$$= 1 + 2x^2 + x^4 - x - 2x^3 - x^5$$
$$= -x^5 + x^4 - 2x^3 + 2x^2 - x + 1,$$

so this is a polynomial of degree 5.

The Standard Form of a Polynomial

So far, the standard forms for the expressions we have discussed in this book have involved two or, in the case of quadratics, three letters for the constants:

Family	Standard form
Linear	$b + mx$
Quadratic	$ax^2 + bx + c$
Power	kx^p
Exponential	ab^x

[2]Note that a constant, like the number 5, is considered a polynomial of degree 0, because it can be written $5x^0$.

Polynomials are different, because the higher the degree, the more terms they may have, and thus the more letters it takes to write them. For this reason, it becomes increasingly difficult to write the general form of a polynomial as the degree gets larger. Therefore, we must take a different approach to writing the general formula for polynomials. Instead of using different letters like $a, b, c \ldots$ for the coefficients, we number them using subscripts, like this:

$$a_0, a_1, a_2, \ldots.$$

This way, the coefficient of x^4 is written a_4, the coefficient of x^3 is written a_3, and so on. The constant term, which we can think of as the coefficient of $x^0 = 1$, is written a_0. Using this notation, the standard forms for first- through fifth-degree polynomials are written like this:

Degree of polynomial	Standard form
1	$a_1 x + a_0$
2	$a_2 x^2 + a_1 x + a_0$
3	$a_3 x^3 + a_2 x^2 + a_1 x + a_0$
4	$a_4 x^4 + a_3 x^3 + a_2 x^2 + a_1 x + a_0$
5	$a_5 x^5 + a_4 x^4 + a_3 x^3 + a_2 x^2 + a_1 x + a_0$

Example 2   Identify the coefficients of (a) $2x^3 + 3x^2 + 4x - 1$ (b) $4x^5 - x^3 + 3x^2 + x + 1$ (c) x^2.

Solution   (a) The subscript of the coefficient should match up with the power of x, so

$$2x^3 + 3x^2 + 4x - 1 = a_3 x^3 + a_2 x^2 + a_1 x + a_0.$$

Thus, $a_3 = 2$, $a_2 = 3$, $a_1 = 4$, and $a_0 = -1$.
(b) Writing this as

$$4x^5 - x^3 + 3x^2 + x + 1 = 4x^5 + 0 \cdot x^4 + (-1)x^3 + 3x^2 + 1 \cdot x^1 + 1 \cdot x^0,$$

we see that it has coefficients

$$a_5 = 4, \quad a_4 = 0, \quad a_3 = -1, \quad a_2 = 3, \quad a_1 = 1, \quad a_0 = 1.$$

(c) Writing this as
$$x^2 = 1 \cdot x^2 + 0 \cdot x + 0,$$

we get $a_2 = 1$, $a_1 = 0$, and $a_0 = 0$.

Polynomials of Degree n

If we do not know the degree of a polynomial, we use a letter, such as n, to represent it. So n represents the highest power of the variable that appears in the polynomial, and the coefficient of this term is written a_n.

Example 3   Consider the polynomial
$$13x^{17} - 9x^{16} + 4x^{14} + 3x^2 + 7.$$

(a) What is the degree, n?
(b) Write out the terms $a_n x^n$ and $a_{n-1} x^{n-1}$.
(c) What is the value of a_{n-3}?

Solution

(a) The degree is $n = 17$.

(b) The term $13x^{17}$ is $a_n x^n$ where $a_n = 13$. The term $-9x^{16}$ is $a_{n-1} x^{n-1}$ where $a_{n-1} = -9$.

(c) The degree is $n = 17$, so $a_{n-3} = a_{14}$ is the coefficient of x^{14}, which means that $a_{n-3} = 4$.

In general, we have:

The Standard Form of a Polynomial

The standard form of a polynomial in x is

$$a_n x^n + a_{n-1} x^{n-1} + \cdots + a_1 x + a_0,$$

where n is a non-negative integer and $a_0, a_1, \ldots, a_n$ are constants.
- The constants $a_0, a_1, \ldots, a_n$ are the *coefficients*.
- If $a_n \neq 0$, then n is the *degree* of the polynomial.

Since the expression for a general polynomial is cumbersome, we sometimes use the function notation $p(x)$ to refer to it.

The Constant Term

In the polynomial

$$p(x) = a_n x^n + a_{n-1} x^{n-1} + \cdots + a_1 x + a_0,$$

the term a_0 is the only term not multiplied by a positive power of x, and we call it the *constant term* of the polynomial. The constant term gives the value of $p(0)$:

$$p(0) = a_n \cdot 0^n + a_{n-1} \cdot 0^{n-1} + \cdots + a_1 \cdot 0 + a_0 = a_0.$$

Thus the constant term is the vertical intercept of the graph of p.

Example 4   Give the constant term of   (a) $4 - x^3 - x^2$ (b) $z^3 + z^2$ (c) $(q - 1)(q - 2)(q - 3)$.

Solution

(a) Writing this as $-x^3 - x^2 + 4$, we see that its constant term is $a_0 = 4$.

(b) Since the only two terms in this polynomial involve positive powers of z, it has constant term $a_0 = 0$.

(c) We could calculate the constant term of this polynomial by expanding it out. However, since we know that the constant term is the value of the polynomial when $q = 0$, we can say that

$$a_0 = (0 - 1)(0 - 2)(0 - 3) = (-1)(-2)(-3) = -6.$$

Example 5   Which of Figures 8.1 and 8.2 is the graph of $y = x^4 - 4x^3 + 16x - 16$?

Solution   Since the graph in Figure 8.1 passes through the origin, it is the graph of a function that has the value 0 at $x = 0$. But $x^4 - 4x^3 + 16x - 16$ has the value -16 at $x = 0$, because its constant term is -16. So it must correspond to the graph in Figure 8.2. This makes sense, since the graph appears to intercept the y-axis at around $y = -16$.

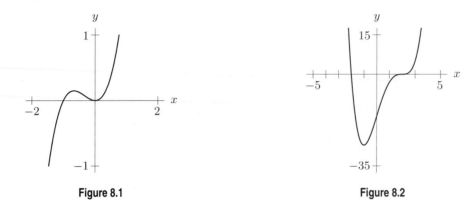

Figure 8.1   Figure 8.2

Example 6   Find the constant term of the polynomial in Example 4 on page 299, and interpret it terms of interest rates.

Solution   The polynomial is $800(1 + r)^4$, and gives the bank balance 4 years after an initial deposit of $800. Although we have not expanded this polynomial out, we know that the constant term is the value when $r = 0$, so

$$\text{Constant term} = 800(1 + 0)^4 = 800.$$

This is the value of the bank balance after 4 years if the interest rate is 0. It makes sense that this is the same as the initial deposit, since if the interest rate is 0 the balance does not grow.

The Leading Term

We already know that all first-degree polynomials have graphs that are lines, and all second-degree polynomials have graphs that are parabolas. This suggests that polynomials of the same degree have certain features in common and that the term with the highest power plays an important role in the overall behavior of a polynomial function. In a polynomial $a_n x^n + a_{n-1} x^{n-1} + \cdots + a_1 x + a_0$, with $a_n \neq 0$, the highest power term $a_n x^n$ is called the *leading term* of the polynomial.

Example 7   Give the degree and leading term of

(a) $s - 3s^3 + s^2$ (b) $2137w^4 + 0.003w^5$ (c) $1 - \dfrac{m^9 \sqrt{2}}{7} + 22m^4$

Solution   (a) The degree is $n = 3$ and the leading term is $-3s^3$.
(b) The degree is $n = 5$ and the leading term is $0.003w^5$.
(c) The degree is $n = 9$ and the leading term is $-\dfrac{\sqrt{2}}{7}m^9$.

Example 8 If $p(x)$ has degree 4 and constant term -2, and $q(x)$ has degree 5 and constant term 7, what can you say about

(a) The degree of $p(x)q(x)$?
(b) The coefficient of x in $p(x)q(x)$?
(c) The degree of $p(x) + q(x)$?
(d) The constant term of $3p(x) - 2q(x)$?
(e) The degree of $p(x)^2$?

Solution

(a) When we multiply two polynomials together, the leading term in the product comes from multiplying together the leading terms in each factor. The leading term in $p(x)$ is a constant times x^4 and the leading term in $q(x)$ is a constant times x^5, so the leading term in $p(x)q(x)$ is a constant times $x^4 \cdot x^5 = x^9$. Thus the degree of $p(x)q(x)$ is 9.

(b) We do not have enough information: to know the coefficient of x in $p(x)q(x)$ we would need to know more about the coefficients in $p(x)$ and $q(x)$.

(c) Since $p(x)$ has terms only up to degree 4, the leading term in $p(x) + q(x)$ is the term in x^5 coming from $q(x)$, so the degree is 5.

(d) The constant terms in $p(x)$ and $q(x)$ are the only terms contributing to the constant term in $3p(x) - 2q(x)$, so that term is $3(-2) - 2 \cdot 7 = -20$.

(e) When we multiply $p(x)$ by itself, the highest degree terms comes from the highest degree of $p(x)$, which is a constant times x^4, times itself, giving a constant times x^8. Thus the degree of $p(x)^2$ is 8.

Example 9 For what values of the constant a is the degree of $(ax^3 + 1)^2 - x^6$ less than 6?

Solution One approach would be to place this polynomial in standard form. However, we can determine the leading term with fewer steps as follows:

$$(ax^3 + 1)^2 - x^6 = (ax^3)^2 + \text{(lower order terms)} - x^6$$
$$= a^2 x^6 - x^6 + \text{(lower order terms)}$$
$$= (a^2 - 1)x^6 + \text{(lower order terms)}.$$

Notice that for most values of a, the degree of the polynomial is 6. However, the term $(a^2 - 1)x^6$ disappears if $a^2 - 1 = 0$, that is, if $a = \pm 1$. Therefore, for these values of a, the degree is less than 6.

Summary

- The standard form for a polynomial in x is
$$a_n x^n + a_{n-1} x^{n-1} + \cdots + a_1 x + a_0, \quad a_n \neq 0.$$

- Here, n is called the *degree* of the polynomial and is the power of the highest-power term or *leading term*.

- The *coefficients* are $a_0, a_1, \ldots, a_{n-1}$, and a_n.

- The coefficient a_0 is also called the *constant term*. It determines the value of the expression at $x = 0$.

- Special categories of polynomials include (in order of increasing degree) *linear, quadratic, cubic, quartic*, and *quintic*.

Exercises and Problems for Section 8.2

Exercises

In Exercises 1–3 find the degree.

1. $2x^6 - 3x^5 - 6x^4 - 4x + 1$

2. $2x^3 - x^2 + 1 - x^3 + 2x^2 - x + 3x^3$

3. $3x^2 + 2x^4 + x - x^4 + 2x^3 - 1 - x^4 + 3x$

In Exercises 4–7, write the polynomials in standard form.

4. $3x - 2x^2 + 5x^7 + 4x^5$

5. $3x^2 + 2x + 2x^7 - 5x^2 - 3x^7$

6. $x(x - 2) + x^2(3 - x)$ **7.** $\dfrac{x^4 - 2x - 14x^3}{7}$

In Exercises 8–13, give the leading term.

8. $3x^5 - 2x^3 + 4$ **9.** $2x^7 - 4x^{11} + 6$

10. $12 - 3x^5 - 15x^3$ **11.** $12x^{13} + 4x^5 - 11x^{13}$

12. $13x^4(2x^2 + 1)$ **13.** x^8

In Exercises 14–18 give the leading coefficient.

14. $5x^6 - 4x^5 + 3x^4 - 2x^3 + x^2 + 1$

15. $1 - 6x^2 + 40x - \frac{1}{2}x^3 + 16x$

16. $100 - \sqrt{6}x + 15x^2$

17. $\sqrt{7}x^3 + 12x - 4 + 6x^2$

18. $x^3 - 2x^2 - \sqrt[3]{9}x^3 + 1$

In Exercises 19–24, give the constant term.

19. $4t^3 - 2t^2 + 17$ **20.** $12t - 2t^3 + 6$

21. $15 - 11t^9 - 8t^4$ **22.** $7t^3 + 2t^2 + 5t$

23. $(3t + 1)(2t - 1)$ **24.** $t(t - 1)(t - 2)$

In Exercises 25–30, is the expression a polynomi[al? For] those that are, give the degree and the leading coeff[icient. As]sume a is a positive integer.)

25. $8x^5 - x^3 + 11x^2 + 50$ **26.** $5 - \sqrt{2}x^2$

27. $ax^{-1} + 2$ **28.** a^{-1}

29. $x^a + a^x, a \neq 1$ **30.** $\dfrac{1 - x(x + 1)}{3}$

List the nonzero coefficients of the polynomials in Exercises 31–34.

31. $3x^4 + 6x^3 - 3x^2 + 8x + 1$ **32.** $2x^5 - 3x^3 + x^7 + 1$

33. $\dfrac{x^{13}}{3}$ **34.** πx

In Exercises 35–38, write polynomials in standard form in x based on the coefficients given assuming all other coefficients equal zero.

35. $a_3 = 2, a_2 = 4, a_1 = -3, a_0 = 5$

36. $a_6 = 17, a_5 = 4, a_2 = 9$

37. $a_{20} = 33, a_{14} = 20, a_{33} = 7, a_7 = 14$

38. $a_1 = 2$

Problems

Write the polynomials in Problems 39–48 in standard form

$$a_n x^n + a_{n-1} x^{n-1} + \cdots + a_1 x + a_0.$$

What are the values of the coefficients $a_0, a_1, \ldots, a_n$? Give the degree of the polynomial.

39. $2x^3 + x - 2$ **40.** $5 - 3x^7$

41. $20 - 5x$ **42.** $\sqrt{7}$

43. $\dfrac{x^2 \sqrt[3]{5}}{7}$ **44.** $3 - 2(x - 5)^2$

45. $15x - 4x^3 + 12x - 5x^4 + 9x - 6x^5$

46. $(x - 3)(2x - 1)(x - 2)$ **47.** $(x + 1)^3$

48. $1 + x + \dfrac{x^2}{2} + \dfrac{x^3}{6} + \dfrac{x^4}{24} + \dfrac{x^5}{120}$

49. What is the constant term of

$$(x + 2)(x + 3)(x + 4)(x + 5)(x + 6)?$$

50. What is the coefficient of x^4 in

$$(x+2)(x+3)(x+4)(x+5)(x+6)?$$

51. What is the degree and leading coefficient of the polynomial $r(x) = 4$?

In Problems 52–56, state the given quantities if $p(x)$ is a polynomial of degree 5 with constant term 3, and $q(x)$ is a polynomial of degree 8 with constant term -2.

52. The constant term of $p(x)q(x)$

53. The degree of $p(x) + q(x)$

54. The degree of $p(x)q(x)$

55. The constant term of $p(x) - 2q(x)$

56. The degree of $p(x)^3 q(x)^2$

In Problems 57–59, what is the degree of the resulting polynomial?

57. The product of a quadratic with a linear polynomial.

58. The product of two linear polynomials.

59. The sum of a degree 8 polynomial and a degree 4 polynomial.

Evaluate the expressions in Problems 60–63 given that

$$f(x) = 2x^3 + 3x - 3, \quad g(x) = 3x^2 - 2x - 4,$$

$$h(x) = f(x)g(x) = a_n x^n + a_{n-1}x^{n-1} + \cdots + a_0.$$

60. n **61.** a_{n-1}

62. a_0 **63.** $h(1)$

64. If the following product of two polynomials,

$$(3x^2 - 7x - 2)(4x^3 - 3x^2 + 5),$$

is written in standard form, what are the constant and leading terms?

65. Given that $(3x^3 + a)(2x^b + 3) = 6x^7 + cx^4 + dx^3 + 3$, find possible values for the constants a, b, c, and d.

66. Find the constants r, s, p, and q if multiplying out the polynomial $(rx^5 + 2x^4 + 3)(2x^3 - sx^2 + p)$ gives

$$6x^8 - 11x^7 - 10x^6 - 12x^5 - 8x^4 + qx^3 - 15x^2 - 12.$$

In Problems 67–71, give polynomials satisfying the given conditions if possible, or say why it is impossible to do so.

67. Two polynomials of degree 5 whose sum has degree 4.

68. Two polynomials of degree 5 whose sum has degree 8.

69. Two polynomials of degree 5 whose product has degree 8.

70. Two polynomials whose product has degree 9.

71. A polynomial whose product with itself has degree 9.

In Problems 72–75, give the value of a that makes the statement true.

72. The degree of $(t-1)^3 + a(t+1)^3$ is less than 3.

73. The constant term of $(t-1)^5 - (t-a)$ is zero.

74. The coefficient of t in $t(a + (t+1)^{10})$ is zero.

75. The constant term of $(t+2)^2(t-a)^2$ is 9.

76. Suppose that two polynomials $p(x)$ and $q(x)$ have constant term 1, the coefficient of x in $p(x)$ is a and the coefficient of x in $q(x)$ is b. What is the coefficient of x in $p(x)q(x)$?

77. Find the product of $5x^2 - 3x + 1$ and $10x^3 - 3x^2 - 1$.

78. Find the leading term and constant term of **(a)** $(x-1)^2$ **(b)** $(x-1)^3$ **(c)** $(x-1)^4$

79. What is the coefficient of x^{n-1} in $(x+1)^n$ for $n = 2, 3$ and 4?

8.3 POLYNOMIAL EQUATIONS

A *zero* of a polynomial $p(x)$ is a real number a such that $p(a) = 0$. In other words, it is a solution to the equation

$$p(x) = 0.$$

In the case of quadratic polynomials, we have seen that we can sometimes solve this kind of equation by factoring the left-hand side. The same approach can also work for polynomials of higher degree.

Example 1   Find the zeros of the polynomial $p(x) = x^3 - x^2 - 6x$ by factoring.

Solution The zeros of the polynomial $p(x) = x^3 - x^2 - 6x$ are the solutions of the equation $x^3 - x^2 - 6x = 0$. By factoring out an x and then factoring the quadratic, $x^2 - x - 6$, we can rewrite the expression on the left of this equation as

$$x^3 - x^2 - 6x = x(x^2 - x - 6) = x(x - 3)(x + 2).$$

This gives

$$x(x - 3)(x + 2) = 0.$$

For the expression on the left to be zero, one of the three linear factors, x, $x - 3$, and $x + 2$, must be zero, giving

$$x = 0, \quad \text{or} \quad x - 3 = 0, \quad \text{or} \quad x + 2 = 0,$$

so the possible zeros of $p(x)$ are

$$x = 0, \quad \text{or} \quad x = 3, \quad \text{or} \quad x = -2.$$

To check that they are zeros, we evaluate p at these values of x:

$$\begin{aligned} p(0) &= 0^3 - 0^2 - 6 \cdot 0 & &= 0 - 0 - 0 & &= 0 \\ p(3) &= 3^3 - 3^2 - 6 \cdot 3 & &= 27 - 9 - 18 & &= 0 \\ p(-2) &= (-2)^3 - (-2)^2 - 6(-2) & &= -8 - 4 + 12 = 0. \end{aligned}$$

Solving Polynomial Equations by Factoring

A *polynomial equation* is an equation of the form

$$p(x) = q(x)$$

where $p(x)$ and $q(x)$ are polynomials. We have seen how factoring a polynomial can help us to find its zeros. Similarly, factoring can also help us to solve polynomial equations.

Example 2 Solve $w^3 - 2w - 1 = (w + 1)^3$.

Solution We write the equation in the form $p(w) = 0$ where p is a polynomial in standard form:

$$\begin{aligned} w^3 - 2w - 1 &= w^3 + 3w^2 + 3w + 1 \\ w^3 - 2w - 1 - w^3 - 3w^2 - 3w - 1 &= 0 && \text{subtract right side from both sides} \\ -3w^2 - 5w - 2 &= 0 && \text{collect like terms (w^3 terms cancel)} \\ 3w^2 + 5w + 2 &= 0 && \text{multiply both sides by } -1. \end{aligned}$$

Since the w^3 terms cancel, we are left with a quadratic equation, whose solutions we can find by factoring or the quadratic formula. Using factoring, we get

$$(3w + 2)(w + 1) = 0,$$

so the solutions are $w = -2/3$, $w = -1$.

Example 3 Solve $x^2 = x^3$.

Solution We subtract x^3 from both sides and factor:

$$\begin{aligned} x^2 - x^3 &= 0 \\ x^2(1 - x) &= 0. \end{aligned}$$

For the expression on the left to be zero, one of the factors x^2 or $1 - x$ must be zero, so

$$x = 0 \quad \text{or} \quad x = 1.$$

A common mistake is to begin by dividing by x^2, obtaining

$$x = 1.$$

This equation is not equivalent to the original equation because it does not have the same solutions. The mistake in dividing by x^2 is that if $x = 0$ then you are dividing by zero.

Example 4   Show that 2 and 3 are the only consecutive positive integers such that the cube of the first is 1 less than the square of the second.

Solution   If we have two consecutive numbers, say n and $n + 1$, such that the cube of the first, n^3, is one less than the square of the second, $(n + 1)^2$, then we have the polynomial equation

$$n^3 = (n + 1)^2 - 1.$$

Solving this equation gives

$$n^3 = n^2 + 2n + 1 - 1$$
$$n^3 - n^2 - 2n = 0$$
$$n(n + 1)(n - 2) = 0.$$

The solutions are $n = 0$, $n = -1$, and $n = 2$. The only positive integer solution is $n = 2$, so the only pair of positive consecutive integers with this property is $n = 2, n + 1 = 3$.

Finding Factors From Zeros

There are two important facts relating factors to zeros:

> (i) If a polynomial has a factor $(x - k)$, then it has a zero at $x = k$.
>
> (ii) If a polynomial has a zero at $x = k$, then it has a factor $(x - k)$.

Although these two facts may sound similar, they tell us different things. Fact (i) is how we know that the polynomial $x^3 - x^2 - 6x$ from Example 1 has zeros at $x = -2, 0, 3$: we determined the zeros by looking at the factors. Fact (ii) lets us go in the other direction, allowing us to determine the factors by looking at the zeros.

Example 5   Find a polynomial that has degree 4 and has zeros at $t = -1, t = 0, t = 1$ and $t = 2$.

Solution   For each zero there is a factor. The zero $t = -1$ tells us that the polynomial has a factor $t - (-1)$, or $t + 1$. The zero $t = 0$ tells us that the polynomial has a factor $t - 0$, or t. Similarly, the other zeros, $t = 1$ and $t = 2$, tell us that the polynomial has factors $t - 1$ and $t - 2$. Since the polynomial has degree 4, and these four factors, when multiplied together, give a polynomial of degree 4, a possible polynomial is

$$t(t + 1)(t - 1)(t - 2).$$

Notice that we could have chosen other polynomials as well, such as

$$2t(t+1)(t-1)(t-2) \quad \text{or} \quad 3t(t+1)(t-1)(t-2).$$

In fact, we could choose any polynomial of the form

$$kt(t+1)(t-1)(t-2), \quad \text{where } k \text{ is any nonzero constant.}$$

Example 6   (a) Use the fact that

$$3000 + 700 + 20 = 3720$$

to find a solution to the equation

$$3x^3 + 7x^2 + 2x = 3720.$$

(b) What does your answer tell you about the factors of the polynomial $3x^3 + 7x^2 + 2x - 3720$?

Solution   (a) We can rewrite $3000 + 700 + 20 = 3720$ as

$$3 \cdot 10^3 + 7 \cdot 10^2 + 2 \cdot 10^1 + 0 \cdot 10^0 = 3720.$$

This means that the equation
$$3x^3 + 7x^2 + 2x = 3720$$

has a solution at $x = 10$. Consequently, the polynomial

$$3x^3 + 7x^2 + 2x - 3720$$

has a zero at $x = 10$.

(b) According to fact (ii) on page 309, we know $(x - 10)$ must be a factor. In fact,

$$3x^3 + 7x^2 + 2x - 3720 = (x - 10)(3x^2 + 37x + 372).$$

Later we see how to find the factors on the right-hand side.

Why Are These Facts True?

Fact (i) on page 309 is true because, if a polynomial $p(x)$ has a factor $(x - k)$, then

$$p(x) = (x - k) \cdot (\text{Other factors}).$$

Putting $x = k$ into this equation we get

$$p(k) = 0 \cdot (\text{Another number}).$$

Since zero times any number is zero, this says that $p(k) = 0$, so $x = k$ is a zero of $p(x)$.
    We account for fact (ii) on page 345.

How Many Zeros Can a Polynomial Have?

The number of factors of a polynomial is less than or equal to the degree of that polynomial. For example, a fourth degree polynomial can have no more than four factors. Otherwise, when we multiply out, the highest power of x would be greater than four. Since each zero of a polynomial corresponds to a factor, the number of zeros is less than or equal to the degree of the polynomial.

Example 7 Without solving, say how many solutions the equation could have:

(a) $x^4 + x^3 - 3x^2 - 2x + 2 = 0$ (b) $3(x^3 - x) = 1 + 3x^3$

(c) $(s^2 + 5)(s - 3) = 0$

Solution (a) Since the degree of the polynomial on the left is 4, the equation can have at most 4 solutions.

(b) This looks like it could simplify to an equation of the form $p(x) = 0$, where $p(x)$ is a cubic polynomial, so there would be up to 3 solutions. However, the leading term on both sides is $3x^3$, so when we simplify, the leading terms cancel, leaving only the linear term. Thus the equation can only have one solution.

(c) The polynomial on the left is a cubic, so there could be up to 3 solutions. However, the quadratic factor, $s^2 + 5$, cannot have any zeros because it is 5 plus a square, so it is always positive. Thus, there is only one solution, coming from the linear factor.

Multiple Zeros

Sometimes the factor of a polynomial is repeated, as in

$$(x - 4)^2 = \underbrace{(x - 4)(x - 4)}_{\text{Occurs twice}}$$

$$\text{or} \quad (x + 1)^3 = \underbrace{(x + 1)(x + 1)(x + 1)}_{\text{Occurs three times}}.$$

In this case we call the zeros *multiple zeros*, because the factor contributing the zero is repeated. For instance, we say that $x = 4$ is a *double zero* of $(x - 4)^2$, and $x = -1$ is a *triple zero* of $(x + 1)^3$. If a zero is not repeated, we call it a *simple zero*.

Visualizing the Zeros of a Polynomial

Since the zeros of a polynomial $p(x)$ are the values $x = a$ where $p(a) = 0$, they are also the x-intercepts of the graph. In Example 1 on page 307 we found that the polynomial $x^3 - x^2 - 6x$ has zeros at $x = -2$, $x = 0$, and $x = 3$ by factoring. Figure 8.3 shows the graph of $y = x^3 - x^2 - 6x$ crossing the x-axis at $x = -2$, $x = 0$, and $x = 3$.

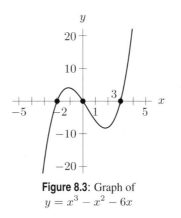

Figure 8.3: Graph of
$y = x^3 - x^2 - 6x$

What Does the Graph Look Like at Multiple Zeros?

In the polynomial $(x-4)^2$, the factor $(x-4)$ occurs twice, so we say the zero at $x = 4$ is a double zero. Likewise, the factor $(x+1)$ occurs three times in the polynomial $(x+1)^3$, so we say the zero at $x = -1$ is a triple zero. The graphs of $(x-4)^2$ and $(x+1)^3$ in Figures 8.4 and 8.5 show typical behavior near multiple zeros.

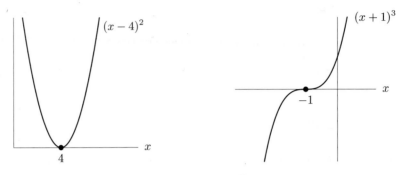

Figure 8.4: Double zero at $x = 4$ **Figure 8.5**: Triple zero at $x = -1$

In general:

The Appearance of the Graph at a Multiple Zero

Suppose $p(x)$ has a multiple zero at $x = k$ that occurs exactly n times, so

$$p(x) = (x-k)^n \cdot (\text{Other factors}), \quad n \geq 2.$$

- If n is even, the graph of the polynomial does not cross the x-axis at $x = k$, but "bounces" off the x-axis at $x = k$. (See Figure 8.4.)
- If n is odd, the graph of the polynomial crosses the x-axis at $x = k$, but it looks flattened there. (See Figure 8.5.)

Example 8   Describe in words the zeros of the polynomials of degree four graphed in Figure 8.6.

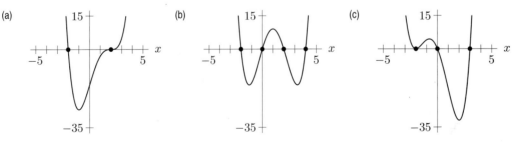

Figure 8.6: Polynomials of degree four

Solution
(a) The first graph has two x-intercepts, one at $x = -2$ and one at $x = 2$. The flattened appearance near $x = 2$ suggests that the polynomial has a multiple zero there. Since the graph crosses the x-axis at $x = 2$ (instead of bouncing off it), the corresponding factor must occur an odd number of times. Since the polynomial is of degree 4, we conclude there is a triple zero at $x = 2$ and a simple zero at $x = -2$.

(b) The second graph has four x-intercepts, corresponding to four simple zeros. We know none of the zeros are repeated because the polynomial is degree 4.

(c) The third graph has three x-intercepts at $x = 0$, $x = 3$, and $x = -2$. The graph bounces at $x = -2$, so this is a double zero of the polynomial. It is not a higher-order zero (such as a quadruple zero) because the polynomial is only of degree 4.

Exercises and Problems for Section 8.3

Exercises

Find the zeros of the polynomials in Exercises 1–6.

1. $(x - 3)(x - 4)(x + 2)$ **2.** $x(x + 5)(x - 7)^2$

3. $x^2 - x - 6$ **4.** $x^4 - 4x^3 + 4x^2$

5. $x^2 + 1$ **6.** $x^4 - 1$

Give all the solutions of the equations in Exercises 7–12.

7. $(x - 1)(x + 2)(x - 3) = 0$

8. $(x + 3)(1 - x^2) = 0$

9. $x^3 + 3x^2 + 2x = 0$

10. $x^3 - 2x^2 + 2^2 x - 2^3 = 0$

11. $(x - 1)x(x + 3) = 0$

12. $x^4 + x^2 - 2 = 0$

Find possible formulas for the polynomials described in Exercises 13–18.

13. The degree is $n = 2$ and the zeros are $x = 2, -3$.

14. The degree is 5 and the zeros are $x = -4, -1, 0, 3, 9$.

15. The degree is $n = 3$ and there is one zero at $x = 5$ and one double zero at $x = -13$.

16. The degree is $n = 3$ and the only zeros are $x = 2, -3$.

17. The degree is $n = 5$ and there is one zero at $x = -6$.

18. The degree is $n = 6$ and there is one simple zero at $x = -1$, one double zero at $x = 3$, and one multiple zero at $x = 5$.

In Exercises 19–25, $p(z) = 4z^3 - z$. Find the given values and simplify if possible.

19. $p(0)$ **20.** $p(\sqrt{5})$ **21.** $p(-1)$

22. $p(4^t)$ **23.** $p(t + 1)$ **24.** $p(3x)$

25. The values of z such that $p(z) = 0$

Problems

26. If $p(x) = x^4 - 2x^2 + 1$, find

  (a) $p(0)$
  (b) $p(2)$
  (c) The values of x such that $p(x) = 0$
  (d) $p(t^2)$.

In Problems 27–34, without solving the equation, decide how many solutions it has.

27. $(x - 1)(x - 2) = 0$

28. $(x^2 + 1)(x - 2) = 0$

29. $(x^2 + 2x)(x - 3) = 0$

30. $(x^2 - 4)(x + 5) = 0$

31. $(x - 2)x = 3(x - 2)$

32. $(2 - x^2)(x - 4)(5 - x) = 0$

33. $(2 + x^2)(x - 4)(5 - x) = 0$

34. $(x^4 + 2)(3 + x^2) = 0$

In Problems 35–43, for what values of a does the equation have a solution in x?

35. $x^2 - a = 0$

36. $2x^2 + a = 0$

37. $ax^2 - 5 = 0$

38. $(ax^2 + 1)(x - a) = 0$

39. $a^2 + ax^2 = 0$

40. $(x^2 + a)(x^2 - a) = 0$

41. $x^3 + a = 0$

42. $x^4 + 5a = 0$

43. $a - x^5 = 0$

44. Find two different polynomials of degree 3 with zeros 1, 2, and 3.

45. **(a)** Find two different polynomials with zeros $x = -1$ and $x = 5/2$.
(b) Find a polynomial with zeros $x = -1$ and $x = 5/2$ and leading coefficient 4.

46. What is the value of

$$5(x - 1)(x - 2) + 2(x - 1)(x - 3) - 4(x - 2)(x - 3)$$

when $x = 3$?

47. What values of the constants A, B, and C, will make

$$A(x - 1)(x - 2) + B(x - 1)(x - 3) - C(x - 2)(x - 3)$$

have the value 7 when $x = 3$?

48. Find the solutions of

$$(x^2 - a^2)(x + 1) = 0, \quad a \text{ a constant}.$$

49. For what value(s) of the constant a does $\left(x^2 - a^2\right)(x + 1) = 0$ have exactly two solutions?

50. Consider the polynomial $x^5 - 3x^4 + 4x^3 - 2x + 1$.

(a) What is the value of the polynomial when $x = 4$?
(b) If a is the answer you found in part (a), show that $x - 4$ is a factor of $x^5 - 3x^4 + 4x^3 - 2x + 1 - a$.

51. Use the identity $(x - 1)(x^4 + x^3 + x^2 + x + 1) = x^5 - 1$ to show that $8^5 - 1$ is divisible by 7.

52. The profit from selling q items of a certain product is $P(q) = 36q - 0.0001q^3$ dollars. Find the values of q such that $P(q) = 0$. Which of these values make sense in the context of the problem? Interpret the values that make sense.

53. American Airlines limits the size of carry-on baggage to 45 linear inches (length + width + height), with a weight of no more than 40 pounds.[3]

(a) If the length and width of a piece of luggage both measure x inches, express the maximum height of the luggage in terms of x.

_____
[3]www.aa.com

(b) Express the volume of the piece of luggage in part (a) in terms of x.
(c) Find the zeroes of your equation from part (b). What does this tell you about the dimensions of the piece of luggage?

In Problems 54–56, Michael receives a different sum of money each birthday starting from his 15th, and invests it in a bank account at an annual interest rate of $r\%$. If

$$x = 1 + \frac{r}{100},$$

then x is the annual growth factor. For example, if he receives \$750 on his 15th birthday, it has grown to $\$750x^5$ five years after that, just before his 20th birthday. If he receives \$500 on his 18th birthday, it has grown to $\$500x^2$ by the same time, since there were two years of growth.

54. The gift on Michael's 15th birthday is \$500 and goes up by \$100 each year after that.

(a) How much money in total has he received just before his 21st birthday?
(b) If $r = 4\%$, what is his investment worth at that time?
(c) Write a polynomial in $x = 1 + r/100$ for the value of his investment just before his 21st birthday.
(d) Graph the polynomial in part (c) and estimate the interest rate if Michael has \$5000 just before his 21st birthday.

55. The first three gifts were \$1000, \$500, and \$750, and the interest rate is $r = 5\%$.

(a) What is the total value of Michael's investments on his 17th birthday, immediately after making the third investment? What is the total value just before his 18th birthday?
(b) Using $x = 1 + r/100$, write polynomial expressions for the value of the investments just after his 17th and just before his 18th birthday.
(c) The total value of his investments just after his 20th birthday is given by

$$1000x^5 + 500x^4 + 750x^3 + 1200x + 650.$$

What were the gifts on his 18th, 19th and 20th birthdays?
(d) Evaluate the polynomial in part (c) for $x = 1.05, 1.06, 1.07$. What do these values tell you about the investment?

56. If $x = 1 + r/100$, the value of his investments just after Michael's 20th birthday is

$$800x^5 + 900x^4 + 300x^2 + 500x + 1200.$$

How much money did he receive on each birthday?

REVIEW EXERCISES AND PROBLEMS FOR CHAPTER EIGHT

Exercises

In Exercises 1–5, give the degree of the polynomial without expanding out fully.

1. $-x + 2x^2 + 1$

2. $x(x - 1)(x + 3)$

3. $(x + 1)^5 - x^5$

4. $(x^2 - 1)(x^2 - 2)(x^2 - x)$

5. $p(x^2)$, where $p(x) = x^4 - x^3 + 3x + 1$.

In Exercises 6–10, give the constant term of the polynomial in x without expanding fully.

6. $(x - 1)(x - 2)(x - 3)(x - 4)(x - 5)$

7. $(x - a)^3(x + 1/a)^3$, $a \neq 0$

8. $(x - 1)^5 - (x + 1)^5$

9. $(x - 1)^6 - (x + 1)^6$

10. $(x - x_0)^3(x^4 + 3x^2 - 2x + 10)$

In Exercises 11–13, for the given polynomials $p(x)$, find:

(a) $p(0)$ **(b)** $p(1)$ **(c)** $p(-1)$ **(d)** $p(-t)$

11. $p(x) = x^3 + x^2 + x + 1$

12. $p(x) = x^5 + 2x^3 - x$

13. $p(x) = x^4 + 3x^2 + 1$

Give all the solutions of the equations in Exercises 14–18.

14. $(x + 3)(1 - x)^2 = 0$

15. $(x + 3^2)(1 - x^2) = 0$

16. $x^3 - x^2 + x - 1 = 0$

17. $x^4 = 2x^2 - 1$

18. $4x^2 - 3x = x^3$

Problems

In Problems 19–24, find the given term or coefficient for $n = 2, 3$ and 4, and describe the pattern.

19. The coefficient of x in $(x + 2)^n$

20. The coefficient of x^{n-1} in $(2x + 1)^n$

21. The leading term of $(3x - 2)^n$

22. The coefficient of x^{n-1} in $(x - 1)^n$

23. The coefficient of x in $(1 - x)^n$

24. The constant term of $(x + a)^n$

In Problems 25–27 let $p(x)$ be a polynomial in x with degree m, not equal to zero, and leading coefficient a. For each polynomial given, find the degree and leading coefficient.

25. $5p(x)$ **26.** $xp(x)$ **27.** $p(x) + 5$

In Problems 28–29, the polynomial $p(x)$ has leading coefficient a and degree m, and the polynomial $q(x)$ has leading coefficient b and degree n. Suppose that $m > n$.

28. What is the leading coefficient and degree of $p(x)q(x)$?

29. What is the leading coefficient and degree of $p(x) + q(x)$?

30. Without using a calculator or computer, sketch $p(x) = x^2(x + 2)^3(x - 1)$.

31. Consider the polynomial $g(t) = (t^2 + 5)(t - a)$, where a is a constant. For which value(s) of t is $g(t)$ positive?

32. Using a calculator or computer, sketch the graph of

$$y = (x - a)^3 + 3a(x - a)^2 + 3a^2(x - a) + a^3$$

for $a = 0$, $a = 1$, and $a = 2$. What do you observe? Use algebra to confirm your observation.

Without solving them, decide how many solutions there are to the equations in Problems 33–38.

33. $x^2 - 10^3 = 0$ **34.** $x^3 - 7^4 = 0$

35. $x^2 + 10^{2.4} = 0$ **36.** $x^3 - 10^{-4.52} = 0$

37. $x^2 - \log 3 = 0$ **38.** $\log(0.2) - x^2 = 0$

39. (a) Find the zeros of the polynomials

$$P(x) = (x - 3)\left(x - \frac{1}{2}\right)$$

and

$$Q(x) = (x - 3)(2x - 1).$$

(b) Are the two polynomials equal?

40. If the cubic $x^3 + ax^2 + bx + c$ has zeros at $x = 1$, $x = 2$, and $x = 3$, what are the values, of a, b, and c?

41. (a) Show that $(1 + x + x^2)(1 - x + x^2) = 1 + x^2 + x^4$.
   (b) Use part (a) to show that neither $1 + x + x^2$ nor $1 - x + x^2$ have zeros.

42. If a is a constant, rewrite the equation $x^3 - ax^2 + a^2x - a^3 = 0$ in a form that clearly shows its solutions. What are the solutions?

43. If a is a non-zero constant, for what values of x is $x^3 - ax^2 + a^2x - a^3$ positive?

44. If the roots of $x^3 + 2x^2 - x - 2 = 0$ are -1, 1, and 2, what are the roots of $(x-3)^3 + 2(x-3)^2 - (x-3) - 2 = 0$? [Hint: Consider what values of $x - 3$ satisfy the second equation.]

45. Find viewing windows on which the graph of $f(x) = x^3 + x^2$ resembles the plots in (a)–(d).

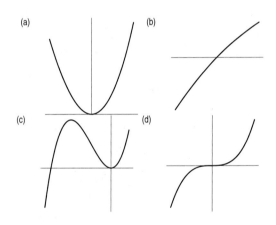

(a)
(b)
(c)
(d)

46. Someone tells you that $x^6 + x^2 + 1$ has no zeros. How do you know the person is right?

47. Someone tells you that $-x^6 + x^2 + 1$ has at least two zeros. How do you know the person is right?

48. Someone tells you that $x^5 + x^2 + 1$ has at least one zero. How do you know the person is right?

49. Find a cubic polynomial with all of the following properties:

   (i) The graph crosses the x-axis at -7, -2, and 3.
   (ii) From left to right, the graph rises, falls, and rises.
   (iii) The graph crosses the y-axis at -3.

Pascal's triangle[4] is an arrangement of integers that begins

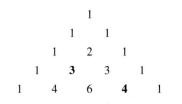

with n entries in each row formed by adding the two entries above it. For instance, the boldface 3 above is formed by adding the preceding entries 1 and 2, and the boldface 4 is formed by adding the preceding entries 3 and 1. Given this information, answer Problems 50–53.

50. Find the next 3 rows of Pascal's triangle.

51. Write the following polynomials in standard form:

$$(x + 1)^0, \quad (x + 1)^1, \quad (x + 1)^2, \quad (x + 1)^3.$$

What do you notice?

52. Without multiplying it out, write $(x + 1)^7$ in standard form. Hint: Make an educated guess based on your answers to Problems 50 and 51.

53. Consider the following pattern:

$$(2x + 3)^0 = 1$$
$$(2x + 3)^1 = 2x + 3$$
$$(2x + 3)^2 = (2x)^2 + 2(2x)3 + 3$$
$$(2x + 3)^3 = (2x)^3 + 3(2x)^2 3 + 3(2x)3^2 + 3^3.$$

Based on this pattern, write $(2x + 3)^4$ in standard form.

In Problems 54–56, Gloria has a credit card balance and decides to pay it off by making monthly payments. For example, suppose her balance is $b_0 = \$2000$ in month $t = 0$ and she makes no more purchases using the card. Each month thereafter, she is charged 1% interest on the debt, and then she makes a $25 payment. To find her balance in month $t = 1$, we multiply $2000 by 1.01 (because the balance goes up by 1%) and then subtract the $25 payment. This gives $b_1 = 2000.00(1.01) - 25 = 1995.00$. Likewise, her balance in months $t = 2, 3$ is given by

$$b_2 = 1995.00(1.01) - 25 = \$1989.95$$
$$b_3 = 1989.95(1.01) - 25 = \$1984.85.$$

Let $x = 1 + r$ where $r = 1\%$ is the monthly interest rate.

---
[4]Though named for Blaise Pascal, who described it in 1655, it was known to medieval Chinese and Islamic scholars and may have been known as early as 450 BC. See http://en.wikipedia.org/wiki/Pascal%27s_triangle.

54. (a) What do b_4 and b_5 represent in terms of Gloria's credit card balance? Evaluate b_4 and b_5.

(b) We can write an expression for b_1 in terms of $x = 1.01$:

$$b_1 = 2000 \underbrace{(1.01)}_{x} - 25 = 2000x - 25.$$

Notice that $b_2 = b_1(1.01) - 25$, that $b_3 = b_2(1.01) - 25$, and so on. Given this, write simplified polynomial expressions for b_2, b_3, b_4, and b_5 in terms of x.

(c) Evaluate your expression for b_5 for $r = 1.5\%$, that is, for $x = 1.015$. What does you answer tell you about interest rates and credit card debt?

55. (a) Gloria's first two monthly payments (in months $t = 1$ and 2) are \$25 and \$60, respectively. With $x = 1.01$, show that b_2, her balance in month $t = 2$, is given by the polynomial

$$2000x^2 - 25x - 60.$$

(b) Her next three monthly payments (in months $t = 3$, 4, and 5) are \$45, \$80, and \$25, respectively. Write a simplified polynomial expression for b_5.

56. In some months, Gloria makes expenditures on her credit card, rather than payments. If $x = 1.01$, after six months her balance is given by the polynomial

$$3000x^6 - 120x^5 - 90x^4 + 50x^3 + 100x^2 - 200x - 250.$$

Describe in words her credit card debt, giving her initial balance and each of her monthly payments or expenditures.

Chapter Nine

Polynomial and Rational Functions

Contents

9.1 LONG-RUN BEHAVIOR OF POLYNOMIAL FUNCTIONS

The graph of a polynomial function

$$y = p(x)$$

can have numerous x-intercepts and bumps. However, if we zoom out and consider the appearance of the graph from a distance and ignore fine details the picture becomes much simpler.

By "zooming out," we mean turning our attention to values of the polynomial for large (positive or negative) values of x. For instance, provided the value of x is large enough, the value of the term $4x^5$ in the polynomial $p(x) = 4x^5 - x^3 + 3x^2 + x + 1$ is far larger than the value of the other terms combined. To see this, we let $x = 100$. The value of the term $4x^5$ is

$$4x^5 = 4(100)^5 = 40,000,000,000,$$

whereas the value of the other terms combined is

$$-x^3 + 3x^2 + x + 1 = -(100)^3 + 3(100)^2 + 100 + 1$$
$$= -1,000,000 + 30,000 + 100 + 1 = -969,899.$$

Therefore, at $x = 100$, the combined value is

$$p(100) = \underbrace{40,000,000,000}_{\text{largest contribution}} + \underbrace{-969,899}_{\text{relatively small correction}}$$
$$= 39,999,030,101.$$

Thinking in terms of money, the value of $4x^5$ by itself is $40 billion, whereas if we include the other terms the combined value is $39.999 billion. Comparing these two values, the difference is negligible, even though (by itself) $969,899 is a lot of money.

Hence, for large values of x,

$$p(x) = 4x^5 + \underbrace{-x^3 + 3x^2 + x + 1}_{\text{relatively small correction}}$$

$$p(x) \approx 4x^5.$$

In general, if the value of x is large enough, we can give a reasonable approximation for a polynomial's value by ignoring or neglecting or the lower degree terms. Consequently, when viewed on a large enough scale, the graph of the polynomial function

$$y = a_n x^n + a_{n-1} x^{n-1} + \cdots + a_1 x + a_0$$

strongly resembles the graph of the power function

$$y = a_n x^n,$$

and we say that the *long-run behavior* of the polynomial is given by its leading term.

The Long-Run Behavior of a Polynomial

In a polynomial $a_n x^n + a_{n-1} x^{n-1} + \cdots + a_1 x + a_0$, with $a_n \neq 0$, the leading term $a_n x^n$ determines the polynomial's long-run behavior.

Example 1   Find a window in which the graph of $y = x^3 + x^2$ resembles the graph of its leading term $y = x^3$.

Solution   Figure 9.1 gives the graphs of $y = x^3 + x^2$ and $y = x^3$. On this scale, $y = x^3 + x^2$ does not look like a power. On the larger scale in Figure 9.2, the graph of $y = x^3 + x^2$ resembles the graph of $y = x^3$. On this larger scale, the "bumps" in the graph of $y = x^3 + x^2$ are too small to be seen. On an even larger scale, as in Figure 9.3, the graph of $y = x^3 + x^2$ is nearly indistinguishable from the graph of $y = x^3$.

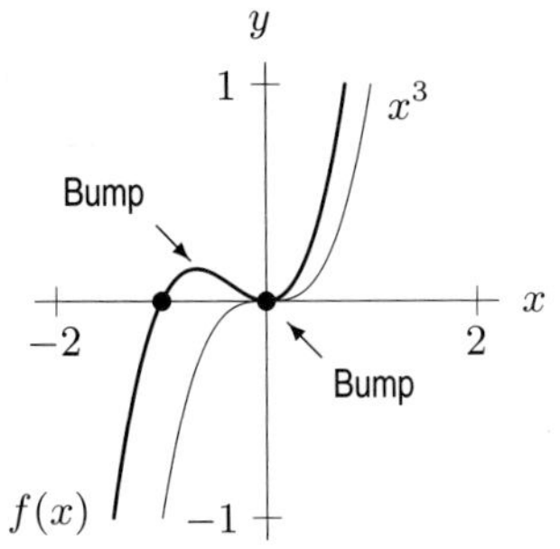

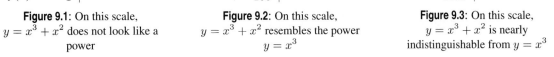

Figure 9.1: On this scale, $y = x^3 + x^2$ does not look like a power

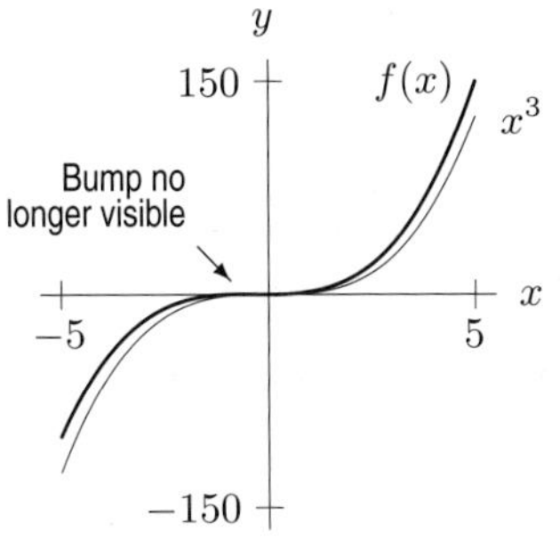

Figure 9.2: On this scale, $y = x^3 + x^2$ resembles the power $y = x^3$

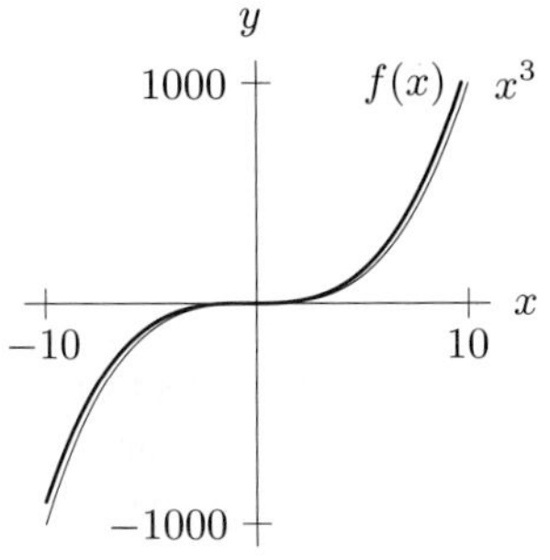

Figure 9.3: On this scale, $y = x^3 + x^2$ is nearly indistinguishable from $y = x^3$

Example 2   Compare the graphs of the polynomial functions

$$f(x) = x^4 - 4x^3 + 16x - 16, \quad g(x) = x^4 - 4x^3 - 4x^2 + 16x, \quad h(x) = x^4 + x^3 - 8x^2 - 12x.$$

Solution   Each of these polynomials is of degree 4, and each has x^4 as its leading term. Thus, all their graphs resemble the graph of x^4 on a large scale. See Figure 9.4.

However, on a smaller scale, the polynomials look different. See Figure 9.5. Two of the graphs go through the origin while the third does not. The graphs also differ from one another in the number of bumps each one has and in the number of times each one crosses the x-axis. Thus, polynomials with the same leading term look similar on a large scale, but may look dissimilar on a small scale.

Figure 9.4: On a large scale, the three polynomials resemble the power $y = x^4$

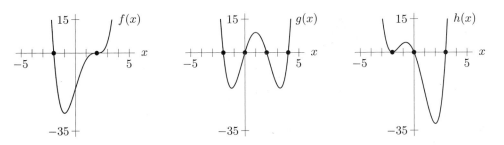

Figure 9.5: On a smaller scale, the three polynomials look quite different from each other

Example 3 Find the leading term of the polynomial $2w^3 + 8w^2$ in Example 3 on page 299 and give a practical interpretation of it.

Solution The polynomial gives the volume of a box whose width is w, length is 4 more than the width, and height is twice the width. The leading term is $2w^3$. This represents the approximate volume of the box when w is large. In that case, for example if $w = 1000$, then $w + 4$ and w are not much different from each other, so the base of the box is approximately a square with side-length w, so

$$\text{Volume} = (\text{Area of base})(\text{Height}) \approx (w^2)(2w) = 2w^3.$$

Long-Run Behavior: Using Polynomials to Estimate Other Functions

In Section 8.1, we saw that we can use a polynomial to estimate the values of another function for certain input values. On the other hand, for large input values, the graph of a polynomial pulls away from the graph of the function being estimated. In the long run, a polynomial behaves like the power function given by its leading term, not like the function it is being used to estimate.

Finding the Formula for a Polynomial from its Graph

Often, we can determine a formula for a polynomial based on its graph.

Example 4 Find a possible formula for the polynomial $f(x)$ graphed in Figure 9.6.

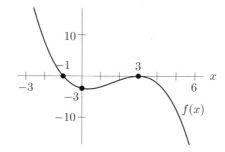

Figure 9.6: Find a formula for this polynomial

Solution Based on its large-scale shape, the degree of $f(x)$ is odd and is larger than 1. The graph has x-intercepts at $x = -1$ and $x = 3$. We see that $x = 3$ is a multiple zero of even power, because the graph bounces off the x-axis here instead of crossing it. (See page 312.) Therefore, a possible formula is

$$f(x) = k(x + 1)(x - 3)^2,$$

where k is a constant.

To find k, we use the fact that $f(x) = -3$ at $x = 0$, so

$$f(0) = k(0 + 1)(0 - 3)^2 = -3,$$

which gives

$$9k = -3 \qquad \text{so} \qquad k = -\frac{1}{3}.$$

Thus, $f(x) = -\frac{1}{3}(x + 1)(x - 3)^2$ is a possible formula for this polynomial.

The formula for $f(x)$ we found in Example 4 is the polynomial of least degree we could have chosen. However, there are other polynomials, such as $y = -(1/27)(x+1)(x-3)^4$, with the same overall behavior as that shown in Figure 9.6.

Example 5   The graph of

$$y = 3x^6 - 2x^5 + 4x^2 - 1$$

crosses the y-axis at $y = -1$. Is there a reason to expect the equation $3x^6 - 2x^5 + 4x^2 - 1 = 0$ to have a solution? If not, explain why not. If so, how do you know?

Solution   The equation $3x^6 - 2x^5 + 4x^2 - 1 = 0$ must have at least two solutions. We know this because on a large scale, the graph of $y = 3x^6 - 2x^5 + 4x^2 - 1$ looks like $y = 3x^6$. (See Figure 9.7.) This means that this function takes on large positive values as x grows large (either positive or negative). Since the graph of $y = 3x^6 - 2x^5 + 4x^2 - 1$ is smooth and unbroken, it must cross the x-axis at least twice to get from its y-intercept of $y = -1$ to the positive values it attains as $x \to \infty$ and $x \to -\infty$.[1]

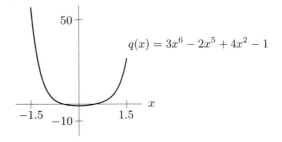

Figure 9.7: Graph must cross x-axis at least twice since $3x^6 - 2x^5 + 4x^2 - 1 = -1$ when $x = 0$, and $3x^6 - 2x^5 + 4x^2 - 1$ looks like $3x^6$ for large x

Note that a sixth degree polynomial such as $3x^6 - 2x^5 + 4x^2 - 1$ in Example 5 can have as many as six real zeros, or none at all. For instance, the sixth degree polynomial $y = x^6 + 1$ has no zeros, because the value of y is nowhere less than 1.

Exercises and Problems for Section 9.1

Exercises

Find possible formulas for the polynomial functions described in Exercises 1–4.

1. The graph crosses the x-axis at $x = -2$ and $x = 3$ and its long-run behavior is like $y = x^2$.

2. The graph crosses the x-axis at $x = -2$ and $x = 3$ and its long-run behavior is like $y = -2x^2$.

3. The graph bounces off the x-axis at $x = -2$, crosses the x-axis at $x = 3$, and has long-run behavior like $y = x^3$.

4. The graph bounces off the x-axis at $x = -2$, crosses the x-axis at $x = 3$, and has long-run behavior like $y = x^5$.

---

[1] We use the symbols $x \to \infty$ to mean x is getting very far from zero in the positive direction and $x \to -\infty$ to mean it is getting very far from 0 in the negative direction.

Find possible formulas for the polynomials shown in Exercises 5–7.

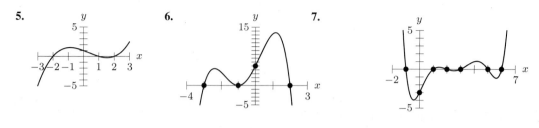

5.

6.

7.

Problems

Answer Problems 8–9 based on Figure 9.8, which shows the same polynomial on 4 different viewing windows.

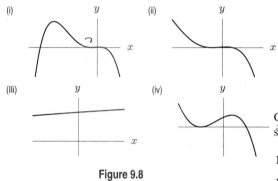

Figure 9.8

8. Rank the windows in order from smallest to largest.

9. One of the following equations describes the polynomial shown in Figure 9.8. Which is it?

(a) $y = -(x + 1)^2(x - 1)(x + 10)$
(b) $y = (x + 1)^2(x - 1)(x + 10)$
(c) $y = (x + 1)(x - 1)(x + 10)$
(d) $y = -(x - 1)^2(x + 1)(x - 10)$

10. The polynomial $p(x)$ can be written in two forms:

I. $p(x) = 2x^3 - 3x^2 - 11x + 6$
II. $p(x) = (x - 3)(x + 2)(2x - 1)$

Which form most readily shows

(a) The zeros of $p(x)$? What are they?
(b) The vertical intercept? What is it?
(c) The sign of $p(x)$ as x gets large, either positive or negative? What is the sign?
(d) The number of time $p(x)$ changes sign as x increases from large negative to large positive x? How many times is this?

11. A polynomial $p(x)$ can be written in two forms:

I. $p(x) = (x^2 + 4)(4 - x^2)$
II. $p(x) = 16 - x^4$

Which form most readily shows

(a) The number of zeros of $p(x)$? Find them.
(b) The vertical intercept? What is it?
(c) The sign of $p(x)$ as x gets large, either positive or negative. What are the signs?

Give the degree, leading term, leading coefficient, and constant term of the polynomials in Problems 12–15.

12. $-3x^2 + x + 5$

13. $x(x + 1)^2$

14. $(2x^3 - 3x^2)^2$

15. $(2x^3 - 3x^2)^2 - (x^2 + 1)^2$

Find possible formulas for the polynomials shown in Problems 16–19.

16.

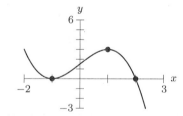

17.

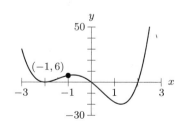

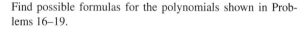

18.

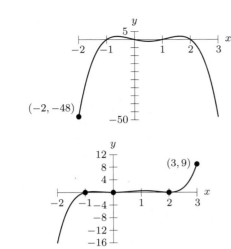

19.

20. Graph $y = 2(x - 3)^2(x + 1)$. Label all axis intercepts.

21. A cubic polynomial, $p(x)$, has $p(1) = -4$ and $p(2) = 7$. What can you say about the zeros of $p(x)$?

22. Explain why $p(x) = x^5 + 3x^3 + 2$ must have at least one zero.

23. Find two different formulas for a polynomial $p(x)$ of degree 3 with $p(0) = 6$ and $p(-2) = 0$, and graph them.

24. By calculating $p(0)$ and $p(1)$ show that

$$p(x) = -x^5 + 3x^2 - 1$$

has at least three zeros.

9.2 RATIONAL FUNCTIONS

If we add, subtract, or multiply two polynomials, the result is always a polynomial. However, when we divide one polynomial by another the result is not necessarily a polynomial, since there could be a remainder.

Example 1 Graph

$$y = \frac{10x^3 - 3x^2 + 1}{10x^3 - 3x^2 - 1}.$$

Solution

See Figure 9.9. Notice that the graph does not resemble the graph of a polynomial. It levels off to a horizontal line as the value of x gets large (either positive or negative). This line is called a *horizontal asymptote*. Furthermore, there is a vertical line on either side of which the value of y gets large (either positive or negative). This line is called a *vertical asymptote*. We learn more about horizontal and vertical asymptotes in the next section.

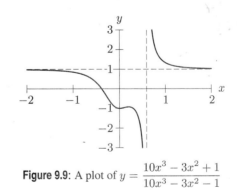

Figure 9.9: A plot of $y = \dfrac{10x^3 - 3x^2 + 1}{10x^3 - 3x^2 - 1}$

In general:

> A **rational function** is a function that can be put in the form
>
> $$f(x) = \frac{a(x)}{b(x)}, \quad \text{where } a(x) \text{ and } b(x) \text{ are polynomials in } x.$$
>
> Here we assume that $b(x)$ is not the zero polynomial.

We have seen that, for integers,

$$\frac{a}{b} = q \text{ with remainder } r \qquad \text{means} \qquad a = bq + r.$$

An alternative to working with remainders is to rewrite

$$a = bq + r \qquad \text{as} \qquad \frac{a}{b} = q + \frac{r}{b}.$$

For instance, instead of dividing 4 into 13 by writing

$$13 = 4 \cdot 3 + 1,$$

we write

$$\frac{13}{4} = 3 + \frac{1}{4}.$$

Notice that, under this approach, we must use fractions (rational numbers), whereas before we could work entirely with integers.[2]

The same approach works with polynomials. Instead of dividing $a(x)$ by $b(x)$ to obtain

$$a(x) = b(x)q(x) + r(x),$$

we write

$$\frac{a(x)}{b(x)} = q(x) + \frac{r(x)}{b(x)}.$$

For instance, instead of dividing $10x^3 - 3x^2 - 1$ into $10x^3 - 3x^2 + 1$ by writing

$$10x^3 - 3x^2 + 1 = (10x^3 - 3x^2 - 1)1 + 2,$$

we write

$$\frac{10x^3 - 3x^2 + 1}{10x^3 - 3x^2 - 1} = 1 + \frac{2}{10x^3 - 3x^2 - 1}.$$

Expressing Rational Functions in Different Forms

A rational number is a ratio of two integers, and by analogy a rational function is a ratio of two polynomials. Just like rational numbers, rational functions can be expressed in different forms. For example, we can cancel a common factor from the numerator and denominator of a rational function.

Example 2   Cancel the common factor in the expression

$$\frac{x^2 - 5x + 6}{x^2 - 4}.$$

Is the resulting expression equivalent to the original expression?

---

[2] A common shorthand for the expression $3 + 1/4$ is the *mixed number* $3\frac{1}{4}$.

Solution   We have

$$\frac{x^2 - 5x + 6}{x^2 - 4} = \frac{(x-2)(x-3)}{(x-2)(x+2)} \qquad \text{common factor of } x - 2$$

$$= \frac{\cancel{(x-2)}(x-3)}{\cancel{(x-2)}(x+2)} \qquad \text{canceling}$$

$$= \frac{x-3}{x+2} \qquad \text{provided } x \neq 2.$$

The expression $(x^2 - 5x + 6)/(x^2 - 4)$ has the same value as the expression $(x-3)/(x+2)$, except at $x = 2$. This is because the latter expression is defined at $x = 2$, whereas the former is not. Thus, the two expressions are not equivalent. This means that the rational functions

$$f(x) = \frac{x^2 - 5x + 6}{x^2 - 4} \quad \text{and} \quad g(x) = \frac{x-3}{x+2}$$

are not exactly the same, although their graphs are almost identical. See Figures 9.10 and 9.11.

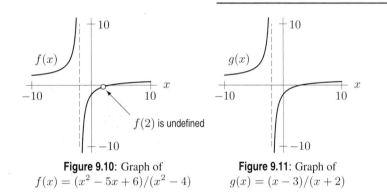

Figure 9.10: Graph of $f(x) = (x^2 - 5x + 6)/(x^2 - 4)$

Figure 9.11: Graph of $g(x) = (x - 3)/(x + 2)$

Example 3   Cancel common factors in each of the following rational expressions.

(a) $\dfrac{x^2 - 2x + 1}{x - 1}$
(b) $\dfrac{2x^2 + 4x}{2x + 6}$
(c) $\dfrac{x^4 - 1}{(x^2 + 1)(x - 3)}$

Solution   (a) We have

$$\frac{x^2 - 2x + 1}{x - 1} = \frac{(x-1)(x-1)}{x-1} = x - 1 \quad \text{provided } x \neq 1.$$

(b) We have

$$\frac{2x^2 + 4x}{2x + 6} = \frac{2(x^2 + 2x)}{2(x + 3)} = \frac{x^2 + 2x}{x + 3}.$$

Canceling the 2 is valid no matter what the value of x, so the two expressions are equivalent.

(c) We have

$$\frac{x^4 - 1}{(x^2 + 1)(x - 3)} = \frac{(x^2 - 1)(x^2 + 1)}{(x^2 + 1)(x - 3)} = \frac{x^2 - 1}{x - 3}.$$

Since $x^2 + 1$ is positive for all values of x, cancelation does not involve possible division by 0, so the two expressions are equivalent.

In working with rational expressions it is common to ignore the few values of x where a particular operation is invalid. When you reduce a rational expression to a simpler form it is often understood that the two forms might not be equivalent at values of x where either of them is undefined. However, when we use the expressions to define functions we must remember that the two functions might not have the same values everywhere.

Example 4   Write each expression in the form $a(x)/b(x)$, for polynomials $a(x)$ and $b(x)$.

(a) $\dfrac{x^2 + 2}{x - 3} + 2$ (b) $\dfrac{x + \frac{1}{x+1}}{x - \frac{1}{x-1}}$ (c) $5x^3 - 2x^2 + 7$

Solution   (a) We put each term over a common denominator and add:

$$\frac{x^2 + 2}{x - 3} + 2 = \frac{x^2 + 2}{x - 3} + 2 \cdot \frac{x - 3}{x - 3}$$

$$= \frac{x^2 + 2}{x - 3} + \frac{2x - 6}{x - 3}$$

$$= \frac{x^2 + 2 + 2x - 6}{x - 3}$$

$$= \frac{x^2 + 2x - 4}{x - 3},$$

so $a(x) = x^2 + 2x - 4$ and $b(x) = x - 3$.

(b) We multiply the numerator and denominator in this expression by $(x + 1)(x - 1)$ to clear the denominators.

$$\frac{x + \frac{1}{x+1}}{x - \frac{1}{x-1}} = \frac{x + \frac{1}{x+1}}{x - \frac{1}{x-1}} \cdot \frac{(x+1)(x-1)}{(x+1)(x-1)} \qquad \text{clearing the denominators}$$

$$= \frac{x(x+1)(x-1) + \frac{(x+1)(x-1)}{x+1}}{x(x+1)(x-1) - \frac{(x+1)(x-1)}{x-1}} \qquad \text{after distributing}$$

$$= \frac{x(x+1)(x-1) + (x-1)}{x(x+1)(x-1) - (x+1)} \qquad \text{after canceling}$$

$$= \frac{x^3 - 1}{x^3 - 2x - 1}, \qquad \text{after simplifying}$$

so we have $a(x) = x^3 - 1$ and $b(x) = x^3 - 2x - 1$.

(c) We have $a(x) = 5x^3 - 2x^2 + 7$ and $b(x) = 1$ (a constant polynomial). In fact, any polynomial is necessarily rational, because we can think of it as a ratio over the constant polynomial $b(x) = 1$.

Remainders

If we divide 12 by 4, the answer is an integer, with no remainders. However, if we divide 13 by 4, we have a remainder of 1:

$$\frac{13}{4} = 3 \quad \text{with remainder 1.}$$

For example, if we distribute 13 cookies among 4 children, each child will get 3 cookies, with one cookie remaining (left over):

Total number of cookies = 4 children × 3 cookies per child + 1 cookie remaining.

In general,

$$\frac{a}{b} = q \quad \text{with remainder } r$$

says, for instance, that a cookies divided among b children will give each child q cookies with r cookies remaining. In symbols:

$$a = bq + r.$$

Notice that r must be less than b, for otherwise there would be enough cookies remaining for each child to have more.

Rational Expressions and Remainders

As is often the case, what works for integers also works for polynomials. If we divide one polynomial by another, sometimes the answer comes out to be a whole polynomial, with no remainder:

$$\frac{x^2 - 5x + 6}{x - 2} = \frac{(x - 2)(x - 3)}{x - 2} = x - 3 \quad \text{(provided } x \neq 2\text{)}.$$

However, if increase the numerator by 1 (as we did when we went from 12 to 13), the answer no longer comes out evenly—the extra 1 will be "left over":

$$\frac{x^2 - 5x + 7}{x - 2} = x - 3 \quad \text{with remainder } r = 1.$$

We can apply the pattern we used for integers:

$$\frac{a(x)}{b(x)} = q(x) \text{ with remainder } r(x) \quad \text{means} \quad a(x) = b(x)q(x) + r(x).$$

Here, $a(x)$ stands for $x^2 - 5x + 7$, the polynomial in the numerator; $b(x)$ stands for $x - 2$, the polynomial in the denominator; $q(x)$ stands for $x - 3$, the quotient; and $r(x)$ stands for 1, the remainder. So

$$\underbrace{x^2 - 5x + 7}_{a(x)} = \underbrace{(x - 2)}_{b(x)}\underbrace{(x - 3)}_{q(x)} + \underbrace{1}_{r(x)}.$$

Example 5   Find the quotient and remainder on dividing $10x^3 - 3x^2 + 1$ by $10x^3 - 3x^2 - 1$.

Solution   One way to work this problem is to rewrite the numerator so that it includes the denominator, as follows:

$$\frac{10x^3 - 3x^2 + 1}{10x^3 - 3x^2 - 1} = \frac{\overbrace{(10x^3 - 3x^2 - 1)}^{\text{denominator}} + 2}{10x^3 - 3x^2 - 1} \quad \text{rewriting numerator.}$$

Since the numerator equals the denominator plus 2, the denominator goes into the numerator one time, and the 2 is left over:

$$\frac{10x^3 - 3x^2 + 1}{10x^3 - 3x^2 - 1} = \frac{\overbrace{(10x^3 - 3x^2 - 1)}^{\text{denominator}} + 2}{10x^3 - 3x^2 - 1} = 1 \quad \text{with remainder } r(x) = 2.$$

Thus, the quotient is $q(x) = 1$ and the remainder is $r(x) = 2$. So

$$\underbrace{10x^3 - 3x^2 + 1}_{a(x)} = \underbrace{(10x^3 - 3x^2 - 1)}_{b(x)} \cdot \underbrace{1}_{q(x)} + \underbrace{2}_{r(x)}.$$

Example 6   Find the quotient and remainder on dividing $30x^3 - 9x^2 + 1$ by $10x^3 - 3x^2 - 1$.

Solution   Notice that the leading term of the numerator is 3 times the leading term of the denominator. We write

$$\frac{30x^3 - 9x^2 + 1}{10x^3 - 3x^2 - 1} = \frac{3\overbrace{\left(10x^3 - 3x^2 - 1\right)}^{\text{denominator}} + 4}{10x^3 - 3x^2 - 1} \qquad \text{rewriting numerator}$$

$$= 3 \quad \text{with remainder 4.}$$

The numerator is 3 times the denominator with 4 left over. Thus, the quotient is $q(x) = 3$ and the remainder is $r(x) = 4$. So

$$\underbrace{30x^3 - 9x^2 + 1}_{a(x)} = \underbrace{(10x^3 - 3x^2 - 1)}_{b(x)} \cdot \underbrace{3}_{q(x)} + \underbrace{4}_{r(x)}.$$

So far, the quotient $q(x)$ and remainder $r(x)$ in our examples have all been constants. This is not always the case.

Example 7   Find the quotient and remainder on dividing $30x^4 - 9x^3 + x + 1$ by $10x^3 - 3x^2 - 1$.

Solution   Notice that the leading term of the numerator is $3x$ times the leading term of the denominator. We write

$$\frac{30x^4 - 9x^3 + x + 1}{10x^3 - 3x^2 - 1} = \frac{3x\overbrace{\left(10x^3 - 3x^2 - 1\right)}^{\text{denominator}} + 4x + 1}{10x^3 - 3x^2 - 1} \qquad \text{rewriting numerator}$$

$$= 3x \quad \text{with remainder } 4x + 1.$$

The numerator is $3x$ times the denominator with $4x + 1$ left over. Thus, the quotient is $q(x) = 3x$ and the remainder is $r(x) = 4x + 1$. So

$$\underbrace{30x^4 - 9x^3 + x + 1}_{a(x)} = \underbrace{(10x^3 - 3x^2 - 1)}_{b(x)} \cdot \underbrace{3x}_{q(x)} + \underbrace{4x + 1}_{r(x)}.$$

Example 8   Use the division algorithm to express the following rational functions in the form

$$q(x) + \frac{r(x)}{b(x)}, \quad \text{where } r(x) = 0 \text{ or degree of } r(x) \text{ less than degree of } b(x)$$

(a) $\dfrac{x^3 - 2x + 1}{x - 2}$ (b) $\dfrac{2x^3 + 3x^2}{x^2 + 1}$

Solution   The long divisions necessary for this example are carried out in Example 1 on page 345.

(a) Polynomial long division gives

$$(x^3 - 2x + 1) = (x - 2)(x^2 + 2x + 2) + 5,$$

so

$$\frac{x^3 - 2x + 1}{x - 2} = x^2 + 2x + 2 + \frac{5}{x - 2}.$$

(b) We have

$$2x^3 + 3x^2 = (2x+3)(x^2+1) + (-2x-3),$$

so

$$\frac{2x^3+3x^2}{x^2+1} = 2x+3 - \frac{2x+3}{x^2+1}.$$

Exercises and Problems for Section 9.2

Exercises

In Exercises 1–6, find the values of x (if any) at which the rational expression is undefined.

1. $\dfrac{9-x}{2x+5}$ **2.** $\dfrac{x^3-8x+1}{x^4+1}$ **3.** $\dfrac{1}{x-\sqrt{2}}$

4. $\dfrac{11x}{x^3+27}$ **5.** $\dfrac{x^2+x}{15x}$ **6.** $\dfrac{4-x}{2}$

In Exercises 7–9, simplify the rational expressions.

7. $\dfrac{11+x^2}{11x+x^3}$ **8.** $\dfrac{1-x}{x^2+x-2}$

9. $\dfrac{4x^2-a^2}{2ax-a^2}$, a a nonzero constant

In Exercises 10–13, solve for x.

10. $5 = \dfrac{3x-5}{2x+3}$ **11.** $12 = \dfrac{6x-3}{5x+2}$

12. $13 = \dfrac{2x^2+5x-1}{3+x}$ **13.** $7 = \dfrac{5x^2-2x+23}{2x^2+2}$

In Exercises 14–19, find the zeros.

14. $y = \dfrac{x^3+4x^2+3x}{2x-1}$ **15.** $y = \dfrac{5x+3}{2x+2}$

16. $y = \dfrac{10x+4}{7x-10}$ **17.** $y = \dfrac{5x^2-4x-1}{10+3x}$

18. $y = \dfrac{-5x^2+1}{3x^2+4}$ **19.** $y = \dfrac{x^3+x^2+12x}{5x^2-2}$

Cancel any common factors in the rational expressions in Exercises 20–25. Are the resulting expressions equivalent to the original expressions?

20. $\dfrac{x^2-6x+5}{x-5}$ **21.** $\dfrac{x^3+6x^2+9x}{x^2+10x+16}$

22. $\dfrac{2x^2+4}{(3x^2+5)(x^2+2)}$ **23.** $\dfrac{x^4}{x^3-x^2}$

24. $\dfrac{x^4-1}{x^3+x}$ **25.** $\dfrac{x^2+9x+20}{x^2+x-6}$

Reduce each expression in Exercises 26–29 to the form $a(x)/b(x)$ for polynomials $a(x)$ and $b(x)$.

26. $\dfrac{3}{2x+6}+5$ **27.** $\dfrac{1+\frac{1}{x}}{2-\frac{3}{x}}$

28. $\dfrac{1}{1+\frac{1}{1+\frac{1}{x}}}$ **29.** $\dfrac{2}{x^5-3x^2+7}$

The rational expressions in Exercises 30–33 are in the form $a(x)/b(x)$ where $a(x)$ and $b(x)$ are polynomials. Find the quotient $q(x)$ and the remainder $r(x)$ so that $a(x) = q(x)b(x) + r(x)$.

30. $\dfrac{x+3}{x+2}$ **31.** $\dfrac{2x+3}{x+1}$

32. $\dfrac{x^3+x+5}{x^2+1}$ **33.** $\dfrac{x^3+5x^2+6x+7}{x^2+3x}$

Problems

34. Let $R(z) = \dfrac{22 - z}{(z - 11)(1 - z)}$. Determine which of the following forms is equal to $R(z)$ and which is equal to $-R(z)$.

(a) $\dfrac{z - 22}{(z - 11)(z - 1)}$

(b) $\dfrac{z - 22}{(z - 11)(1 - z)}$

(c) $\dfrac{22 - z}{(11 - z)(z - 1)}$

(d) $\dfrac{22 - z}{(11 - z)(1 - z)}$

In Problems 35–38, imagine solving the equation by multiplying by the denominator to convert it to a polynomial equation. What is the degree of the polynomial equation?

35. $\dfrac{2x - 5}{x - 7} = 7$

36. $\dfrac{2x + x^2}{2x^2 - 5} = 7$

37. $\dfrac{x^2 + 5x}{2x^2 + 3} = \dfrac{1}{2}$

38. $\dfrac{x^2 - 5}{x^3 + 2} = \dfrac{1}{x + 2}$

Use the Remainder Theorem to find the value of the constant r making the equations in Problems 39–42 identities.

39. $x^2 + 1 = (x - 1)(x + 1) + r$

40. $x^3 + 2x + 3 = (x - 4)(x^2 + 4x + 18) + r$

41.

$$p(x) = (x + 1)(x^4 - 2x^3 + 4x^2 - 5x + 4) + r$$

where $p(x) = x^5 - x^4 + 2x^3 - x^2 - x - 2$.

42.

$$\frac{x^7 - 2x^3 + 1}{x - 2} = q(x) + \frac{r}{x - 2}$$

where $q(x) = x^6 + 2x^5 + 4x^4 + 8x^3 + 14x^2 + 28x + 56$.

9.3 LONG-RUN BEHAVIOR OF RATIONAL FUNCTIONS

The Long-Run Behavior of Rational Functions

In the long-run, every rational function behaves like a power function. For example, consider

$$y = \frac{6x^4 + x^3 + 1}{-5x + 2x^2}.$$

Since the long-run behavior of a polynomial is determined by its highest power term, for large x the numerator behaves like $6x^4$ and the denominator behaves like $2x^2$. Thus, the long-run behavior of this function is

$$y = \frac{6x^4 + x^3 + 1}{-5x + 2x^2} \approx \frac{6x^4}{2x^2} = 3x^2.$$

See Figure 9.12.

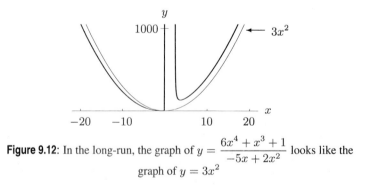

Figure 9.12: In the long-run, the graph of $y = \dfrac{6x^4 + x^3 + 1}{-5x + 2x^2}$ looks like the graph of $y = 3x^2$

In general, given any rational function, for large enough values of x we have:

$$y = \frac{a_n x^n + a_{n-1} x^{n-1} + \cdots + a_0}{b_m x^m + b_{m-1} x^{m-1} + \cdots + b_0} \approx \frac{a_n x^n}{b_m x^m} = \frac{a_n}{b_m} x^{n-m}.$$

This means that on a large scale the rational function resembles the function $y = \left(\dfrac{a_n}{b_m}\right) x^{n-m}$, which is a power function of the form $y = kx^p$, where $k = a_n/b_m$ and $p = n - m$. In summary:

For large enough values of x (either positive or negative), the graph of a rational function looks like the graph of a power function. If $y = \dfrac{p(x)}{q(x)}$, where $p(x)$ and $q(x)$ are polynomials, then the **long-run behavior** is given by

$$y = \frac{\text{Leading term of } p(x)}{\text{Leading term of } q(x)}.$$

Example 1   For positive x, describe the long-run behavior of the rational function

$$y = \frac{x+3}{x+2}.$$

Solution   If x is a large positive number, then

$$y = \frac{\text{Big number} + 3}{\text{Same big number} + 2} \approx \frac{\text{Big number}}{\text{Same big number}} = 1.$$

For example, if $x = 100$, we have

$$y = \frac{103}{102} = 1.0098\ldots \approx 1.$$

If $x = 10,000$, we have

$$y = \frac{10,003}{10,002} = 1.00009998\ldots \approx 1.$$

For large positive x-values, $y \approx 1$. Thus, for large enough values of x, the graph looks like the line $y = 1$, its horizontal asymptote. See Figure 9.13. However, for $x > 0$, the graph is above the line since the numerator is larger than the denominator.

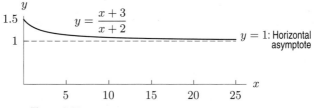

Figure 9.13: For large positive values of x, the graph of $y = (x+3)/(x+2)$ looks like the horizontal line $y = 1$

Example 2   For positive x, describe the long-run behavior of the rational function

$$y = \frac{3x + 1}{x^2 + x - 2}.$$

Solution   The leading term in the numerator is $3x$ and the leading term in the denominator is x^2. Thus for large enough values of x,

$$y \approx \frac{3x}{x^2} = \frac{3}{x}.$$

Figure 9.14 shows the graphs of $y = (3x + 1)/(x^2 + x - 2)$ and $y = 3/x$. For large values of x, the two graphs are nearly indistinguishable. Both graphs have a horizontal asymptote at $y = 0$.

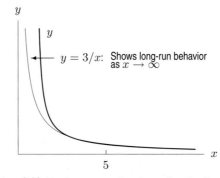

Figure 9.14: For large enough values of x, the function
$y = (3x + 1)/(x^2 + x - 2)$ looks like the function $y = 3/x$

Example 3   Use the fact that

$$\frac{10x^3 - 3x^2 + 1}{10x^3 - 3x^2 - 1} = 1 + \frac{2}{10x^3 - 3x^2 - 1}$$

to explain the horizontal asymptote in Figure 9.9 on page 325.

Solution   If x is very large, the polynomial $10x^3 - 3x^2 - 1$ behaves like its leading term, $10x^3$, so the ratio

$$\frac{2}{10x^3 - 3x^2 - 1}$$

is approximately equal to $2/10x^3$. If x is large and positive, then $2/10x^3$ is the ratio of 2 and a large positive number, so it is a small positive number, and if x is large and negative, then $2/10x^3$ is the ratio of 2 and a large negative number, so it is a small negative number. Either way, the value of

$$y = 1 + \frac{2}{10x^3 - 3x^2 - 1}$$

is close to 1. It is slightly greater than 1 when x is large and positive, and slightly less than 1 when x is large and negative. This is reflected in Figure 9.9, where the graph is slightly above the line $y = 1$ when x is large and positive, and slightly below it when x is large and negative.

Vertical Asymptotes

From Figure 9.14, we see that the graph of

$$y = \frac{3x + 1}{x^2 + x - 2}$$

has a horizontal asymptote at $y = 0$. This means that as we zoom out—that is, as we focus our attention on increasingly large (positive or negative) values of x—the graph increasingly resembles the horizontal line $y = 0$. From the same figure, we see that near $x = 1$, the graph grows very steep, resembling the vertical line $x = 1$. This means the graph has a vertical asymptote at $x = 1$.

The rapid rise (or fall) of the graph of a rational function near a vertical asymptote is due to the denominator's becoming small (close to zero). For this particular function, as the value of x gets close to 1, two things happen: the value of the numerator gets close to $3 \cdot 1 + 1 = 4$, and the value of the denominator gets close to $1^2 + 1 - 2 = 0$. Thus, the value of y grows larger and larger (either positive or negative), since we are dividing a number nearly equal to 4 by a number nearly equal to 0. The rapidly rising (or falling) y-values result in the near-vertical appearance of the graph.

Short-Run Behavior: The Zeros and Vertical Asymptotes of a Rational Function

The short-run behavior of a polynomial can often be determined from its factored form. The same is true of rational functions. If a rational function is given by

$$y = \frac{p(x)}{q(x)}, \qquad p(x), q(x) \text{ polynomials,}$$

then the short-run behavior of $p(x)$ and $q(x)$ tell us about the short-run behavior of the rational function.

A fraction is equal to zero if and only if its numerator equals zero (and its denominator does not equal zero). Thus, the rational function $y = p(x)/q(x)$ has a zero wherever $p(x)$ has a zero, provided $q(x)$ does not have the same zero.

Just as we can find the zeros of a rational function by looking at its numerator, we can find the vertical asymptotes by looking at its denominator. A rational function is large wherever its denominator is small. This means that the rational function has a vertical asymptote wherever its denominator has a zero, provided its numerator does not also have the same zero.

Example 4   Find the zeros and vertical asymptotes of the rational function $y = \dfrac{x + 3}{x + 2}$.

Solution   We see that $y = 0$ if

$$\frac{x + 3}{x + 2} = 0.$$

This ratio equals zero only if the numerator is zero (and the denominator is not zero), so

$$x + 3 = 0$$
$$x = -3.$$

The only zero is $x = -3$. To check, note that at $x = -3$, we have $y = 0/(-1) = 0$. The denominator has a zero at $x = -2$, so the graph has a vertical asymptote there. See Figure 9.15.

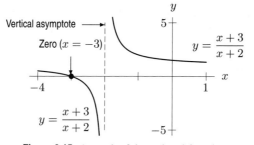

Figure 9.15: A graph of the rational function
$y = \dfrac{x+3}{x+2}$ showing its vertical asymptote and zero

To understand what happens near the asymptotes, we can evaluate at each asymptote the factors of the rational expression that are not zero there. Using the values we find, we approximate the function by one we can understand more easily.

Example 5   Graph $y = \dfrac{25}{(x+2)(x-3)^2}$, showing all the important features.

Solution   Since the numerator of this function is nowhere zero, this rational function has no zeros, which means that the graph nowhere crosses the x-axis. The graph has vertical asymptotes at $x = -2$ and $x = 3$ because this is where the denominator is zero. What does the graph look like near its asymptote at $x = -2$? At $x = -2$, the numerator is 25 and the value of the factor $(x-3)^2$ is $(-2-3)^2 = 25$. Thus, near $x = -2$,

$$y = \frac{25}{(x+2)(x-3)^2} \approx \frac{25}{(x+2)(25)} = \frac{1}{x+2}.$$

So, near $x = -2$, the graph looks like the graph of $y = 1/(x+2)$.

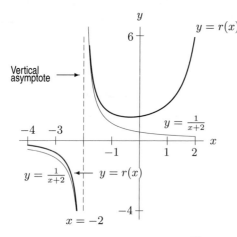

Figure 9.16: The graph of $y = \dfrac{25}{(x+2)(x-3)^2}$
resembles the graph of $1/(x+2)$ near the
asymptote at $x = -2$

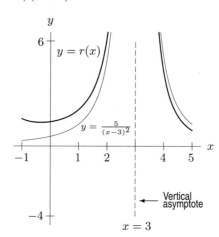

Figure 9.17: The graph of
$y = \dfrac{25}{(x+2)(x-3)^2}$ resembles the graph
of $5/(x-3)^2$ near the asymptote at $x = 3$

What does the graph of $y = 25/((x+2)(x-3)^2)$ look like near its vertical asymptote at $x = 3$? Near $x = 3$, the numerator is 25 and value of the factor $(x+2)$ is approximately $(3+2) = 5$. Thus, near $x = 3$,

$$y \approx \frac{25}{(5)(x-3)^2} = \frac{5}{(x-3)^2}.$$

Near $x = 3$, the graph looks like the graph of $y = 5/(x-3)^2$.

At $x = 0$,

$$y = \frac{25}{(0+2)(0-3)^2} = \frac{25}{18},$$

so the graph crosses the y-axis at $25/18$. The long-run behavior of this rational function is given by the ratio of the leading term in the numerator to the leading term in the denominator. The numerator is 25, and if we multiply out the denominator, we see that its leading term is x^3. Thus, the long-run behavior is given by $y = 25/x^3$, which has a horizontal asymptote at $y = 0$. See Figure 9.18.

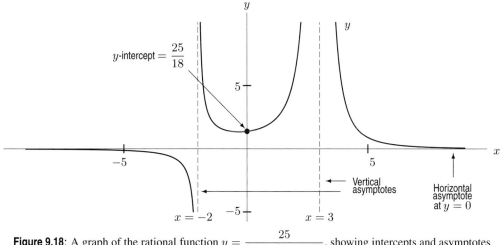

Figure 9.18: A graph of the rational function $y = \dfrac{25}{(x+2)(x-3)^2}$, showing intercepts and asymptotes

The Graph of a Rational Function

We can now summarize what we have learned about the graphs of rational functions.

For a rational function given by $y = \dfrac{p(x)}{q(x)}$, where p and q are polynomials with different zeros, then:

- The **long-run behavior** is given by the ratio of the leading terms of $p(x)$ and $q(x)$.
- The **zeros** are the same as the zeros of the numerator, $p(x)$.
- A **vertical asymptote** occurs at each of the zeros of the denominator, $q(x)$.

If $p(x)$ and $q(x)$ have zeros at the same x-values, the rational function may behave differently. See page 338.

Can a Graph Cross an Asymptote?

The graph of a rational function never crosses a vertical asymptote. However, the graphs of some rational functions cross their horizontal asymptotes. The difference is that a vertical asymptote occurs where the function is undefined, so there can be no y-value there, whereas a horizontal asymptote represents the limiting value of the function as $x \to \pm\infty$. There is no reason that the function cannot take on this limiting y-value at a finite x-value. For example, the graph of $y = \dfrac{x^2 + 2x - 3}{x^2}$ crosses the line $y = 1$, its horizontal asymptote; the graph does not cross the vertical asymptote, the y-axis. See Figure 9.19.

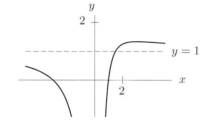

Figure 9.19: A rational function can cross its horizontal asymptote

When Numerator and Denominator Have the Same Zeros: Holes

The rational function $y = \dfrac{x^2 + x - 2}{x - 1}$ is undefined at $x = 1$ because the denominator equals zero at $x = 1$. However, the graph of y does not have a vertical asymptote at $x = 1$ because the numerator of y also equals zero at $x = 1$. At $x = 1$,

$$y = \frac{x^2 + x - 2}{x - 1} = \frac{1^2 + 1 - 2}{1 - 1} = \frac{0}{0},$$

and this ratio is undefined. What does the graph of this function look like? Factoring the numerator gives

$$y = \frac{(x - 1)(x + 2)}{x - 1} = \frac{x - 1}{x - 1}(x + 2).$$

For any $x \neq 1$, we can cancel $(x - 1)$ and rewrite the formula for y as

$$y = x + 2, \qquad \text{provided } x \neq 1.$$

Thus, the graph is the line $y = x + 2$ except at $x = 1$, where the function is undefined. The line $y = x + 2$ contains the point $(1, 3)$, but the graph of this rational function does not. Therefore, we say that the graph has a *hole* in it at the point $(1, 3)$. See Figure 9.20.

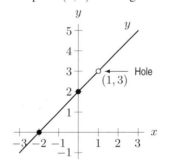

Figure 9.20: The graph of $y = (x^2 + x - 1)/(x - 1)$ is the line $y = x + 2$, except at the point $(1, 3)$, where it has a hole

Not All Rational Functions Have Vertical Asymptotes

It is tempting to assume that the graph of any rational function has a vertical asymptote. However, this is not always the case. The function shown in Figure 9.20 gives an example of a rational function whose graph has no asymptote: its graph has a hole. The function

$$y = \frac{1}{x^2 + 3}.$$

gives another example. Here, the denominator is always greater than 3; it is never 0. We see in Figure 9.21 that this function does not have a vertical asymptote.

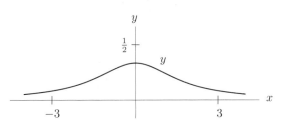

Figure 9.21: The rational function $y = 1/(x^2 + 3)$ has no vertical asymptote

Exercises and Problems for Section 9.3

Exercises

Match the following rational functions to the statements in Exercises 1–4. A statement may match none, one, or several of the given functions. You are not required to draw any graphs to answer these questions.

(a) $y = \dfrac{1}{x^2 + 1}$ (b) $y = \dfrac{x - 1}{x + 1}$

(c) $y = \dfrac{x - 2}{(x - 3)(x + 1)}$ (d) $y = \dfrac{(x - 2)(x - 3)}{x^2 - 1}$

1. This function has no zeros (x-intercepts).
2. The graph of this function has a vertical asymptote at $x = 1$.
3. This function has long-run behavior $y \to 0$ as $x \to \pm\infty$, that is, its graph has a horizontal asymptote at $y = 0$.
4. The graph of this function has no vertical asymptotes.

Match the following rational functions to the statements in Exercises 5–8. A statement may match none, one, or several of the given functions. You are not required to draw any graphs to answer these questions.

(a) $y = \dfrac{x^2 + 1}{x^2 - 1}$ (b) $y = \dfrac{x - 1}{x^2 - x - 2}$

(c) $y = \dfrac{x}{x + 1}$ (d) $y = \dfrac{1 + x^2}{2 + x^2}$

5. This function has no zeros (x-intercepts).

6. This function has long-run behavior $y \to 1$ as $x \to \pm\infty$, that is, its graph has a horizontal asymptote at $y = 1$.

7. The graph of this function has a vertical asymptote at $x = -1$.

8. The graph of this function contains the point $(0, 1/2)$.

In Exercises 9–11 find the vertical asymptotes of the functions given.

9. $f(x) = \dfrac{2x - 2}{x + 1}$

10. $g(r) = \dfrac{r - 6}{r^2 - 3r - 4}$

11. $h(x) = \dfrac{x^2 + x - 6}{x^2 - 7x + 10}$

In Exercises 12–14 find the horizontal asymptotes of the functions given.

12. $r = \dfrac{3t^4 - 3t^3 + t - 500}{3t^2 + 500t - 12}$

13. $g = \dfrac{x - 4}{x^2 + 16}$

14. $s = \dfrac{4z^3 - z + 9}{1000 + 3z^3}$

Problems

15. A rental car is driven d miles per day. The cost c is

$$c = \frac{2500 + 8d}{d} \text{ cents per mile.}$$

(a) Find the cost per mile for a 100 mile per day trip.
(b) How many miles must be driven per day to bring the cost down to 13 cents or less per mile?
(c) Rewrite the formula for c to make it clear that c decreases as d increases. What is the limiting value of c as d gets very large?

16. The cost of printing n posters is $10 + 0.02n$ dollars.

(a) Write a formula for the average cost a per poster of a printing of n posters.
(b) How many posters must be printed to make the average cost 4¢ per poster?
(c) Rewrite the formula for the average cost to show why it is impossible to print posters for less than 2¢ per poster.

17. When doctors test a patient for a virus, such as HIV, the results are not always accurate. For a particular test, the proportion, P, of the people who test positive and who really have the disease is given by

$$P = \frac{0.9x}{0.1 + 0.8x},$$

where x is the fraction of the whole population that has the disease.

(a) If 1% of the population has the disease, what is P?
(b) What value of x leads to $P = 0.9$?
(c) If doctors do not want to use the test unless at least 99% of the people who test positive really have the disease, for what values of x should it be used?

18. The rational function $r(x)$ can be written in two forms:

I. $r(x) = \dfrac{(x-1)(x-2)}{x^2}$

II. $r(x) = 1 - \dfrac{3}{x} + \dfrac{2}{x^2}$

Which form most readily shows

(a) The zeros of $r(x)$? What are they?
(b) The value of $r(x)$ as r gets large, either positive or negative? What is it?

19. The rational function $q(x)$ can be written in two forms:

I. $q(x) = 1 + \dfrac{2x-7}{x^2-1}$

II. $q(x) = \dfrac{(x-2)(x+4)}{(x-1)(x+1)}$

(a) Show that the two forms are equivalent.
(b) Which form most readily shows

  (i) The zeros of $q(x)$? What are they?
  (ii) The vertical asymptotes? What are they?
  (iii) The horizontal asymptotes? What are they?

20. Find the horizontal asymptotes of

$$y = \frac{6x+1}{2x-3}$$

and

$$y = \frac{4x+4}{2x-3} + 1.$$

What do you observe? Use algebra to explain your observation. [Hint: Note that $6x+1 = (4x+4) + (2x-3)$.]

21. Let $p(x) = \dfrac{x+3}{2x-5}$ and let $q(x) = \dfrac{3x+1}{4x+4}$.

(a) Find the horizontal and vertical asymptotes of $p(x)$ and $q(x)$.
(b) Let $f(x) = p(x) + q(x)$. Write $f(x)$ as a single rational expression.
(c) Find the horizontal and vertical asymptotes of $f(x)$. Describe the relationship between the asymptotes of $p(x)$ and $q(x)$ and the asymptotes of $f(x)$.

22. Let $R(x) = \dfrac{5}{(x-3)(x-2)}$.

(a) Show that $R(x) = \dfrac{5}{x-3} - \dfrac{5}{x-2}$.
(b) Which of the two forms of $R(x)$ helps to easily determine the zeros and the vertical asymptotes of $R(x)$? Find them.
(c) If $\dfrac{k}{(x-a)(x-b)} = \dfrac{k}{x-a} - \dfrac{k}{x-b}$ is an identity in x, what can you say about k, a, and b?

REVIEW EXERCISES AND PROBLEMS FOR CHAPTER NINE

Exercises

In Exercises 1–5, find the values of x at which the rational function is undefined.

1. $\dfrac{3x + 4}{3x^2 + 7x + 4}$

2. $\dfrac{x + 7}{2\pi}$

3. $\dfrac{2x + 1}{2x^2 - 5x - 3}$

4. $\dfrac{32}{32 - x^5}$

5. $\dfrac{2 - x^2}{x^4 - x^2 - 6}$

In Exercises 6–8, simplify the rational functions. Assume a is a nonzero constant.

6. $\dfrac{x^2 - 2x - 15}{6 - x - x^2}$

7. $\dfrac{ax^2 + 4x}{2ax}$

8. $\dfrac{3x^2 - 2ax - a^2}{x - a}$

Problems

9. Express as ratio of two polynomials:

$$\dfrac{\dfrac{2x^2 + 1}{x - 3} - \dfrac{x - 1}{x - 4}}{\dfrac{x + 5}{x - 3} \cdot \dfrac{x^2 + 2}{x - 4}}.$$

10. Figure 9.22 gives the graph of the rational function

$$y = \dfrac{k(x - p)(x - q)}{(x - r)(x - s)}.$$

Based on the graph, find values of k, p, q, r, s given that $p < q$ and $r < s$.

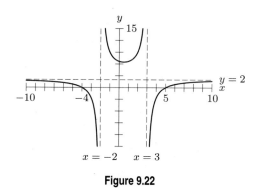

$x = -2 \qquad x = 3$

Figure 9.22

11. Give the zeros, holes, vertical asymptotes, long-run behavior, and the y-intercept of

$$y = \dfrac{x^2 - 4x + 3}{x^2 - x - 6}.$$

You are not required to graph this function.

12. The rational function $r(x)$ can be written in two forms:

I. $r(x) = \dfrac{x^2 - 9}{x^2 - 4}$

II. $r(x) = \dfrac{(x - 3)(x + 3)}{(x - 2)(x + 2)}$

Which form most readily shows

(a) The zeros of $r(x)$? What are they?
(b) The vertical intercept? What is it?
(c) The vertical asymptotes? What are they?

13. (a) Show that the rational function

$$r(x) = \dfrac{3x + 10}{4x + 8}$$

can also be written in the form

$$r(x) = \dfrac{3}{4} + \dfrac{1}{x + 2}.$$

(b) Explain how you can determine the horizontal asymptote from the second form, and find the asymptote.

14. At a fund raiser, Greg buys a raffle ticket. If he is the only one to buy a raffle ticket, he is sure to win the prize. If only one other person buys a ticket, his chance of winning is $1/2$. In general, if x tickets are sold, Greg's chance of winning is given by $w = 1/x$.

(a) Find the horizontal asymptote of $w = 1/x$.
(b) What is the practical interpretation of the horizontal asymptote?

Find formulas for the functions described in Problems 15–20. Some problems may have more than one possible answer.

15. The graph of h is a parabola with vertex $(h, k) = (3, 4)$ and with y-intercept 12.

16. The domain of u is all $x, x \neq 2, x \neq -3$.

17. The graph of w is given in Figure 9.23.

18. The graph of h is a parabola with vertex $(-2, 5)$ and y-intercept -1.

19. The domain of u is $2 \leq x \leq 8$.

20. The graph of w is given in Figure 9.24.

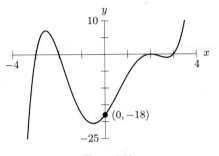

Figure 9.23

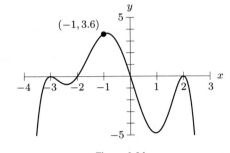

Figure 9.24

TOOLS FOR CHAPTER 9

In Example 6 on page 310 we saw that

$$3x^3 + 7x^2 + 2x - 3720 = (x - 10)(\text{Another polynomial}),$$

because the left hand side is zero when $x = 10$. To find the other polynomial, we must divide $x - 10$ into $3x^3 + 7x^2 + 2x - 3720$. We start by look at the leading term $3x^3$. Since $3x^3 = 3x^2 \cdot x$, we can write

$$3x^3 + 7x^2 + 2x - 3720 = (x - 10)\left(3x^2 \qquad\right).$$

Notice we have left space on the right for the lower degree terms in the second factor on the right. We set this up in more traditional format as

$$\begin{array}{r} 3x^2 \qquad\qquad \\ x-10\overline{)\ 3x^3 + 7x^2 + 2x - 3720} \end{array}.$$

To find the next term after $3x^3$, we see what is left to divide after we subtract $(x-10)(3x^2)$:

$$\begin{array}{r} 3x^3 + 7x^2 + 2x - 3720 = (x-10)\left(3x^2 \qquad\right). \\ -3x^3 + 30x^2 \qquad\qquad\qquad\qquad\qquad \\ \hline 37x^2 + 2x \qquad\qquad\qquad\qquad\qquad \end{array}$$

Or, in more traditional style:

$$\begin{array}{r} 3x^2 \qquad\qquad \\ x-10\overline{)\ 3x^3 + 7x^2 + 2x - 3720} \\ -3x^3 + 30x^2 \qquad\qquad\qquad \\ \hline 37x^2 + 2x \qquad\qquad \end{array}.$$

Now we can find the next term in the quotient by comparing the divisor, $x - 10$, with the new dividend $37x^2 + 2x - 3720$. The new leading term is $37x^2 = 37x \cdot x$, so the next term is $37x$, giving

$$\begin{array}{r} 3x^2 + 37x \qquad \\ x-10\overline{)\ 3x^3 + 7x^2 + 2x - 3720} \\ -3x^3 + 30x^2 \qquad\qquad\qquad \\ \hline 37x^2 + 2x \qquad\qquad \end{array}.$$

Continuing in this way, we get the completed calculation

$$\begin{array}{r} 3x^2 + 37x + 372. \\ x-10\overline{)\ 3x^3 + 7x^2 + 2x - 3720} \\ -3x^3 + 30x^2 \qquad\qquad\qquad \\ \hline 37x^2 + 2x \qquad\qquad \\ -37x^2 + 370x \qquad\qquad \\ \hline 372x - 3720 \\ -372x + 3720 \\ \hline 0 \end{array}$$

Notice that the remainder is zero, as we expected.

In general, we have to allow for the possibility that the remainder might not be zero. For example, to divide $x^2 + x + 1$ into $3x^4 - x^3 + x^2 + 2x + 1$, we write

$$3x^4 - x^3 + x^2 + 2x + 1 = (x^2 + x + 1)\left(3x^2 \qquad\right) \qquad,$$

leaving space for the remainder as well as the quotient. Or, in traditional format

$$
\begin{array}{r}
3x^2 \qquad\qquad\qquad \\[2pt]
x^2 + x + 1 \overline{\smash{\big)}\ 3x^4\ -x^3\ +x^2 + 2x + 1}
\end{array}.
$$

We subtract $(x^2 + x + 1)(3x^2) = 3x^4 + 3x^3 + 3x^2$:

$$
\begin{array}{r}
3x^2 \qquad\qquad\qquad \\[2pt]
x^2 + x + 1 \overline{\smash{\big)}\ 3x^4\ -x^3\ +x^2 + 2x + 1} \\[2pt]
-3x^4 - 3x^3 - 3x^2 \qquad\quad \\ \hline
-4x^3 - 2x^2 + 2x
\end{array}
$$

Again, we find the next term in the quotient by comparing the divisor, $x^2 + x + 1$, with the new dividend $-4x^3 - 2x^2 + 2x$. The ratio of the leading terms is now $-4x^3/x^2 = -4x$, so the next term is $-4x$, giving

$$
\begin{array}{r}
3x^2 - 4x \qquad\qquad\quad \\[2pt]
x^2 + x + 1 \overline{\smash{\big)}\ 3x^4\ -x^3\ +x^2 + 2x + 1} \\[2pt]
-3x^4 - 3x^3 - 3x^2 \qquad\quad \\ \hline
-4x^3 - 2x^2 + 2x
\end{array}
$$

Continuing in this way, we get the completed calculation

$$
\begin{array}{r}
3x^2 - 4x + 2 \qquad\qquad \\[2pt]
x^2 + x + 1 \overline{\smash{\big)}\ 3x^4\ -x^3\ +x^2 + 2x + 1} \\[2pt]
-3x^4 - 3x^3 - 3x^2 \qquad\quad \\ \hline
-4x^3 - 2x^2 + 2x \\
4x^3 + 4x^2 + 4x \\ \hline
2x^2 + 6x + 1 \\
-2x^2 - 2x - 2 \\ \hline
4x - 1
\end{array}
$$

This means that $4x - 1$ is the remainder, so

$$
3x^4 - x^3 + x^2 + 2x + 1 = (x^2 + x + 1)(3x^2 - 4x + 2) + 4x - 1.
$$

The process for dividing one polynomial into another is called *the division algorithm.*

The Division Algorithm

Given two polynomials $a(x)$ and $b(x)$, where $b(x)$ is not the zero polynomial, we can divide $b(x)$ into $a(x)$ to get a polynomial $q(x)$ (the quotient) and a polynomial $r(x)$ (the remainder) such that

$$
a(x) = q(x)b(x) + r(x)
$$

and either $r(x)$ is the zero polynomial or it has degree less than that of $b(x)$.

When we divide a positive integer b into another integer a, we get a remainder which is less than the divisor, b. The analogous fact for polynomial division is the degree of the remainder being less than the degree of the divisor, $b(x)$.

Example 1   Divide

(a) $x - 2$ into $x^3 - 2x + 1$ (b) $x^2 + 1$ into $2x^3 + 3x^2$

Solution   (a) We have

$$
\begin{array}{r}
x^2 + 2x + 2 \\
x - 2 \overline{)\ x^3 \qquad\quad - 2x + 1} \\
-x^3 + 2x^2 \\
\hline
2x^2 - 2x \\
-2x^2 + 4x \\
\hline
2x + 1 \\
-2x + 4 \\
\hline
5
\end{array}
$$

so

$$(x^3 - 2x + 1) = (x - 2)(x^2 + 2x + 2) + 5.$$

(b) We have

$$
\begin{array}{r}
2x + 3 \\
x^2 + 1 \overline{)\ 2x^3 + 3x^2} \\
-2x^3 \qquad\quad - 2x \\
\hline
3x^2 - 2x \\
-3x^2 \qquad - 3 \\
\hline
-2x - 3
\end{array}
$$

so

$$2x^3 + 3x^2 = (2x + 3)(x^2 + 1) + (-2x - 3).$$

The Remainder Theorem

A particularly simple division is when we divide the polynomial $x - a$, for a constant a, into some other polynomial $p(x)$. In that case the remainder is a constant, since it must have degree less than the degree of $x - a$. Thus

$$p(x) = (x - a)q(x) + \text{Constant}.$$

What is the constant? If we put $x = a$ in the equation above, we get

$$p(a) = 0 \cdot q(a) + \text{Constant}.$$

Since 0 times anything is 0, this tells us that the constant is $p(a)$. We have shown

---

The Remainder Theorem

For any polynomial $p(x)$ and any constant a, we have

$$p(x) = (x - a)q(x) + p(a),$$

where $q(x)$ is the quotient of $p(x)/(x - a)$.

Notice that if $p(a) = 0$ this tells us that $x - a$ is a factor of $p(x)$. This proves fact (ii) discussed on page 309.

Exercises on the Division Algorithm

The rational expressions in Exercises 1–4 are in the form $a(x)/b(x)$ where $a(x)$ and $b(x)$ are polynomials. Use the division algorithm to find the quotient $q(x)$ and the remainder $r(x)$ so that $a(x) = q(x)b(x) + r(x)$.

1. $\dfrac{2x^2 - 3x + 11}{x + 7}$

2. $\dfrac{2x^2 - 3x + 11}{x^2 + 7}$

3. $\dfrac{x^4 - 2}{x - 1}$

4. $\dfrac{x^4 - 2}{x^2 - 1}$

Use what you know about polynomial division to answer Problems 5–7.

5. Given that
$$\frac{p(x)}{x^2 - 5x + 6} = x + 3,$$
find
$$\frac{p(x)}{x - 3}.$$

6. Given that
$$\frac{p(x)}{x - 4} = x^2 + 7x + 10,$$
find
$$\frac{p(x)}{x^2 + x - 20}.$$

7. Given that
$$\frac{x^4 + x + 1}{x + 1} = x^3 - x^2 + x \quad \text{with remainder } r = 1,$$
find
$$\frac{x^4 + x + 2}{x + 1}.$$

Without actually dividing (which is a lot of trouble), state values of $q(x)$ and $r(x)$ for Problems 8–13 such that
$$\text{If} \quad \frac{a(x)}{b(x)} = q(x) \text{ with remainder } r(x)$$
$$\text{then} \quad a(x) = b(x)q(x) + r(x).$$

Explain your reasoning. *Example.* By writing
$$\frac{8x^5 + 2x + 1}{8x^5 + 2x + 2} = \frac{8x^5 + 2x + 2 - 1}{8x^5 + 2x + 2},$$

we see that $q(x) = 1$ and $r(x) = -1$, because
$$\underbrace{8x^5 + 2x + 2 - 1}_{a(x)} = \underbrace{(8x^5 + 2x + 2)}_{b(x)} \cdot \underbrace{1}_{q(x)} + \underbrace{(-1)}_{r(x)}.$$

8. $\dfrac{8x^5 - 3x^2 + 2x + 3}{8x^5 - 3x^2 + 2x + 2}$

9. $\dfrac{8x^5 - 3x^2 + 3x + 3}{8x^5 - 3x^2 + 2x + 2}$

10. $\dfrac{8x^5 + 2x^4 - 3x^2 + 2x + 2}{8x^5 - 3x^2 + 2x + 2}$

11. $\dfrac{16x^5 - 6x^2 + 4x + 7}{8x^5 - 3x^2 + 2x + 2}$

12. $\dfrac{8x^6 - 3x^3 + 2x^2 + 2x + 5}{8x^5 - 3x^2 + 2x + 2}$

13. $\dfrac{(8x^5 - 3x^2 + 2x + 2)^2 + 3}{8x^5 - 3x^2 + 2x + 2}$

Refer to the Remainder Theorem to answer Problems 14–17.

14. Given that
$$\frac{x^3 + 2x + 3}{x - 2} = x^2 + 2x + 6 \quad \text{with remainder } r,$$
find r.

15. Given that
$$\frac{p(x)}{x - 3} = x^2 + 5 \quad \text{with remainder } 2,$$
find $p(x)$.

16. Given that
$$\frac{p(x)}{x - 4}$$
has remainder $r = 7$, find $p(4)$.

17. Given that
$$\frac{x^2 + 3x + 5}{x - a}$$
has remainder $r = 15$, find a.

ANSWERS TO ODD NUMBERED PROBLEMS

Section 0.1

1. $0.07p$
3. $0.06(p - 1000)$
5. Add 1, multiply by 2
9. $2(1 - x) + 3$
11. $2(1 - (x + 3))$
13. $12x$
15. $n - 6$
17. $t + 4$
19. $5q + 2$
21. 176
23. 3
25. 70
27. -8
29. -12
31. $75s + 15c$ dollars
33. $704.5w/h^2$
35. $p + n + 10d + 5q$
37. Production twice that of well 1
39. Production 80 barrels/day less than well 1
41. Production 3 times wells 1 and 2 combined
47. Yes; $a = b^2, x = \theta^2$
49. Yes; $a = 2, x = 4y$
51. 150π ft^2
53. 108π ft^2
55. $V_0 + r_1 t + r_2 t$
57. $(g_1 + g_2 + g_3 + g_4 + 2f)/6$
59. $x/z + (L + 1)(y + 2k)$
63. $w = \pm 3$

Section 0.2

1. (a) $\dfrac{1}{5}b; \dfrac{b}{5}$
   (b) Yes
3. (a) $0.8b; \dfrac{b}{8/10}$
   (b) No
5. Not equivalent
7. Not equivalent
9. Equivalent
11. $-3y^3 + x$
13. a
15. $2z^4 + 5z^3$
17. $2t - 3t$ and $-t$ equivalent

-11	-7	0	7	11
-22	-14	0	14	22
33	21	0	-21	-33
11	7	0	-7	-11
11	7	0	-7	-11

19. I and $-(-I)$ are equivalent

I	-2	-1	0	1	2
$-I$	2	1	0	-1	-2
$-(-I)$	-2	-1	0	1	2

21. Incorrect
23. Correct
27. Yes
29. 85
33. $1 - (1 - x)$ equiv. to x
35. $x \cdot x \cdot x + x \cdot x \cdot x + x \cdot x + x \cdot x + x \cdot x +$ $x + x + x + x$
37. Always even

Section 0.3

1. (ii)
3. Yes
5. $5m/6$
7. $(2x + 1)/(x + 1)$
9. $e/8$
11. $3/j$
13. $(b + a)/(ab)$
15. $-1/((x - 2)(x - 3))$
17. $2a/((a - b)(a + b))$
19. $1/x$
21. $1/5$
23. $(a + b)/ab$
25. 1
27. $-c/(2d)$
29. $(3(a + 2))/(2(b + 3))$
33. $-(1/2)(a + 1) + 1$ and $-(1/2)a + (1/2)$ are equivalent
35. $a = 2, b = x, c = u$
37. $a = 1/r, b = r^2 s^3, c = -rs^2$
39. (c)
41. (b)
43. None
45. (a) Not equivalent
    (b) Not equivalent
    (c) Equivalent
    (d) Not equivalent
    (e) Equivalent
    (f) Equivalent
47. $AB/(B + A)$
49. $n + n + n + f + f + f$, other answers possible

Section 0.5

1. Not a solution
3. Solution
5. Solution
7. (c)
9. $x + 2 = 2x$
11. $x - 6 = 30$
13. A doubled number is 16
15. Ten more than a number is twice the number
17. Four less than number is 3 times number
19. $t = 15$
21. $y = 25$
23. (a)

x	0	1	2	3	4
$1 + 5x$	1	6	11	16	21

(b) $x = 3$
25. Not an identity
27. Not an identity
29. $1 \le s \le 240$, s an integer
31. $30 \le V \le 200$
33. $n, 16 \le n < 24$, n an integer
35. $20 = 6 + 88t - 16t^2$
37. $20 = -58 + 2v$
39. (a) 3500
    (b) $3500 - 700t = 0$
41. Yes
43. No
45. $t = 0, 4$
51. Identity
53. $-1, 1, 2$
55. 1
59. $0 < x < 7$
61. (a) $R_T = R_1 R_2/(R_1 + R_2)$
    (b) $10/7$ ohms

Section 0.6

1. Subtract 7; $x = 3$
3. Multiply by 9; $x = 153$
5. Subtract 3 then divide by 2; $x = 5$
7. Subtract 5 then multiply by 3; $x = 45$
9. Valid, add $2x$ to both sides
11. Valid, subtract 1 from both sides and rearrange
13. Invalid
15. Valid, divide both sides by 5
17. Subtract 0.1, -0.2
19. Add t, 8
21. Mult. by -1, 4
23. Mult. by -5, -20
25. No
27. (e)
29. (b)
31. (b)
33. (e)
35. Decreases
37. Increases
39. (a) Yes
    (b) No
41. No; $x = 0$ is solution

Chapter 0 Review

1 $2(x+2) - 4$

3 $(x-1)^2 + 1$

5 $-18/25$

7 $1/4$

9 $A = 2r, B = 3s$

11 $A = n, B = m + z^2$

13 $A = 5, B = 7$

15 One plus three times a number is 15

17 Fifteen more than number is three times the number

19 Seven less than a twice a number is 5 times the number

21 $a = 14$

23 $b = 40$

25 1 and 2

27 0

29 -8

31 Neither value

33 Subtract 5 from both sides

35 Multiply both sides by 2

37 Multiply both sides by 4

41 Equation, $2n + (n+3) = 21$

43 Equation, $2(J+S) = 140,000$

45 Expression, $3w - 50$

47 (a) Total number of widgets company receives
   (b) Proportion of widgets supplied by A
   (c) Nothing
   (d) Number of defective widgets supplied by A
   (e) Total number of defective widgets company receives
   (f) Proportion of defective widgets company receives
   (g) Proportion of non-defective widgets company receives

49 $a = b$ and $x = z$

51 $a = 1/7$ and $x = t$

53 $9w - 15$

55 (a) and (c); (b) and (f); (d) and (e)

57 $r_1 + r_2 - 100$

59 (a) $(2p + 3q)/5$
   (b) q
   (c) p
   (d) $(Ap + Cq)/(A + C)$

61 (a) $4p + 3e + 9c$ dollars
   (b) $4p/(4p + 3e + 9c)$
   (c) $Pp + Ee + Cc$; $Pp/(Pp + Ee + Cc)$

63 $400p + 200q + 120r + 30s$

65 (D)

67 (b) 40
   (c) 2 pm

69 (a) No
   (b) Yes

71 (c)

73 (c)

75 max of $E - 0.075I$ and 0

77 $50,000,000 = 3f + 5,000,000$

79 True

81 True

Section 1.1

1 $w = f(c)$

3 Dep: N, Ind: C

5 $f(-7) = -9/2$

7 $1/3$

9 $1/\sqrt{2} = \sqrt{2}/2$

11 4

13 (a) 99
   (b) -15
   (c) 145
   (d) 177

15 (a) 4
   (b) 3
   (c) 2
   (d) 2 and 4

17 $C = 1.06P$

19 $10 - 6u$

21 $-2 + 3n$

23 $-2 + 3t^3$

25 $n + 1$

27 Average daily downloads at current price

29 Change in daily downloads if price drops 10 cents

31 Average daily downloads 20% price drop

33 Average daily revenue

37 Mileage difference if 5 lbs/in^2 underinflated

39 (a) $0, 9, 8, 3, 0, 6$
   (b) 3
   (c) -8
   (d) 16
   (e) 6

Section 1.2

1 (a) $f(1) = 8, f(3) = 6, f(1)$ greater

3 (a) $C(100) = -20, C(200) = -40, C(100)$ greater

5 (a) 40: $^\circ$C
      3: liters
   (b) 3 liters

7 Equivalent

9 Not equivalent

11 $Q = (1/4)t; k = 1/4$

13 $Q = (b+r)t; k = b + r$

15 $Q = (\alpha - \beta)t/\gamma; k = (\alpha - \beta)/\gamma$

17 0.00591 meters/year

19 22

21 34

23 Increases

25 Decreases

27 Increases

29 (a) 5 children
   (b) $40
   (c) $250
   (d) $c/3$

31 $y = 5 + 3(x - (-1)); m = 3, x_0 = -1$

33 $a = 5, b = 4, c = -2, d = 3$

35 Value increased by $30,000

37 Value decreased at average rate of $1,000/year

39 $a \neq 3$

Section 1.3

1 $x = 1$

3 $x = \pm 4$

5 (a) $t = 6$
   (b) $t = 1, t = 2$

7 (a) 75
   (b) 5

9 20

11 3 cm

13 (a) $15 - d/20 = 10$
   (b) gallons of gas

10 gallons when $d = 100$

15 (a) $s = 0.5, s = 0.75, s = 1$
   (b) $s = 0$
   (c) $s = 0.25$

17 (a) $p = 50$

19 (a) Year 2
   (b) $t = 3$

Section 1.4

1 Yes, 5

3 No

5 Yes, 1/9

7 No

9 Yes, 42

11 $q = 8$

13 $t = 10$

15 (a) $m = kr$
   (b) 76.125 ft
   (c) 4.091 in

17 $k = 0.1$; a 10% off sale

19 (a) $k = 4.25$
   (b) $k = 5.15$
   (c) $k = 4.38$
   (d) $109.50

21 (a) 2006: $83,333.33
      1967:$1451.86
   (b) About 57

23 $D = 30t, k = 30, 150$ miles

25 (a) $C = kx$
   (b) $C = 9.50x$
   (c) Dollars per yard
   (d) $52.25

27 Blood mass $= 0.05$ (Body mass);
    3.5 kilograms

29 $29.52

31 Yes

33 No

Chapter 1 Review

1 $w(8000) = 6000$

3 (a) 1.25
  (b) 0.5
  (c) $(a^2 + 1)/(5 + a)$.

5 0

7 Undefined

9 $(\pi + 1)/(2\pi + 1)$

11 (a) -4
  (b) ± 2

13 (a) w
  (b) $(-4, 10)$
  (c) $(6, 1)$

15

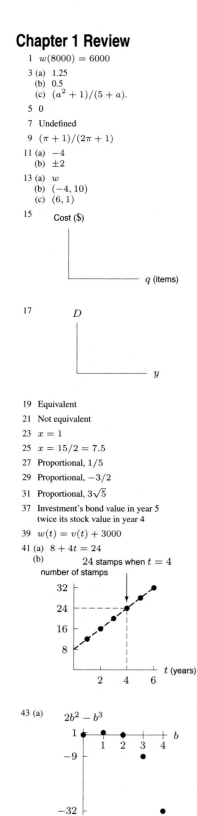

Cost ($)

q (items)

17

D

y

19 Equivalent

21 Not equivalent

23 $x = 1$

25 $x = 15/2 = 7.5$

27 Proportional, $1/5$

29 Proportional, $-3/2$

31 Proportional, $3\sqrt{5}$

37 Investment's bond value in year 5 twice its stock value in year 4

39 $w(t) = v(t) + 3000$

41 (a) $8 + 4t = 24$
  (b)

24 stamps when $t = 4$
number of stamps

43 (a)

$2b^2 - b^3$

(b) $b = 0$ and $b = 2$

45 (a) $100,000 - 10,000t = 70,000$
  (b)

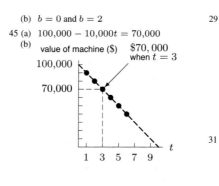

value of machine ($) $70,000 when $t = 3$

100,000

70,000

1 3 5 7 9   t

47 (a) $t = 4$ days

49 (a) $a = 0, a = 1$
  (b) $a = 3, a = 4, a = 5, a = 6$
  (c) $a = 2$

51 (a) $M = 167H$
  (b) $M = 20b$
  (c) $B = 8.35H$, yes

53 $C = 81.667b$,
  326.667 calories

55 $(f(5) - f(3))/(5 - 3)$

57 $10,374.04$

59 $q(t_2) < q(t_1)$

Section 2.1

1 (a) (i) $50,005

      (ii) $55,000

      (iii) $50,000
  (b) Start up costs

3 $h = 60 + 0.2t$ inches

5 $T = 30 - 0.04d$

7 $b = -5300, m = 250$

9 $b = 0.007, m = -0.003$

11 $b = 0, m = 0.5$

13 (a) Initial rental cost
   (b) $100 per hour

15 $600, 5$

17 $b = 25, m = 0.06$

19 $b = 50, m = 1.2$

21 $b = 100, m = -3$

23 Vertical intercept: 20,000, initial cost;
   Slope: 1500, annual cost

25 Vertical intercept: 120, initial distance;
   Slope: -5, speed toward shore

27

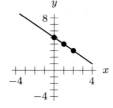

29

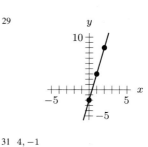

31 $4, -1$

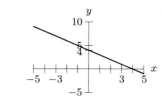

33 $-2, 3$

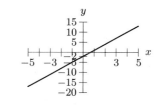

35 $-0.2, -0.5$

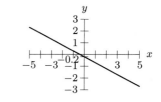

39 (a) (iii)
   (b) (i)
   (c) (ii)
   (d) (iv)
   (e) (v)

41 Yes, slope a, vertical intercept $5a$

43 Yes

45 $m = 1/3; b = -11$

47 $m = -2/3; b = 20/3$

49 $m = \pi; b = 0$

59 Time the journey takes (in hours)

61 Total fuel cost

63 $V/(nT)$ in^3 of seed

Section 2.2

1 Linear

3 Linear

5 Not linear

7 Linear

9 Linear

11 Linear

13 Not linear

15 Linear

17 Linear

19 Linear

21 Not linear

23 Not linear

25 $x^2 + 5x - 2$, not linear

27 $4, 3$

29 $w + 1, w$

31 $mn + m + 7, m + 5$

33 $y = 9 + 3x$

35 $g(n) = 22 - 2/3n$

37 $(1, 5), 3$

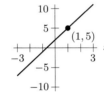

39 $(1, 3), 1/2$

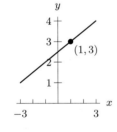

45 (a) Cost of gas; expenses; car rental

   (b) Dollars; miles

   (c) Yes

47 (iii)

49 $B = 10 + 2x$

53 $350 + 95d$ dollars

55 $40,000 + 5000d$ dollars

57 Yes

59 $\frac{1}{1/2}\left(1 - \left(-\frac{5}{2}\right)x\right)$

61 (a) $N = 550 - 100t$

   (b) $0 \le t \le 5.5$

Section 2.3

1 (a) $1900

  (b) 7 credits

3 Not possible

5 500

7 $x = 6$

9 $z = 22/3 = 7.333$

11 $y = 28$

13 $x = 4$

15 $x = 28$

17 $a = 2$

19 $p = -1.54$

21 $r = -14/5 = -2.8$

23 $m = 15/8 = 1.875$

25 $k = -9/2 = -4.5$

27 $n = -23/4 = -5.75$

29 $B = -3$

31 Positive

33 Negative

35 Positive

37 Negative

39 Positive

41 Negative

43 One

45 Infinite

47 One

49 Infinite

51 $t = y/(3\pi)$

53 $a = 2(s - v_0 t)/t^2$

55 $w = (c - ab)/(2a)$

57 $m = (3w - 2u - z)/(u + w - z)$

59 $y = z - x$

61 $m = (p - n)/2$

63 $w = 3x$

65 $t = 1/11$

67 (a) $62

   (b) $211

   (c) 800 miles

69 1.238 pounds

71 18 ft

73 (a) $A = 0$

   (b) $A > 0$

   (c) None

75 (a) $A = 5$

   (b) $A > 5$

   (c) None

77 (a) Any value of A

   (b) No value of A

   (c) None

79 (a) None

   (b) $A > 0$

   (c) $A = 0$

81 One

83 Infinite

85 One

87 Infinite

89 $x = 3.2$

91 2.6

93 $x = 7/3$

95 $x = -5$

Section 2.4

1 $y = 160 - 3x, m = -3, b = 160$

3 $y = 300 - 3x, m = -3, b = 300$

5 (a) (V)

  (b) (IV)

  (c) (I)

  (d) (VI)

  (e) (II)

  (f) (III)

  (g) (VII)

7

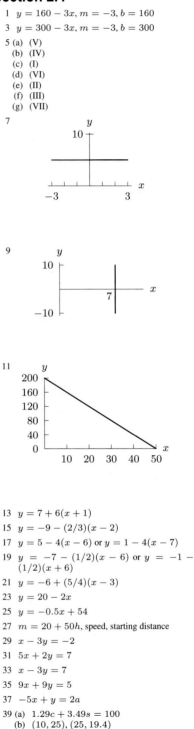

9

11

13 $y = 7 + 6(x + 1)$

15 $y = -9 - (2/3)(x - 2)$

17 $y = 5 - 4(x - 6)$ or $y = 1 - 4(x - 7)$

19 $y = -7 - (1/2)(x - 6)$ or $y = -1 - (1/2)(x + 6)$

21 $y = -6 + (5/4)(x - 3)$

23 $y = 20 - 2x$

25 $y = -0.5x + 54$

27 $m = 20 + 50h$, speed, starting distance

29 $x - 3y = -2$

31 $5x + 2y = 7$

33 $x - 3y = 7$

35 $9x + 9y = 5$

37 $-5x + y = 2a$

39 (a) $1.29c + 3.49s = 100$

   (b) $(10, 25), (25, 19.4)$

(c)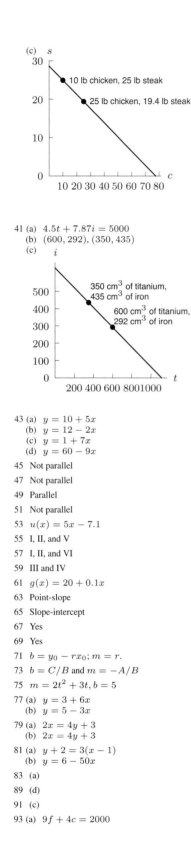

41 (a) $4.5t + 7.87i = 5000$
(b) $(600, 292), (350, 435)$
(c)

43 (a) $y = 10 + 5x$
(b) $y = 12 - 2x$
(c) $y = 1 + 7x$
(d) $y = 60 - 9x$

45 Not parallel

47 Not parallel

49 Parallel

51 Not parallel

53 $u(x) = 5x - 7.1$

55 I, II, and V

57 I, II, and VI

59 III and IV

61 $g(x) = 20 + 0.1x$

63 Point-slope

65 Slope-intercept

67 Yes

69 Yes

71 $b = y_0 - rx_0; m = r.$

73 $b = C/B$ and $m = -A/B$

75 $m = 2t^2 + 3t, b = 5$

77 (a) $y = 3 + 6x$
(b) $y = 5 - 3x$

79 (a) $2x = 4y + 3$
(b) $2x = 4y + 3$

81 (a) $y + 2 = 3(x - 1)$
(b) $y = 6 - 50x$

83 (a)

89 (d)

91 (c)

93 (a) $9f + 4c = 2000$

(b) $f = 66.7$

95 (a) 50
(b) 48
(c) $h = 100 - (5/4)a$

97 $750a + 1200r = 1,000,000$

Section 2.5

1 (a) $707
(b) $35 + 112 \cdot c$ dollars

3 $50 + 45h$ dollars

5 $2400 - 500y$ dollars

7 $1.03p$

9 $80 + 2p$

11 $200 - 50h$ miles

13 (a) $C = 11{,}835 - 1035(t - 1)$ dollars
(b) $4590
(c) $0

19 $N = 40{,}000 - 1400t$

21 $y = 1400 + 6x$

23 $d = 77 - 3.2t$

25 Yes

27 No

29 Yes

31 $N = 800 - 60t$

33 III

35 (a) $62.00
(b) $37.00 + 0.25m$ dollars
(c) $37.00d + 25$ dollars

37 1477 dollars per capita in circulation in 1995; Annual increase of 94.5 dollars per capita; -15.6, in 1979 no dollars per capita in circulation; no

39 (a) $I = 145 + 12.5(t - 2)$ thousand dollars
(b) (i) $182,500
(ii) $970,000
(iii) $95,000
(c) Reliable; not reliable; possibly reliable

41 $100 + 4t$ gallons for $0 \le t \le 25$; filling; 25 minutes; 200 gallons for $t > 25$

43 (a) $P = 3.25 + 0.1t$
(b) $P = 3.25 + 0.1(t - 10)$

45 Linear, $Q = 5.13 + 0.22t$

47 (a) Yes, $I = 46 + 453.6w$
(b) Number of grams in a pound (instrument weighs in grams), bad calibration of instrument

Section 2.6

1 $x = 6, y = -1$

3 $x = 1, y = -1$

5 $a = 5, b = -2$

7 $p = -1, r = 4$

9 $(x, y) = (2, 3)$

11 $(m, n) = (0.1, 0.5)$

13 $(x, y) = (10, -20)$

15 $(x, y) = (-1, 3)$

17 $(x, y) = (3, -7)$

19 $(e, f) = (6, -7)$

21 $x = 2, y = 3$

23 $(2.8, 2.1)$

25 $x = 1, y = 5$

27 $x = 1.5, y = -4$

29 $(x, y) = (3, -1)$

31 $(\alpha, \beta) = (6, 5)$

33 $(x, y) = (-2, -5)$

35 $(\kappa, \psi) = (2, -1)$

37 $x = 5, y = -5$

39 $x = 1, y = b$

41 If $0 < x < 5$ use A, if $5 < x \le 12$ use B, if $x = 5$ use either company

43 $29/2$ and $5/2$

45 (a) $4.05
(d) Yes

47 $n = 10$

49 $u = 6/5, d = 1/20$; UK cup is $6/5$ US cup, UK dessertspoon is $1/20$ UK cup

Chapter 2 Review

1 4.29, 3.99

3 250, $1/36$

5 $220 - 5t$

7 $b = 200, m = 14$

9 $b = 0, m = 1/3$

11 $b = 7/3, m = 2/3$

13 -7

15 $4h$

17 Positive

19 Positive

21 Negative

23 Positive

25 Negative

27 No solution

29 Linear

31 Linear

33 Not linear

35 Linear

37 Linear

39 Linear

41 $y = 8$

43 $t = 14/11 = 1.273$

45 $x = 9$

47 $x = 12/7$

49 Initial value, annual depreciation

51 $C = \frac{5}{9}(F - 32)$

53 $x = (3C - 2B)/(A - 6B)$

55 (a) $y = 7 + 3x$
(b) $y = 8 - 10x$

57

59 $y = -7 + 3x; b = -7, m = 3.$

61 $y = -12/7 + (4/7)x; b = -12/7, = 4/7$

63 $y = 0 + \sqrt{8}x; b = 0, m = \sqrt{8}$

65 $3x - 2y = -1$

69 $y = 5 + 2x$

71 $y = 8 - (3/4)x$

73 $y = -16 - 3x$

75 $y = 13 - (2/3)x$

77 $y = 6 + (1/3)x$

79 $y = 9 - (3/5)x$

81 $y = -13 - (4/5)x$

83 $y = -60 - 3x$

85 $s(t) = 8200 + 150t$

87 $w(x) = -3x + 32$

89 $g(x) = -x + 55$

91 $f(x) = -5x - 1$

93 $p = 125,000 + 5000h$

95 $s = 9 - 0.5q$

97 $q = 2500 - 2000p$

99 Yes

101 No

103 No

105 $600 + 24P$ dollars

107 Starting amount, rate of increase

109 Number of FDIC-insured banks in 1993; decline of 480 FDIC-insured banks a year; 2020

111 (iv)

113 (ii)

115 (v)

117 (i)

119 (vi)

121 (iv)

123 (iii)

125 (v)

127 (vi)

129 (iv)

131 $y = r + 3 + st$

133 $y = 2r + st$

135 $y = 0.5s$

137 $v(x) = 7x + 7$

139 $3.45d$ dollars, linear

141 $0.25I - 1250$ dollars, linear

143 (a) $65t - 75(t - 1)$ miles
(b) $1 \le t \le 7.5$

145 $100 - 4t$ gallons for $0 \le t \le 25$; emptying; 25 minutes

147 (a) $T = 53 - 3/1000(E - 7000)$
(b) (i) $45°F$

(ii) $74°F$

149 (a) $C = 320 + 0.2(n - 100)$
(b) $m = 0.2; b = 300$
(c) $450
(d) 1000 CDs

151 (a) $0.39; $0.11; $y = 0.39 - 0.0255x$
(b) 15.3; No
(c) $0.44; completely
(d) 1966; no agreement

(e) No

157 (a) $F = 2C$
(b) $320°F = 160°C$

159 $f(t) = 4500 - 0.25t$

161 2.5

Section 3.1

1 Power function

3 Not a power function

5 Power function

7 (a) $f(0) = 0$
(b) $f(2) = 12$
(c) $f(a) = 3a^2$
(d) $f(x + 1) = 3x^2 + 6x + 3$

9 (a) $f(0) = 0$
(b) $f(2) = 40$
(c) $f(a) = 5a^3$
(d) $f(x + 1) = 5x^3 + 15x^2 + 15x + 5$

11 2; 1

13 1/3; 1

15 2; 4π

17 (a) Coef: $\pi/5$; exp: 3
(b) $8\pi/5$ cm^3
(c) $64\pi/5$ cm^3

19 (a) Proportional
(c) Increases

21 (a) Inversely
(b)

x	1	10	100	1000
y	1	0.1	0.01	0.001

(c) Decreases

23 $p > 1$

25 $p = 1$

27 $p > 1$

29 (a) Odd
(b) Positive

31 a^4

33 b^{-4}

35 (a) 600 hours, 240 hours
(b) Decreases
(c) $w = 2400/h$, inversely

37 144 feet; 400 feet; larger after five seconds

39 Larger for body mass 70 kilograms

41 $x^2 \cdot x^3 = x^5 = x^{2+3}$

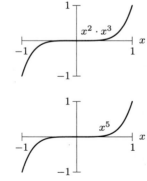

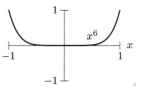

Section 3.2

1 Coefficient: 3; exponent: 1/2

3 Coefficient: 4; exponent: $-1/4$

5 Coefficient: 1/4; exponent: 3

7 Coefficient: 2; exponent: -2

9 Coefficient: $3^{1/4}/2$; exponent: 1/2

11 $k = 2/3, p = -1/2$

13 $k = 49, p = 6$

15 $k = 27, p = 6$

17 $k = 1/5, p = -1$

19 Not possible

21 $k = 1/8, p = -3/2$

23 $k = 64, p = 6$

25 $f(a)$

27 $f(b)$

29 $f(b)$

31 1; (III)

33 3; (I)

35 -1; (II)

37 -2; (II)

39 3; (I)

41 0.3; (IV)

43 (a) Not possible
(b) $P = 8w; k = 8; p = 1$

45 $k = 2/3^{1/4}, p = 1/2$

47 (a) 22,289
(b) 58,045
(c) 5,377

49 Unchanged

51 Decreases

53 Increases

55 Remains unchanged

57 Increases

Section 3.3

1 $x = 50^{1/3} = 3.684$

3 $x = 1/16$

5 No solutions

7 $a = 81$

9 $y = 123$

11 $x = 83/3$

13 $c = 2.438$

15 $p = 0, p = 1.871, p = -1.871$

17 $L = \pm 1/4$

19 $x = -1/27$

21 $z = 720\pi$

23 $r = \pm\sqrt{2A/\pi}$

25 $x = 1/(3y^2),\ x = 0$

27 (i)

29 (iv)

31 (vi)

33 (vi)

35 (i)

37 (i)

39 (i)

41 (ii)

43 $0.18

45 (a) $r = C/2\pi$

(b) $r = \sqrt{A/\pi}$

(c) $r = (3V/4\pi)^{1/3}$

(d) $r = \sqrt{V/\pi h}$

(e) $r = \sqrt{3V/\pi h}$

49 (a) $A = 0$
(b) $A > 0$
(c) $A > 0$

51 (a) $A = 0$
(b) $A > 0$
(c) $A > 0$

53 (a) $A = 0$
(b) Any $A \neq 0$
(c) Any $A \neq 0$

55 (a) $A = 0$
(b) No A
(c) Any $A \neq 0$

57 (a) No A
(b) $A < 0$
(c) $A < 0$

59 (a) No A
(b) No A
(c) No A

61 (a) $A = 4$
(b) $A > 4$
(c) No A

63 (a) $A = 4$
(b) $0 < A < 4$
(c) $A \leq 0$

65 (a)

(b) $x = 17.100$

Section 3.4

1 $y = 23.25x^5$

3 $y = 40x^2$

5 $s = 14.1421\sqrt{t}$

7 $k = 2;\ c = 2d^2;\ c = 98$

9 $S = kh^2$

11 $r = kA^{1/2}$

13 Multiplied by a factor of 8

15 Halved

17 (a) 8
(b) 8

19 $y = 2x^2$

21 $y = -x^4$

23 (a) 4
(b) 9
(c) 1/4
(d) 0.01

25 Multiplied by 1/4

27 Increases by 33.1%

29 (a) $R = kr^4$
(k is a constant)
(b) $R = 4.938r^4$
(c) $3086.42\ \text{cm}^3/\text{sec}$

31 (a) 7228 feet
(b) 0.25

33 (a) Life span increases with body size. Bird.
(c) (ii) 9.78×10^{37} kg. Unrealistic.

35 Yes

37 No

Chapter 3 Review

1 $y = 3x^{-2};\ k = 3,\ p = -2$

3 $y = (3/8)x^{-1};\ k = 3/8,\ p = -1$

5 $y = (5/2)x^{-1/2};\ k = 5/2,\ p = -1/2$

7 $y = 0.2x^2;\ k = 0.2,\ p = 2$

9 $y = 125x^3;\ k = 125,\ p = 3$

11 $y = (1/5)x;\ k = 1/5,\ p = 1$

13 Base: w; exponent: -1; coefficient: $-1/7$

15 Base: v; exponent: 2; coefficient: 5

17 Base: x; exponent: 1; coefficient: 12

19 Base: x; exponent: -2; coefficient: 3

21 Base: r; exponent: 2; coefficient: 48

23 Base: a; exponent: 2; coefficient: π^2

25 $x = 2.614$

27 $x = 16$

29 $x = 7$ and $x = -3$

31 (a) $L = (R^2C^4)/(4\pi^2)$
(b) $C = \pm\sqrt{2\pi\sqrt{L}/R}$

33 Two

35 Zero

37 One

39 (i)

41 (iii)

43 (i)

45 (i)

47 (iii)

49 (i)

51 (iii)

57 (c) and (d)

59 (a) 3.172 cm

(b) $r = 0.671R^{1/4}$
(c) Yes, 1/4

61 Inversely proportional to cube root of P

63 (a)

(b) No
(c) Yes
(d) Increasing
(e) Decreasing
(f) $w = 0.086d^{2.4}$

Chapter 3 Tools-1

1 24

3 -8

5 x^8

7 x^5

9 $(a + b)^{10}$

11 $(a + b)^7$

13 $(a/3)^b$

15 2^{n+2}

17 B^{2a+1}

19 $((x^2 + y)(x + y^2))^3$

21 Negative

23 Negative

25 Negative

27 Negative

Chapter 3 Tools-2

1 1

3 2

5 1/9

7 3/2

9 $x^{-1/2};\ n = -1/2$

11 $x^{3/2};\ n = 3/2$

13 $x^{-2},\ n = -2$

15 Positive

17 Negative

19 Negative

21 Positive

Section 4.1

1 (b) $x \neq 3$

3 (b) All x

5 (b) $x \geq 3$

7 All real numbers

9 $y \geq 0$

11 (a) All real numbers
   (b) All real numbers

13 (a) All real numbers
   (b) $y = 7$

15 (a) All real numbers
   (b) All real numbers

17 (a) $x > 4$
   (b) $y > 0$

19 (a) $x \geq -1$
   (b) $y \geq 0$

21 (a) $x \neq -1$
   (b) $y \neq 3$

23 (a) $0 \leq t \leq 12$
   (b) $0 \leq y \leq 200$

25 (a) $1 \leq x \leq 7$
   (b) $2 \leq y \leq 18$

27 (a) $0 \leq a \leq 6$ years
   (b) $0 \leq V \leq \$18{,}000$

29 All k; range is all real numbers

Section 4.2

1 $h(x) = 2x + 1, k = 1/3, p = 5$

3 $u(x) = 5 - x^3, k = 1, p = 1/2$

5 $h(x) = 1 + \sqrt{x}, k = 100, p = 4$

7 $h(x) = 12 - \sqrt[3]{x}, k = 5, p = 1/2$

9 $g(x) = x^3 + 3$

11 $g(x) = (x + 1)^3$

13 $g(x) = (x + 1)^3 - 3$

15 $g(x) = 2 - x$

17 $g(x) = 7 - x$

19 $y = x^2 + 2$

21 $y = (x - 1)^2$

23 $y = (x - 3)^2 + 1$

25

27

29

31 $y = (x + 1)^4$

33 $w = t^6 + 5$

35 $q = 3 + 10s^3$

37 $y = x^4 + x^2 + 1$

39 Inside function: multiply by 2 and add 1

41 $u = x^2 + 1$ and $y = \sqrt{u}$;
   Other answers are possible

43 $u = x^3$ and $y = 3u - 2$;
   Other answers are possible

45 (a) $f(g(x)) = x^{3/2} + 1$
   (b) $g(f(x)) = \sqrt{x^3 + 1}$

47 $f(x) = 5x^3$

49 $y = \sqrt{5x + 2}$

51 $R = 25 - 0.08\sqrt{t}$

53 (a) Erosion starts 30 years earlier
   (b) Height is 50 cm higher

55 Input s, output q

Section 4.4

1 Add 8

3 Raise to the $1/5^{\text{th}}$ power

5 Add 2 then divide by 5

7 Divide by 2 then raise to $1/5^{\text{th}}$ power

13 (a) $y = 8x^5 + 4$
   (b) Subtract 4, divide by 8, then take the 5th root.

17 $g(y) = \sqrt[3]{y/12}$

19 $g(y) = (y/(1 - y))^2$

21 $x = (19 - \sqrt{37})/2$

Chapter 4 Review

1 Domain: All real numbers,
   Range: $y \geq 0$

3 Domain: $x \geq 0$,
   Range: $y \geq 0$

5 Domain: All real numbers,
   Range: $P \geq 0$

7 Domain: $x > 0$
   Range: $y > 0$

9 All real numbers

11 All real numbers

13 $g(x) = 2x^2 - 4$

15 $g(x) = 2(x + 2)^2 - 2$

17 $k \neq 5$; range: all real numbers except 5

19 $0 \leq G \leq 0.75$

21 (a) all real
   (b) $s \geq 2$

23 Inside function: Square, then add 1

25 $u = x - 1$ and $y = 1 + 2u + 5u^2$;
   Other answers are possible

27 (a) $f(g(x)) = 45x^2$
   (b) $g(f(x)) = 15x^2$

29 (a) $f(g(t)) = -3 - 5t^2$
   (b) $g(f(t)) = (2 - 5t)^2 + 1$
   (c) $f(f(t)) = 25t - 8$
   (d) $g(g(t)) = (t^2 + 1)^2 + 1$

31 (a) $y = b + mx + k$; the y-intercept is $b + k$

   (b) $y = b + m(x - k)$; the y-intercept is $b - mk$

33 $y = 2x^5 - 1$

35 (a) $C = f(r) + 5$
   (b) $C = f(r - 3)$

37 Input W, output U

41 $g(y) = (4y^2 - 4)/(y^2 - 7)$

Section 5.1

1 $f(x) = x^2 - 3x$;
   $a = 1, b = -3, c = 0$

3 $q(m) = m^2 - 14m + 49$;
   $a = 1, b = -14, c = 49$

5 $h(x) = x^2 - (r + s)x + rs$;
   $a = 1, b = -(r + s), c = rs$

7 $m(t) = 2t^2 - 4t + 14$;
   $a = 2, b = -4, c = 14$

9 $q(p) = 4p^2 - 5p + 6$;
   $a = 4, b = -5, c = 6$

11 $l(l - 6)$ square feet

13 Quadratic

15 Quadratic

17 Quadratic

19 Not quadratic

21 Positive: $x < -5$ or $x > 4$; negative: $-5 < x < 4$

23 Positive for all values of $x \neq 6$

25 $y = -(x - 6)^2 + 15$

27 $y = (x - 3)^2 - 5$

29 $y = (-7/4)(x - 2)^2 + 3$.

31 $y = -(x - 3)^2 + 9$

33 $h = 0; k = 0$

35 $h > 0; k = 0$

37 $h < 0; k > 0$

39 (a) $a < 0, c < 0$
   (b) $h > 0, k = 0$
   (c) Yes; $r = s, r > 0, s > 0$

41 (a) $a > 0, c > 0$
   (b) $h < 0, k > 0$
   (c) Does not factor

43 (a) $a > 0, c > 0$
   (b) $h < 0, k = 0$
   (c) Yes; $r = s, r < 0, s < 0$

45 (a)

47 No match

49 (e)

51 (c)

53 (a) 960 feet
   (b) 10 seconds

55 (a) Weight at birth
   (b) 22.952 lb

57 $0.01x^2 + 40x$ dollars

59 $5x^2 - 15x + 10$

61 $f(x) = (x - (a + 1))(x - 3a)$;
   $f(x) = x^2 - (4a + 1)x + 3a(a + 1)$

63 h increases, k decreases

Section 5.2

1. $(x+3)(x+5)$
3. $(x-2)(x-6)$
5. $y: 12, x: -4, -1$
7. $y: 2, x: \pm\sqrt{17/4} - 5/2$
9. $(1/2, -1/4)$
11. $(0, 2)$
13. $(x-6)^2$
15. $6(x-4)^2 + 2$
17. $c(x-2d)^2 + (b+5)$
19. -9
21. No minimum
23. $y = 2(x+2)^2 + 12; a = 2, h = -2, k = 12$
25. $y = 2(x+5)^2 - 38; a = 2, h = -5, k = -38$
27. $y = 3(x+7/6)^2 - 11/12; a = 3, h = -7/6, k = 11/12$
29. (a) $x^2 - 4x - 21, (x+3)(x-7)$
    (b) $-21, -21, -21; -24, -24, -24$
31. (a) $y = 6x^2 - 23x + 21$, $a = 6, b = -23, c = 21$
    (b) $y = 6(x-7/3)(x-3/2)$, $a = 6, r = 7/3, s = 3/2$
    (c) $y = (x-23/12)^2 + (-25/24)$, $a = 6, h = 23/12, k = -25/24$
33. $a = -5, b = -2, c = 3$
35. $a = 1/5, b = 4/5, c = 7/5$
37. $a = 2, b = -6, c = -2$
39. $(x+2)(x+5)$
41. $(3t+10)^2$
43. $a = -2, h = 5, k = 5$
45. $a = 4, h = 2, k = 6$
47. $a = 1, h = -6, k = -16$
49.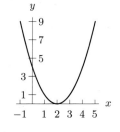
51. $y = 8(x-3/2)(x-(-5/4))$, $a = 8, r = 3/2, s = -5/4$
53. $y = -2x(-4x+1) + (-15); k = -2, v = -4, w = -15$
55. (b) Factored; $p = 3$ or $p = 9$
    (c) Standard; $54,000
    (d) Vertex; profit is $18,000 when $p = 6$
57. (a) $A = (100 - w)w$
    (b) 2500 sq ft

Section 5.3

1. $a = 2, b = -0.3, c = -9$
3. $a = -1, b = 0, c = 4$
5. $a = -4, b = 10, c = -6$
7. $a = 48, b = -307, c = 6$
9. $a = 5, b = -10, c = 0$
11. $x = \pm\sqrt{7} = \pm2.648$
13. $x = 0, x = -4$
15. No solution
17. $x = 3 \pm \sqrt{3}$
19. $x = 1 \pm \sqrt{\frac{5}{2}}$
21. $x = \pm\sqrt{5 \pm \sqrt{5}}$
23. $x = 6 \pm \sqrt{41}$
25. No solutions
27. $x = \frac{1 \pm \sqrt{177}}{4}$
29. (a) $x = 5 \pm \sqrt{40}$
    (b) $x = 5 \pm \sqrt{40}$
31. (a) $x = -7/2 \pm \sqrt{29}/2$
    (b) $x = -7/2 \pm \sqrt{29}/2$
33. (a) $-17/2 \pm \sqrt{321}/2$
    (b) $-17/2 \pm \sqrt{321}/2$
35. (a) $x = 8 \pm \sqrt{121/2}$
    (b) $x = 8 \pm \sqrt{121/2}$
37. (a) $x = -17/10 \pm \sqrt{269}/10$
    (b) $x = -17/10 \pm \sqrt{269}/10$
39. (a) $x = -1/2, x = -3$
    (b) $x = -1/2, x = -3$
41. (a) $x = -1/2$
    (b) $x = -1/2$
43. (a) No solutions
    (b) No solutions
45. $(2 \pm \sqrt{52})/6$
47. $a = -2, h = -3, k = -4$
49. $a = 1, h = -3, k = 3$
51. $a = 2, h = -3, k = 22$
53. Two solutions
55. One solution
57. No solutions
59. (a) 900 feet
    (b) 7.5 seconds
61. $t = \sqrt{k}/4$
63. None
65. $A > 10$
67. $c < 1/3$
71. (a) $A = lw + \pi w^2/8$
    (b) $w = 2.9$ ft, $l = 5.8$ ft, $r = 1.5$ ft
73. 10 in by 12 in
75. (a) No intercepts
    (b) 2
    (c) 3, 1
    (d) $2 \pm \sqrt{2}$

Section 5.4

1. $x = 2, 3$
3. $x = -2, -3$
5. $x = 5, -1$
7. No solution
9. $x = 3, -2, -7$
11. $x = 3, x = -5$
13. $x = 0, x = -2$
15. $x = 4, x = -1$
17. $x = 2, x = 6$
19. $x = (-6 \pm \sqrt{52})/2 = -6.606$ and 0.606
21. $x = 1, x = 7$
23. $x = (1 \pm \sqrt{34})/3$
25. $x = 3/2, 1/3$
27. $(x-2)(x+3) = 0$
29. $(x-2)^2 = 3$
31. $(x-p)^2 = q$
33. $x^2 - 2ax + a^2 = 0$
35. $0, 2, -1$
37. 1
39. (a) 0
    (b) 0 and 6.122 seconds
41. $x = (5 \pm \sqrt{37})/6$
43. One solution
45. Two solutions
47. No solution
49. One solution
51. $t = \pm\sqrt{5}$
53. $t = 9$
55. No solutions
57. (a) $p = (10 - q^2)/(3q)$
    (b) $q = (-3p \pm \sqrt{9p^2 + 40})/2$
59. (a) $p = 0, p = -q/2$
    (b) $q = 0, q = -2p$
61. (a) $x = -3$ and $x = -4$
    (b) $x = -3$ and $x = -4$
63. No

Chapter 5 Review

1.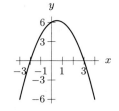
3.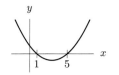
5.

7

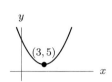

$(3,5)$

9 Not quadratic

11 Quadratic

13 Not quadratic

15 Not quadratic

17 Max: a

19 Min: 2

21 $a = -7, h = 3, k = 11$

23 $(x + 2)(x + 1) = 0; -1, -2$

25 $(3z + 1)(2z + 1) = 0; -1/3, -1/2$

27 (a) $y = 6x^2 - 29x + 35,$
      $a = 6, b = -29, c = 35$
     (b) $6(x - 7/3)(x - 5/2),$
      $a = 6, r = 7/3, s = 5/2$
     (c) $y = 6(x - 29/12)^2 - 1/24,$
      $a = 6, h = 29/12, k = -1/24$

29 $x = \pm 2$

31 $x = 6$

33 $x = 1, x = 2$

35 $x = -59, x = -47$

37 $x = -14, x = -7$

39 $x = -8$

41 $x = -6 \pm \sqrt{29}$

43 $x = -5 \pm \sqrt{22}$

45 $x = 0$

47 $x = -1/2, x = -2$

49 $x = 4, x = -3/2$

51 $y = (-1 \pm \sqrt{109})/6$

53 $t = (5 \pm \sqrt{19})/2$

55 $y = -5 \pm \sqrt{27}$

57 $r = (1 \pm \sqrt{21})/2$

59 $v = 1, v = -4$

61 If x is positive, left-hand side is positive

63 $9t^2 - 3t - 2 = 0$

65 $x^2 + (\sqrt{3} - \sqrt{5})x - \sqrt{15} = 0$

67 10 seconds

69 10

71 (a) $(x - 3)^2 + 11$
     (b) Smallest value is 11, at $x = 3$

73 One x-intercept,

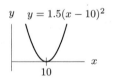

y $y = 1.5(x - 10)^2$

x

10

75 No x-intercepts,

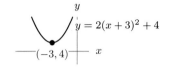

$y = 2(x + 3)^2 + 4$

$(-3, 4)$ x

77 (a) Negative, 12
     (b) $-2, 6$
     (c) $2, 16$

79 $a = 2, r = -5, s = 3$

81 $a = 2, v = -2, w = -30$

83 (a) \$400
     (b) \$200

85 $y = 8$

87 $y = x^2$

89 $A < 0$

91 $A \geq 0$

93 $r = 0$ or $s = 0$

95 $x = a, x = 1$

97 $a = 1$

99 $a \leq 0$

101 $a = -5$

103 $a = 1$

105 (c) $c \leq k$

107 $(a + 1)^2 + 1 = 0$

109 $8, -6$

Chapter 5 Tools

1 $(b + 3)(b - 6)$

3 $2x(x + 4)(2a - 1)$

5 $t = 0, t = 5, t = -5$

7 $t = 3, t = -2$

9 $t = -5$

11 $t = 3/2, t = 3, t = -3$

13 $x = 4, x = 5, x = -5$

15 $x = -3/2, x = -\sqrt[3]{5/2} = -1.357$

17 $x = -a$

19 $y^2 + 2y - 3$

21 $6a^2 + 5a - 6$

23 $r^3 - 3r^2 + 2r$

25 $r^3 - 5r^2 + 13r - 36$

27 $a^2 - b^2 - 2bc - c^2$

29 $-2x - h, h \neq 0$

31 $x^2 - 16x + 64$

33 $x^2 - 26x + 169$

35 $x^2 + 2xy + y^2$

37 $25p^4 - 10p^2q + q^2$

39 $x^3 - 3x^2y + 3xy^2 - y^3$

41 $q^4 + 4q^3r + 6q^2r^2 + 4qr^3 + r^4$

43 $x^2 - 81$

45 $x^2 - 144$

47 $3x^2 - 31x - 22$

49 $15x^3 + 11x^2 - 56x$

51 $x^3 + 9x^2 + 12x - 4$

53 $48x^3 + 4x^2 - 58x - 14$

55 $6x^4 + 12x^3 - 24x^2 - 60x - 30$

57 $x^6 - 2x^5 + 8x^4 - 15x^3 - 2x^2 + 8x - 16$

59 $(a + b)(x - y)$

61 $(2x - y)(4x - 3)$

63 $(3v - 2w)(2v^2 + 3w^3)$

65 $(y - 6)(y + 1)$

67 $(n - 6)(n + 5)$

69 $(g - 10)(g - 2)$

71 $(t - 2)(t - 25)$

73 $(q + 10)(q + 5)$

75 $(b + 6)(b - 4)$

77 $(x + 3y)(x + 8y)$

79 $2(z + 7)(z - 1)$

81 $(z + 4)(4z + 3)$

83 Cannot be factored

85 $3(w + 6)(w - 2)$

87 $(t^2 + 9)(t^2 - 6)$

89 $2(3w - 4)(2w + 1)$

91 Cannot be factored

93 $(x + 4)(x - 4)$

95 $(s - 6t)^2$

97 Cannot be factored

99 $2x(3x^3 + 4z^2)^2$

101 $(r^p + 9)^2$

103 $-3t^3(t + 2v)^2(t - 2v)^2$

105 $x(x^2 + 4)$

107 $(x + 9)(x - 9)$

109 $(x + 12)(x - 12)$

111 $(x - 7)(x - 7)$

113 $(x - 11)(x - 11)$

115 $x^2(x - 9)(x - 9)$

117 $(x - 9)(x + 6)$

119 $x(x + 12)(x + 11)$

121 $(5x + 8)(x - 9)$

123 $(qx - 7)(2x + p)$

125 None

127 None

131 $(x + 7)(x + 8)$

133 $(x + 2b)(x + 2b - 1)$

135 $((a + b)/2)^2 - ((a - b)/2)^2 = ab$

137 False

139 False

147 (a) $1 + x + x^2 + x^3 + x^4 + x^5 + x^6 + x^7$
     (b) $1 + x + x^2 + x^3 + \cdots + x^{14} + x^{15}$
     (c) $1 + x + x^2 + x^3 + \cdots + x^{30} + x^{31}$
      $(1+x)(1+x^2)\left(1 + x^{2^2}\right)\left(1 + x^{2^3}\right)$
      $\ldots \left(1 + x^{2^k}\right) = 1 + x + x^2 + x^3 + \ldots +$
      $x^{2^{k+1} - 1}$

Section 6.1

1 No

3 Yes; Init: 0.75; growth: 0.2

5 $Q = 300 \cdot 3^t$; $a = 300, b = 3$

7 $200 \cdot 9^t$; $a = 200, b = 9$

9 Not exponential

11 Exponential

13 $P = MZ^t$

15 $800 \cdot d^k$

17 $V_0 \left(\frac{2}{3}\right)^n$

19 Nz^t

23 (III)

25

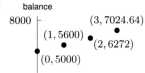

27

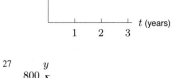

29 (a)   (i)   10,194,900
       (ii)   10,189,803
      (iii)   10,184,708
      (iv)   10,174,525
   (b) $f(t) = 10,200,000(0.9995)^t$

31 220,000; 1.016

33 Increases by factor of 1.3728

35 $P = 400(0.8)^t$

37 $Q = 50(1/25)^t$; $a = 50, b = 1/25$

39 $Q = 40 \cdot 4^t$; $a = 40, b = 4$

41 $Q = 0.2 \cdot 10^t$, $a = 0.2, b = 10$

Section 6.2

1 Growth

3 Decay

5 Decay

7 1.085

9 3.15

11 70%

13 400%

15 $a = 200, b = 1.031, r = 3.1\%$

17 $a = \sqrt{3}, b = \sqrt{2}, r = b - 1 = 41.42\%$

19 $a = 5, b = 2, r = 100\%$

21 Initial value \$2200; growth rate 21.1%/yr

23 Initial value \$8800; decreases by 4.6%/yr

25 $\$5(1.3)^t$ million

27 $2(1 + r)^{10}$ million dollars; r is growth rate

29 Radioactive substance decaying at 3%

31 Machine depreciating at 2%

33 $a = 90, b = 0.1, r = -90\%$

35 $a = 70, b = 1/2, r = -50\%$

37 $a = 2000, b = 1.0617, r = 6.17\%$

39 (a)   Pop 222,000 in 2000
       Growth rate 5.6%/yr
   (b)   382,818

41 (a)   14,251,670
   (b)   14,612,237
   (c)   14,981,927
   (d)   15,749,603 or 15,749,602

43 Austin's growth factor over 3 yrs

45 Phoenix's fractional growth in two years

47 No

49 13.22%

Section 6.3

1 $a = 1000, b = 2, T = 12$ if t in years

3 $a = 50, b = 5, T = 15$ if t in years

5 $a = 250, b = 2, T = 6$ if t in years

7 $a = 200, b = 5/4, T = 90$ if t in minutes

9 $a = 800, b = 1.021, t_0 = 1995$

11 $a = 200,000, b = 1.082, t_0 = 7$

13 $V = 12,000(1.02292)^t$

15 Second account

17 $a = 1/3, b = 2^{1/3} = \sqrt[3]{2}$

19 $a = 7, b = 8$

21 Initially 1200; drops by 1.5% each month

23 (a)

Ja	Fe	Ma	Ap	Ma	Ju
1	1	2	4	8	16

   (b) 2^{n-2} for $2 \leq n \leq 6$
   (c)   32 inches

25 (a)   Earns 0.88% monthly
   (b)   11.09%

27 (a)   0.11% interest daily
   (b)   49.37%

29 (a)   3% interest every quarter
   (b)   12.55%

31 $\$6(1.05)^7$

33 $(1.05)^{25} - 1$

35 $\$20(1.10)^7$

37 $((1.05)^{10} - 1)100\%$

39 $\$250(0.99)(1.05)^{10}$ million

41 (a)   1 week
   (b)   5 days
   (c)   2 weeks

43 Substance ϵ

45 (a) $a = 800, b = 2, \tau = 15$; doubles every 15 years
   (b)   4.73%

47 (a) $a = 75, b = 10, \tau = 30$; grows tenfold every 30 years
   (b)   7.98%

49 (a) $a = 400, b = 2/3, \tau = 14$; decreases by 1/3 every 14 years

   (b) -2.85%

51 $a = 1500, b = 1.1487, r = 14.87\%$

53 $a = 8000, b = 1.08, r = 8\%$

55 24.47%

57 80.25%

Section 6.4

1 $t = 2$

3 $t = 0$

5 $t = -2$

7 $t = 0$

9 $t = 2.6$

11 $2.6 < t < 2.8$

13 Positive

15 Negative

17 (a)   Dom: all real; Range: $Q > 0$
   (b)   Dom: $t \geq 0$; Range: $0 < Q \leq 200$

19 (a)   (IV)
   (b)   (V)
   (c)   (III)

21 (a)

x	0	1	2	3	4
$g(x)$	28	30.8	33.88	37.268	40.995

   (b)   (i) $x = 0, 1$
       (ii) $x = 2, 3, 4$
      (iii) $x = 3$

23 (a)

x	-3.5	-1.5	0.5	2.5
y	8.351	58.648	411.854	2892.246

   (b) $x = -1.5, 0.5, 2.5$

25 (a)   57.2; 52.1; 51.1; 50.2; 49.3; 48.4; 47.5; 39.4
   (b)   About 3.7 hours

27 (b) $t = 30$
   (c) $t = 5, 10$

29 (a)   Years 0, 1, 2
   (b) $t = 1$

31 (a)

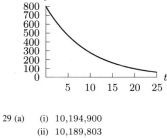

   (b)   (i) $x \leq 2$
       (ii) $x \geq 3$
   (c)   Use more points

33 No solution

35 Positive

37 Positive

39 Negative

41 Negative

43 Positive

45 (a)   (III)

(b) (II)
(c) (IV)

47 (a) $A > 0$
   (b) $A = 1$
   (c) $A > 1$

49 (a) $A > 0$
   (b) $A = 1$
   (c) $0 < A < 1$

51 (a) $A > 0$
   (b) $A = 1/2.1$
   (c) $A > 1/2.1$

53 (a) $A < 0$
   (b) $A = -1$
   (c) $A < -1$

55 $x = 3u$

57 $x = (1/2)(3u - 1)$

59 $x = -2u$

61 $x = 1.176$

63 $x = 1.505$

Section 6.5

1 $f(x) = 80.5347(0.9788)^x$

3 $P = 10,000 \cdot 0.9958^t$

5 $f(t) = 40(1.1214)^t$

7 $p(x) = 16.6537(1.0373)^x$

9 $v(x) = 1.6818(5.6569)^x$

11 7.177%

13 12.94%

15 45 min

17 0.486%

19 10.409%

21 9.05%

23 −9.43%

25 4.56%

27 $g(t) = 8000(0.9215)^t$

29 $V = 3450(1.0475)^t$

31 $f(x) = 503.153(0.9551205)^x$

33 $V = 3500 \cdot 2^{t/8}$

35 (a) $0.5Q_0$ after 62 days; $0.25Q_0$ after 124 days
   (b) After 186 days
   (c) $0.9889Q_0$

37 $Q_0(0.87)^t$

39 (a) $P = 936.7(1.053)^t$ billion dollars
   (b) 1212.7 billion dollars

41 (a) 1990 pop: 396,000
      2000 pop: 541,000
   (b) $396(1.03169)^t$
   (c) 2010 pop: 739,100

43 $f(t) = 150(0.8)^t$

45 $P - 282(1.01)^{20} = 0$

Chapter 6 Review

5 $h = 6(1.05)^t$

7 $Q = (8/5)(5/2)^{t/3}$

9 $Q = (6/3^{1/3})3^{t/9}$

11 $Q = 2(7/2)^{t/3} = 2(1.518)^t$

13 $Q = 7.537(0.937)^t$

15 Yes; $a = 2, b = 5$

17 No

19 Yes; $a = 1, b = 2/9$

21 Exponential, constant -21, base $1/2$

23 Power function, constant 1, exponent -1

25 Decreasing; 4%

27 Decreasing; 47.5%

29 $B \cdot 2^n$

31 $V_0 \cdot k^3$

33 $V_0 \cdot k^{N/h}$

35 1.6

37 2

39 9.5%

41 116%

43 Linear

45 Exponential

47 Exponential

49 $y = \sqrt{2}^x = (1.414)^x$

51 Quadratic

53 Exponential

55 Linear

57 $a = 1700, b = 1.117, r = 11.7\%$

59 $a = 120, b = 3.2, r = 220\%$

61 $f(t) = 8.5859(1.11806)^t$

63 $v(x) = 10.4613(1.06936)^x$

65 $90(0.83)^t$ mg

67 $\$30(1.022)^t$

69 $1500(1.03)^t - 1500(1.02)^t$

71 4.05%

73 $Q = Q_0(0.9)^t$

75 (a) (I), (IV), (V), (VI)
   (b) (II), (III)
   (c) (V), (VI)

77 $g(t) = 172.290(0.90856)^t$

79 (a) (III)
   (b) (II)

81 -6.11% per day

83 16.99% per year

85 Second investment

87 Less than

89 Less than

91 Less than

93 Greater than

95 Greater than

97 (b)

99 (e)

101 (d)

103 (c)

105 (c)

107 Linear: a, b, c if $n = 1$, none if $n \neq 1$
    Exponential: none

109 Linear: A
    Exponential: t, base b^2

111 $64 \cdot 4^x$; $a = 64, b = 4$

113 $500\left(1 - 2^{-\frac{1}{3}t}\right)$; $a = 500, k = 1/3$

115 T is the number of weeks

117 s is number of days since the day after sac opens

Section 7.1

1 $10^{-2} = 0.01$

3 $10^{1.301} = 20$

5 $10^{3.699} = 5000$

7 $10^{3x^2 + 2y^2} = \alpha\beta$

9 $\log 100,000 = 5$

11 $\log 200 = 2.301$

13 $\log 39,994 = 4.602$

15 $\log 97 = a^2 b$

17 4

19 -0.145

21 1.248

23 1.756

25 -3

27 3/2

29 Undefined

31 1

33 7

35 $\sqrt{10}$

37 100,000

39 $x = 100$

41 $x = 316,227.766$

43 12

45 $3 < \log 8991 < 4$

47 $4 < \log(0.99 \cdot 10^5) < 5$

49 $-2 < \log 0.012 < -1$

51 Between 1.42 and 1.44

53 3.5; 3.5051,

x	3.4	3.5	3.6	3.7
10^x	2511.886	3162.278	3981.072	5011.872

55 -0.301,

x	-0.296	-0.298	-0.300	-0.302
10^x	0.506	0.504	0.501	0.499

Section 7.2

1 4.3; 4.3219,

x	4.1	4.2	4.3	4.4
2^x	17.148	18.379	19.698	21.112

3 3.3; 3.3219,

x	3.1	3.2	3.3	3.4
0.5^x	0.117	0.109	0.102	0.095

5 $x = -4$

7 $x = 3/2$

9 $\log 19/\log 2, 4.248$

11 $\log 17/\log 12, 1.140$

13 $2/\log 80, 1.051$

15 $\log 520/\log 1.041, 155.638$

17 $\log 600/\log 1.033, 197.028$

19 12.9104

21 No; for example $a = 0, b = 1$

23 $\log A + 2\log B$

25 Not possible

27 Not possible

29 $3\log A + (1/2)\log B$

31 $u + v$

33 $2u + 3v + 0.5w$

35 $9v^2$

37 $2x + 1$

39 $(x+1)^2$

41 (a) 0.301, 1.301, 2.301, 3.301
   (b) 4.301, −0.699

43 $28 \cdot 10^{t\log 1.121}; a = 28, k = \log 1.121$

45 $420 \cdot 10^{(\log 2)(-0.2)t}, a = 420, k = -0.2\log 2$

Section 7.3

1 1.519

3 −1.284

5 0.799

7 $\log(5.2)/\log 1.041, 41.030$

9 $\log 15/\log 1.033, 83.409$

11 $\log 2.5/\log(2/3), -2.260$

13 $\log(8/70)/\log 0.882, 17.275$

15 $17\log 0.125/\log(2/3), 87.185$

17 $\log(3/7)/\log(0.923/0.891), -24.013$

19 $\log(97/84.2)/\log(0.982/0.891), 1.455$

21 $\log(0.877/0.315)/\log(0.782/0.916), -6.474$

23 $\log(8400/2300)/\log 1.0417$

25 $t = \frac{4\log 3.5}{\log 3}$

27 $x = 200$

29 0.485

31 −0.2

33 1000

35 500,000.5

37 $x = 10,000,000,000$

39 No

41 Yes

43 No

45 Yes

Section 7.4

1 15.850 minutes

3 628 days

5 2206

7 7.258 years

9 16.980 hours

11 30.187 years

13 $t = 3.289$

15 $t = 0.458$

17 14

19 $t = 6.213$ months

21 20 months

23 $t = 32.137$

25 $t = -13.853$

27 (a) 11.786 years
   (b) 20.149 years

29 (a) 3.094 years
   (b) 5.290 years

31 (a) (i) 6.579 years
       (ii) 3.289 years

33 (a) 3.61%
   (b) 19.545 years

35 (a) $4.02
   (b) Faster
   (c) 88 years, 2026

37 (a) False
   (b) 3.322 inches

39 (a) 30 million
   (b) 480 million
   (c) 2.529 hours
   (d) 30 minutes

Chapter 7 Review

1 $10^1 = 10$

3 $10^{1.733} = 54.1$

5 $10^r = w$

7 $\log 1,000,000 = 6$

9 $\log 0.1 = -1$

11 $\log 0.558 = -0.253$

13 (a) $\log 100 = 2$
   (b) $\sqrt{100} = 10$

15 2

17 0

19 Undefined

21 1/3

23 3.68

25 $2n + 1$

27 Undefined

29 5.9

31 $\log 20/\log 5 = 1.861$

33 $\log 30/\log\sqrt{2} = 9.814$

35 −0.572

37 1/3

39 0.6724

41 2.126

43 −1.924

45 15/2

47 0.956

49 $\log 60/\log 1.02$

51 $(\log 800 - \log 250)/(\log 0.8 - \log 1.1)$

53 9

55 $x = -\frac{2}{3}$

57 $x = \log\frac{9}{8}$

59 $\log 1.5$

61 $t = \frac{\log 0.5}{\log 1.117}$

63 $t = \frac{\log\frac{1}{6}}{\log 0.5}$

65 $t = \frac{\log(z-r)}{\log b}$

67 $\log(90/40)/\log(1.118/1.007)$

69 $5x - 3$

71 $6x^3$

73 $-x/3$

75 $3x/2$

77 $x = 3$

79 1000

81 $x = 10^{1/3}$

83 $x = 10^{2.5} = 316.228$

85 $x = 10^{4/3} = 21.544$

87 7.273 years

89 23.449 years

91 7.17 weeks

93 Yes, 32.893 years

95 Between 1.64 and 1.66

101 (VI)

103 (II)

105 (V)

107 $v - u$

109 $(2u + 0.5v)/(3w + u/3)$

111 $-5uw(\log 2 + v)$

Section 8.1

1 Polynomial

3 Polynomial

5 Not polynomial

7 Yes

9 No

11 Yes

13 $800 + 3200r + 4800r^2 + 3200r^3 + 800r^4$

15 Yes,

x	0.1	0.2	0.3
$f(x)$	1.054	1.118	1.195
$g(x)$	1.054	1.118	1.192

17 $g(0.5) = 177/128$

Section 8.2

1 6

3 3

5 $-x^7 - 2x^2 + 2x$

7 $\frac{1}{7}x^4 - 2x^3 - \frac{2}{7}x$

9 $-4x^{11}$

11 x^{13}

13 x^8

15 $-1/2$

17 $\sqrt{7}$

19 $a_0 = 17$

21 $a_0 = 15$

23 $a_0 = -1$

25 Polynomial, degree 5, leading coefficient 8

27 Not a polynomial

29 Not a polynomial

31 $a_4 = 3, a_3 = 6, a_2 = -3, a_1 = 8, a_0 = 1$

33 $a_{13} = 1/3$

35 $2x^3 + 4x^2 - 3x + 5$

37 $7x^{33} + 33x^{20} + 20x^{14} + 14x^7$

39 $2x^3 + x - 2; n = 3;$
    $a_3 = 2, a_2 = 0, a_1 = 1, a_0 = -2$

41 $-5x + 20; n = 1;$
    $a_1 = -5, a_0 = 20$

43 $\frac{\sqrt[3]{5}}{7} \cdot x^2; n = 2;$
    $a_2 = \sqrt[3]{5}/7, a_1 = 0, a_0 = 0$

45 $-6x^5 - 5x^4 - 4x^3 + 36x; n = 5;$
    $a_5 = -6, a_4 = -5, a_3 = -4, a_2 = 0,$
    $a_1 = 36, a_0 = 0$

47 $x^3 + 3x^2 + 3x + 1; n = 3;$
    $a_3 = 1, a_2 = 3, a_1 = 3, a_0 = 1$

49 720

51 $0, 4$

53 8

55 7

57 3

59 8

61 $a_{n-1} = a_3 = -4$

63 $h(1) = -6$

65 $a = 1, b = 4, c = 2, d = 9$

67 Possible answer: $2x^5 + 3x^4$ and $-2x^5 + 7x^4$

69 Impossible

71 Impossible

73 1

75 $\pm 3/2$

77 $50x^5 - 45x^4 + 19x^3 - 8x^2 + 3x - 1$

79 $2, 3, 4$

Section 8.3

1 $x = 3, 4, -2$

3 $x = 3, -2$

5 None

7 $1, -2, 3$

9 $0, -1, -2$

11 $0, 1, -3$

13 $(x - 2)(x + 3)$

15 $(x - 5)(x + 13)^2$

17 $(x + 6)^5$ is one possibility

19 0

21 -3

23 $4t^3 + 12t^2 + 11t + 3$

25 $0, 1/2, -1/2$

27 Two

29 Three

31 Two

33 Two

35 $a \geq 0$

37 $a > 0$

39 $a < 0$

41 All a

43 All a

45 (a) $(x+1)(2x-5)$ and $-3(x+1)(2x-5)$
    (b) $2(x+1)(2x-5)$

47 $A = 7/2; B$ and C anything

49 $1, -1, 0$

53 (a) $45 - 2x$
    (b) $x^2(45 - 2x)$
    (c) $x = 0$ and $x = 22.5$

55 (a) \$2377.50; \$2496.38
    (b) $1000x^2 + 500x + 750; 1000x^3 + 500x^2 + 750x$
    (c) \$0, \$1200, \$650
    (d) \$4662.25; \$4784.73; \$4910.73

Chapter 8 Review

1 2

3 4

5 8

7 -1

9 0

11 (a) 1
    (b) 4
    (c) 0
    (d) $-t^3 + t^2 - t + 1$

13 (a) 1
    (b) 5
    (c) 5
    (d) $t^4 + 3t^2 + 1$

15 $1, -1, -9$

17 $1, -1$

19 $4, 12, 32; n \cdot 2^{n-1}$

21 $9x^2, 27x^3, 81x^4; 3^n x^n$

23 $-2, -3, -4; -n$

25 $m, 5a$

27 $m; a$

29 a, m

31 $t > a$

33 Two

35 None

37 Two

39 (a) $x = 3$ and $x = 1/2$
    (b) No

43 $x > a$

45 (a) $-0.1 \leq x \leq 0.1, 0 \leq y \leq 0.011$
    (b) $-1.1 \leq x \leq -0.9, -0.121 \leq y \leq 0.081$
    (c) $-1.1 \leq x \leq 0.3, -0.121 \leq y \leq 0.117$
    (d) $-20 \leq x \leq 20, -7600 \leq y \leq 8400$

49 $(x+7)(x+2)(x-3)/14 = x^3/14 + 3x^2/7 - 13x/14 - 3$

51 Coefficients of polynomials entries in Pascal's triangle

53 $16x^4 + 96x^3 + 216x^2 + 216x + 81$

55 (b) $2000x^5 - 25x^4 - 60x^3 - 45x^2 - 80x - 25$

Section 9.1

1 $y = (x+2)(x-3)$

3 $y = (x+2)^2(x-3)$

5 $y = \frac{1}{4}(x+2)(x-1)(x-2)$

7 $y = (1/60)(x+1)(x-1)(x-2)(x-3)(x-5)(x-6)$

9 (a)

11 (a) I; $x = 2, -2$
    (b) II; 16
    (c) II; negative

13 Degree 3; leading term x^3;
    leading coefficient 1; constant term 0

15 Degree 6; leading term $4x^6$;
    leading coefficient 4; constant term -1

17 $y = 2x(x-2)(x+2)^2$

19 $y = \frac{1}{4}x^2(x+1)(x-2)^2$

21 One zero between $x = 1$ and $x = 2$;
    possibly others

23 Possible answers: $(x+2)(x+3)(x+1)$,
    $(x+2)(x^2+3)$

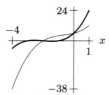

Section 9.2

1 $x = -5/2$

3 $x = \sqrt{2}$

5 $x = 0$

7 $1/x$

9 $(2x+a)/a$ provided $x \neq a/2$

11 $x = -1/2$

13 $x = (-1 \pm \sqrt{82})/9$

15 $x = -3/5$

17 $x = -1/5, 1$

19 $x = 0$

21 No common factors

23 $x^2/(x-1)$; not equivalent

25 No common factors

27 $a(x) = x + 1, b(x) = 2x - 3$

29 $a(x) = 2, b(x) = x^5 - 3x^2 + 7$

31 $q(x) = 2, r(x) = 1$

33 $q(x) = x + 2, r(x) = 7$

35 One

37 One

39 $r = 2$

41 $r = -6$

Section 9.3

1 (a)

3 (a) and (c)

5 (a) and (d)

7 (a), (b), (c)

9 $x = -1$

11 $x = 5$

13 $g = 0$

15 (a) 32 ¢/mile
(b) 500 or more
(c) $c = (2500/d) + 8$; 8 ¢/mile

17 (a) 0.0833
(b) 1/2
(c) 0.917 or more

19 (b) (i) II; $x = 2, -4$

(ii) II; $x = 1, -1$
(iii) I; $y = 1$

21 (a) $p(x)$: vertical asymptote $x = 5/2$, horizontal asymptote $y = 1/2$; $q(x)$: vertical asymptote $x = -1$, horizontal asymptote $y = 3/4$
(b) $y = (10x^2 + 3x + 7)/(8x^2 - 12x - 20)$
(c) Vertical asymptote $x = 5/2, -1$; horizontal asymptote $y = 5/4$

Chapter 9 Review

1 $x = -1, x = -4/3$

3 $x = 3, x = -1/2$

5 $x = \sqrt{3}, x = -\sqrt{3}$

7 $(ax + 4)/(2a)$ provided $x \neq 0$

9 $\dfrac{2x^3 - 9x^2 + 5x - 7}{x^3 + 5x^2 + 2x + 10}$

11 Hole at $x = 3$; zero at $x = 1$; vertical asymptote at $x = -2$; horizontal asymptote at $y = 1$; y-intercept at $y = -1/2$

13 (b) $y = 3/4$

15 $h(x) = \frac{8}{9}(x - 3)^2 + 4$

17 $w(x) = \frac{1}{4}(x + 3)(x + 2)(x - 2)^2(x - 3)$

19 $\sqrt{x - 2} + \sqrt{8 - x}$ or $\sqrt{(x - 2)(8 - x)}$ or $\sqrt{\sqrt{6} - \sqrt{x - 2}}$

Chapter 9 Tools

1 $a(x) = 2x^2 - 3x + 11, b(x) = x + 7, q(x) = 2x - 17, r(x) = 130.$

3 $a(x) = x^4 - 2, b(x) = x - 1, q(x) = x^3 + x^2 + x + 1, r(x) = -1.$

5 $x^2 + x - 6$

7 $\dfrac{x^4 + x + 1}{x + 1} = x^3 - x^2 + x$ with remainder $r = 2$

9 $q(x) = 1; r(x) = x + 1$

11 $q(x) = 2; r(x) = 3$

13 $q(x) = 8x^5 - 3x^2 + 2x + 2; r(x) = 3$

15 $x^3 - 3x^2 + 5x - 13$

17 $a = 2$ or $a = -5$

INDEX